Springer Undergraduate Mathematics Series

T0253817

For other titles published in this series, go to
www.springer.com/series/3423

Jeremy Gray

Worlds Out
of Nothing

A Course in the History
of Geometry in the 19th Century

 Springer

Jeremy Gray
Department of Mathematics
The Open University
Walton Hall
Milton Keynes, MK7 6AA
United Kingdom
j.j.gray@open.ac.uk

The Mathematics Institute
The University of Warwick
Warwick
United Kingdom

Whilst we have made considerable efforts to contact all holders of copyright material contained in this book, we have failed to locate some of them. Should holders wish to contact the Publisher, we will be happy to come to some arrangement with them.

Cover illustration elements reproduced by kind permission of:
Aptech Systems, Inc., Publishers of the GAUSS Mathematical and Statistical System, 23804 S.E. Kent-Kangley Road, Maple Valley, WA 98038, USA. Tel: (206) 432 -7855 Fax (206) 432 -7832 email: info@aptech.com URL: www.aptech.com.
American Statistical Association: Chance Vol 8 No 1, 1995 article by KS and KW Heiner 'Tree Rings of the Northern Shawangunks' page 32 fig 2.
Springer-Verlag: Mathematica in Education and Research Vol 4 Issue 3 1995 article by Roman E Maeder, Beatrice Amrhein and Oliver Gloor 'Illustrated Mathematics: Visualization of Mathematical Objects' page 9 fig 11, originally published as a CD ROM 'Illustrated Mathematics' by TELOS: ISBN 0-387-14222-3, German edition by Birkhauser: ISBN 3-7643-5100-4.
Mathematica in Education and Research Vol 4 Issue 3 1995 article by Richard J Gaylord and Kazume Nishidate 'Traffic Engineering with Cellular Automata' page 35 fig 2. Mathematica in Education and Research Vol 5 Issue 2 1996 article by Michael Trott 'The Implicitization of a Trefoil Knot' page 14.
Mathematica in Education and Research Vol 5 Issue 2 1996 article by Lee de Cola 'Coins, Trees, Bars and Bells: Simulation of the Binomial Process' page 19 fig 3. Mathematica in Education and Research Vol 5 Issue 2 1996 article by Richard Gaylord and Kazume Nishidate 'Contagious Spreading' page 33 fig 1. Mathematica in Education and Research Vol 5 Issue 2 1996 article by Joe Buhler and Stan Wagon 'Secrets of the Madelung Constant' page 50 fig 1.

Springer Undergraduate Mathematics Series ISSN 1615-2085
ISBN 978-0-85729-059-5   e-ISBN 978-0-85729-060-1
DOI 10.1007/978-0-85729-060-1
Springer London Dordrecht Heidelberg New York

British Library Cataloguing in Publication Data
A catalogue record for this book is available from the British Library

Mathematics Subject Classification (2010): 01A05, 01A055, 01A60, 03-03, 14-03, 30-03, 51-03, 53-03

Cover design: VTEX, Vilnius

Printed on acid-free paper

Springer is part of Springer Science+Business Media (www.springer.com)

Preface

In 1789 geometry was at a low ebb. Euler had written on it – he wrote on everything – but his successor, Joseph Louis Lagrange was much more of an algebraist and an analyst. The great flowering of French mathematics that may be said to have started with Lagrange's arrival in Paris in 1787 at the age of 51 and continued with Laplace, Legendre, and then the generation of Cauchy, Fourier and Poisson, accomplished much and innovated widely, but less in geometry than in other areas. Only Gaspard Monge stood out, both as an original geometer and as an inspiring teacher. In 1914, at the other end of what historians call the long 19th century, geometry was not only a major branch of mathematics with several new branches springing from it, it could claim to have been one of the most provocative and challenging in its implications for the nature of mathematics. There were whole new geometries, some contesting the centrality of Euclidean geometry and therefore the framework for all of mechanics, others arguably yet more fundamental, and pervading more and more of the domain of pure mathematics. This transformation of geometry is the theme of this book.

This book discusses the ideas, the people, and the way they (the people and the ideas) fit into larger pictures. It starts with the rediscovery of projective geometry in post-revolutionary France by Poncelet, a student of Monge. It then takes in the discovery of non-Euclidean geometry by Bolyai in Austria–Hungary and Lobachevskii in Russia in the 1820s. The reception of Poncelet's work was poor, that of the work of Bolyai and Lobachevskii dismal, and we consider why this was. With these clear examples in place showing the importance of the social in the reception of mathematical ideas, the book then considers how matters turned around. For projective geometry, the algebraic methods of Möbius and Plücker are prominent because they not only led to the resolution of a major paradox facing Poncelet's work they also opened up the little-studied domain of plane curves of degree 3 or more. For non-Euclidean geometry, the crucial ideas are those of Gauss, of Riemann, however obscurely they were presented, and of Beltrami. By the 1870s, the new geometries were secure. Projective geometry was fast becoming the central topic in geometry, non-Euclidean geometry had been given a presentation acceptable to mathematicians at least, and geometry was a topic of growing importance in mathematics. The book moves towards its conclusion with a consideration of Klein's view of geometry, the subsequent growth of axiomatic projective geometry, first in Italy and soon afterwards in the hands of David Hilbert, and with the work of Henri Poincaré on non-Euclidean geometry, including his conventionalism. The book ends by briefly examining the formal aspect of mathematics, the relation to physics in the work of Einstein, and some thoughts about the truth, if any, of mathematics.

The book offers a number of unfamiliar ways in which mathematics can be thought about. Its relations to philosophy and to physics are noted and at times explicitly discussed, and indeed quite deliberately the book discusses some of those aspects of 19th-century geometry of interest to philosophers of mathematics today. Because the book is about mathematics but is not a mathematics course, original motivations matter, individuals' critical judgements matter and forming an overview matters. There is an important place in a modern mathematics degree for imperfect understanding (as opposed to education exclusively delivered in bite-sized chunks designed to promote technical mastery) because imperfect understanding is all we can bring to most issues. No one understands enough about all the major problems of the day, but all of us must grapple with them on pain of being irresponsible. We can, sometimes, form what we must admit is an imperfect grasp of what the experts say, and base our actions on that. In a much more modest way, a book that invites you to think about how imperfect judgements can most reasonably be formed has a certain justification. It is equally true that writing about mathematics is a valuable skill. Almost every graduate will go on to spend a great deal of time writing: writing reports, bids for money, research papers, policy statements, teaching documents.

We lack a recent history of projective geometry and of many, if not all, of its major protagonists. For too long now, readers have had to fall back on Coolidge's *A history of geometrical methods*, first published in 1940 [43], supplemented by a very small number of more specialised studies. Coolidge's book has its uses – it is mathematically accurate – but it slights the social aspects, rewrites the mathematics, and can degenerate to a list of results. This book therefore offers the first recent treatment of the work of Poncelet, although the reader should consult Bos et al. [22] for a splendidly thorough account of Poncelet's porism. The argument presented here that Plücker's resolution of the duality paradox was decisive for the future of projective geometry seems to be new, and indeed Plücker's contribution has been rather marginalised. The argument that research in projective geometry was partly driven by a desire to make it fundamental is new. I note it here in such matters as the production of a projective definition of conic sections. It should be followed through to a proper study of projective transformations and the axiomatisations of the subject; I hope to complete such an analysis in a forthcoming paper.

I have written in various places on various aspects of the history of non-Euclidean geometry and I did not want to burden this book with too much repetition. I continue to hold a restrained view of what Gauss knew; however, Erhard Scholz urges a different view forcefully [219]. I have tried to draw attention to some of the better recent investigations into Riemann, but much had to be set aside. I was happy to be prodded into rethinking what I knew about Beltrami's papers, which make an unexpected appearance in Klein's study of projective geometry.

For the later period, Toepell's invaluable study of Hilbert's route into projective geometry [236] made that work easier, and Marchisotto's work in progress on Pieri will do much to illuminate the Italian contributions (also helpfully discussed recently by Avellone et al. [5] and Bottazzini [23]). But we are particularly fortunate to have *David Hilbert's lectures on the foundations of geometry, 1891–1902* edited by M. Hallett and U. Majer finally in print [107]. This book not only reproduces the early lecture courses Hilbert gave that led up to his famous *Grundlagen der Geometrie* of 1899 [119], and which are already very informative about how Hilbert worked and reworked his ideas, it has an excellent, instructive commentary. Speculations at the end of the present book about Einstein's ideas, and about the philosophical implications of the new geometries, owe much to the work of other people, but I believe they are fresh in the present context.

However, I must stress that this book grew out of a course in which I lectured on some, even most, of the content of the 30 principal chapters to be found here, including the revision material, and invited students to read the rest. It is my hope that the book can be used in the same way, and various selections of the material are possible. It is not an attempt to write a complete history of geometry in the 19th century, for the simple reason that too much is missing. Quite simply, a manageable course that reflects the coherence of the developments and captures enough of their significance to be worthwhile must leave things out. I could have said more about Monge's mathematical work, for which see Taton's book [235] and the relevant chapter in Lützen's book on Liouville [158]. Research needs to be done on Chasles' presentation of projective geometry, and the way his work eclipsed that of Poncelet. Similarly, Steiner's contribution is slighted here and does not seem to have been studied by historians of mathematics recently. Von Staudt was almost wholly omitted from the course – amends are made here in the Appendix.

Not entirely for reasons of space, I have also omitted Grassmann entirely.[1]
He came to regard his first book, of 1844, as "an utter failure" [95, p. 19]
and radically revised it in 1862. This was both an extension of the former and
a reworking of it in more mathematical, less philosophical dress, because, as
Grassmann put it in his 1862 book, of "the difficulty ... the study of that work
caused the reader on account of what they believed to be its more philosophical
than mathematical form." [96, p. xiii] As he noted in the foreword to the
second edition of his first book in 1878, its reception only began in 1867 with
a favourable notice by Hankel and more importantly by Clebsch.

Almost every social and contextual issue raised could have been discussed
further. When much fuller histories of this material are written, such matters
as the how, when and where of what was done in geometry in the 19th century
will have to be examined, audiences analysed, many gaps filled in.

On a positive note, the 19th century is a remarkable century in the history
of mathematics. It is the century in which our ideas of mathematics changed
most radically, and in which many of the ideas that form the undergraduate
syllabus were created (rigorous analysis, group theory, linear algebra, and much
of the geometry that this book is particularly about). I hope readers will find it
interesting to see some of the ideas which may have occupied them at university
emerge in their historical context.

Mathematics is a great cultural adventure. It's been going for over 5,000
years as a written activity[2], it shows no signs of stopping. This book offers
a chance to examine some of the things that make mathematics important,
and to obtain some kind of an overview of that activity. If we ask why anyone should care about all this old material, one answer takes us to another
question: what is attractive about mathematics? It is notoriously hard to tell
non-mathematicians what the appeal of the subject is. Another answer is that,
up to a point, we live in a society that sells us certain ideas about mathematics in particular and education in general as intrinsically interesting, and
marginalises others. So we can notice that society offers us certain loaded
choices, and ask how, and why. In this way, we start to understand mathematics as a social enterprise, and begin to think about what mathematics is,
and why.

I have taken the opportunity to make a number of changes to the book.
Athanase Papadopoulos is preparing the first English translation of Lobachevskii's *Pangéométrie* of 1855 (see [156]) and he was kind enough to show me
his work in progress. This has led me to reconsider what I have written about

---

[1] Grassmann's major works were published in 1844 [93] and 1862 [94], English translations are in [95] and in [96], respectively.

[2] My thanks to Eleanor Robson for telling me that the earliest mathematics exercises
for trainee scribes are from late fourth-millennium Uruk (a city in southern Iraq).

Lobachevskii, to make a short mention of his path to his great discovery and his publications of 1829, and to make some mention of his *Pangéométrie*. I hope readers will read that work in full, when it becomes available in its new, English, translation with its richly informative commentary. I also thank him for telling me, what I should have known, that the University of Kasan was not a backwater but in fact the second-oldest university in Russia, founded after Moscow and before St Petersburg, and was well staffed from its inception. This, of course, helps explain why Martin Bartels was there and was one of Lobachevskii's teachers.

Marvin Greenberg, whose *Euclidean and non-Euclidean geometries* I warmly recommend for its many insights from the standpoint of modern geometry, has spurred me in numerous ways to reconsider what I wrote, and I thank him for making me look again at Saccheri's arguments, at Janos Bolyai's squaring of the circle, and at several other matters. Readers may also profitably consult his paper Greenberg, M.J. 2010 Old and new results in the foundations of elementary plane Euclidean and non-Euclidean geometries, *American Mathematical Monthly*, March 198–219.

I would also like to draw attention here to Robin Hartshorne's *Geometry: Euclid and beyond* for its particularly lucid explanations of many mathematical topics that lie just beyond this book, including, for example, the subject of non-Archimedean geometries, which relate in fascinating ways to the topic of parallelism.

Conversations with a number of historians of mathematics have sharpened my opinions on a number of topics. It turns out to be harder than I had thought to say when the subject of projective geometry acquired a local habitation and a name. From the time of Poncelet onwards people spoke of projective properties of figures and of their projections, but as these ideas came to constitute a definite subject it was often called "synthetic geometry". I am not aware of the term "projective geometry" in the writings of Chasles, Hesse, Steiner, or von Staudt. Klein spoke of projective geometry in his Erlangen programme as late as 1872; the first to call a book by that name may be Cremona in 1873.

A number of people wrote in to tell me of numerous smaller, mostly typographical, errors in the book. I thank them all and apologise to everyone. I particularly want to thank Dirk Schlimm and Marvin Greenberg, but also, among the students on the MSc course at the Open University in various years, Peter Lush and Dave Sixsmith.

Very amusingly, it has turned out that the picture said to be of Legendre (Figure 7.3 in the first printing) is not of the mathematician Adrien-Marie Legendre at all, but of an entirely unrelated namesake. The whole story, with information about how it was discovered, is told in Peter Duren's article "Changing Faces: The mistaken portrait of Legendre" in the December 2009 issue of the

Notices of the American Mathematical Society, 56, 1440-1444. A newly discovered cartoon of Legendre, shown on the front cover of the journal, is now thought to be the only likeness we have of the mathematician.

How to use this book

This book is the fruit of a secondment to the University of Warwick for one term a year from 2001 to 2004, where I taught it as an option to third- and fourth-year students. To underline the fact that this book can be used as a history course, in which opinions play a major role, and not a dispassionate mathematics course, I have retained the informal, chatty style of the lectures. In it I have tried to present everything the student needs to read except for some primary sources that are available over the Web, and also to give detailed advice (to students and their instructors) on how to tackle the assignments.

A book on the history of geometry studies a body of ideas and practices historically. What is discussed is the production and reception of ideas, and how this was affected by the social context. The ideas are those of mathematics, the practices those of mathematicians. The mathematics involved is a substantial mathematical diet, even at the University of Warwick, where students may study non-Euclidean geometry. However, I make the claim that a course based on this book can be done, and done well, by anyone willing to think about these ideas and that all the necessary mathematics is presented here. This is a defensible claim because the ideas here are to be understood as a historian would treat them, and not as a mathematician would. That is to say, they are to be understood so that arguments about their production and reception can be understood. Insofar as these arguments require that an informed opinion be held about some mathematical argument, that piece of mathematics belongs in a history course, but the whole mathematical package (precise definitions, proper proofs, abundant practice with techniques) does not.

That said, the mathematics in a course on the history of mathematics should not be mysterious, and the assessment was devised to minimise this risk. It is given here much in the form it was given to the Warwick university students, and at about the same time (see Chapters 12, 21 and 31, where my ideas about assessment in general and these assessment questions in particular are set out). The opening question each year asked for a short explanation of the reception of the work of either Poncelet or Bolyai and Lobachevskii. This tests the student's ability to digest some six chapters of material and isolate the key features in about 500 words. The word length in this assignment is deliberately short so that the assignment is hard and a process of analysis and selection must take place. The second assignment (1,500 words) asked that they talk their way through either Salmon's classification of quartic curves, or Cremona's projective study of conics and duality, or Lobachevskii's 1840 presentation of non-Euclidean geometry (students are encouraged to use the Poincaré disc model). Real mathematical understanding is required to do this well. The Salmon material, which closely follows Plücker's original presentation of these ideas, was new to the students, quite tricky and not free of error. The Cremona passage is in a kind of mathematics they cannot have seen anywhere else, and the Lobachevskii question asks them to explain in what way the disc model clarifies what Lobachevskii left obscure. (The students knew by then that this lack of clarity was not a major reason for Lobachevskii's ideas doing so badly in his lifetime.) It is only with some idea of the validity of a disc model that the Lobachevskii passage becomes readable at this level, and even then it is not easy. The final assignment was to summarise the key developments in geometry in the long 19th century and account for them. It is evident that the mathematical ideas have not come from nowhere, and have not succeeded simply on their own merits; the purpose of this assignment was to give the students a chance to demonstrate some overall mastery of a slew of ideas and relate them to contemporary opinions about their worth.

Inevitably, answers to such questions are more like opinions (let us hope, well-supported opinions) than they are facts. There *are* facts in history, a great many of them, but how they are organised, what weight to attach to them, is a matter for each historian to decide. Being a historian is a matter of producing arguments: establishing that these facts are evidence relevant to a certain claim; that the conclusion they support is this and not that; that the confidence one can have in the conclusion is strong or weak. So the answers to historical questions are a mixture of facts and arguments, and one aspect of reading history is sorting out facts from judgements in any argument. Different historians, different people, inevitably and rightly differ among themselves in the judgements they reach.

I have tried to suggest a range of source material, but inevitably language is a problem. The original sources for this material are overwhelmingly German and French, and I was able on occasion to find French students willing to go further with their study of Chasles. There is probably more worth in including Italian, and even Russian or Latin, than English sources, and even allowing for the existence of translations (and providing some more), there is not enough that English students can read. To that extent, even the splendid, and ever growing, Digital Mathematics Library [53] is not enough. On the other hand, the seductive Web ever beckons, with items from the wonderful to the downright odd. Safest to say that used with thought, and in conjunction with other material, it is a fine resource.

The fact that this book can be used as a course explains some other possibly unexpected features. I did not ask the students to read Lobachevskii's booklet of 1840 in detail until quite late, because it is surely too hard without something like a disc model in hand. I could have provided more mathematical examples at every stage, but I have tried to steer a line between providing enough information for it to be possible for a student to make an informed judgement about, say, what Plücker did, and teaching how to dualise a wide range of curves. That would require wrestling with resolvents; but what a historian of mathematics has to do is reach an informed opinion and I judged that eloquent examples could profitably substitute for theorems. Students were invited to familiarise themselves with these examples all the same. For the benefit of users of this book, some mathematics is presented very swiftly in various places so that other approaches can be taken to this material.

The best way to find out about the lives of mathematicians is to consult the *Dictionary of scientific biography* [90] and then to go to the sources mentioned there. Entries vary from the short to almost book length; that for Laplace is a multi-authored work of 130 pages. The Web is another source of information, of uneven merit.

I should like to thank the staff at both The Open University and the University of Warwick who made it possible for me to teach this course at Warwick. Above all, I should like to thank the students at Warwick. Each year, well over a hundred enrolled, completed the course and reported positively on their experiences. Most were taking degrees in mathematics, some in mathematics and statistics, some in physics, some in philosophy. Individually and collectively they demonstrated a hugely encouraging desire to grapple with difficult material. They did so successfully and often impressively, and there are many points in this book that I owe to their persistence and indeed to their insights. I am also happy to thank Nicholas Jackson for his help with everything to do with producing the index.

The manuscript then passed to Jennifer Harding, whose careful and astute editing did much to sharpen this book, and whom I am also happy to thank. I also thank Aaron Wilson for his fine work on the illustrations. At the same time Springer sent it to no less than six referees, and while they all reported positively on the book, which was gratifying, several took the time to identify failings and to suggest improvements, and for those comments I am particularly grateful. My good friend Erhard Scholz also read the book with his customary care and his erudition has saved me from errors I would blush to admit to; I offer him heartfelt thanks.

Acknowledgements

The translations from the work of Chasles, Gergonne, and Poncelet are taken from *The history of mathematics: A reader*, J. G. Fauvel and J. J. Gray (eds.), and appear by kind permission of The Open University. The material from Beltrami's *Saggio* is taken from J. Stillwell, *Sources of hyperbolic geometry*, History of Mathematics series, vol. 11, American Mathematical Society, Providence RI, and London Mathematical Society, London, 1996, and appears by kind permission of the American Mathematical Society. The material from Poincaré's essays "Space and geometry" and "Experiment and geometry" are taken from H. Poincaré, *Science and hypothesis*, Dover Publications, who state that to the best of their knowledge this material is in the public domain. The material from A. Einstein, *Relativity; the special and the general theory, a popular exposition*, translated by R. W. Lawson, is taken from the edition published by Methuen in 1920. To the best of our knowledge this material is in the public domain.

The pictures of Michel Chasles and Henri Poincaré appear by kind permission of the Dibner Institute Collection at the Smithsonian Institution Libraries, Washington, DC.

The picture of Riemann is taken from *Bernhard Riemann 1826–1866: Turning points in the conception of mathematics* by Detlef Laugwitz, translated by Abe Shenitzer, Birkhäuser, 1998, and appears by kind permission of the publishers.

The picture of Oswald Veblen appears by kind permission of the American Mathematical Society.

The remaining pictures, those of Monge, Poncelet, Lambert, Gauss, Lobachevskii, Möbius, Plücker, Hesse, Riemann, Beltrami, Klein, Cremona, Hilbert and Enriques, are taken from sources which to the best of our knowledge are in the public domain.

Every effort has been made to contact all holders of copyright material contained in this book, but if any copyright holder wishes to contact the publisher, the publisher will be happy to come to an arrangement with them.

Contents

List of Figures xxiii

1. Mathematics in the French Revolution 1
   1.1 The French Revolution 1
   1.2 Some mathematicians 4
       1.2.1 Monge 6
   1.3 Descriptive geometry 8

2. Poncelet (and Pole and Polar) 11
   2.1 Poncelet reminisces 11
   2.2 Poncelet's mathematics 16
   2.3 From Poncelet's *Traité* 18
       2.3.1 Commentary 21
   2.4 Pole, polar and duality 22

3. Theorems in Projective Geometry 25
   3.1 The theorems of Pappus, Desargues and Pascal 25
   3.2 Some properties of some transformations 35
   3.3 Cross-ratio 39
       3.3.1 Porismata 40

4. Poncelet's *Traité* 43
   4.1 Poncelet's singular claims 43
       4.1.1 Meeting 44
   4.2 Cauchy responds 47
   4.3 Other responses 51
   4.4 Poncelet's more conventional methods 52

5. Duality and the Duality Controversy 53
   5.1 Pole and polar 53
   5.2 Gergonne versus Poncelet................................ 55
       5.2.1 Curves of higher degree 56
   5.3 Gergonne.. 57

6. Poncelet and Chasles 63
   6.1 What was done – differing opinions 63
   6.2 Institutions and careers 66
   6.3 Chasles 67
   6.4 What was done? 69
   6.5 Chasles, Steiner and cross-ratio 70
   6.6 Extracts from Chasles' *Aperçu historique* 73
       6.6.1 On descriptive geometry 73
       6.6.2 On Monge and his school.......................... 74
       6.6.3 On Monge's work 75
   6.7 A quick introduction to modern projective geometry 76
       6.7.1 The real projective plane 76
       6.7.2 Projective spaces 78

7. Lambert and Legendre.. 79
   7.1 Saccheri 81
   7.2 Lambert.. 84
   7.3 Legendre 86
   7.4 Lambert.. 87

8. Gauss 91
   8.1 Gauss.. 91
   8.2 Schweikart and Taurinus 93
   8.3 What Gauss knew 96
       8.3.1 Gaussian curvature 98

9. János Bolyai... 101
   9.1 János and Wolfgang Bolyai 101
   9.2 János Bolyai's geometry 104
   9.3 János Bolyai's section 32 112

10. Lobachevskii.. 115
   10.1 Lobachevskii and Kasan 115
   10.2 Lobachevskii's new geometry.......................... 118
       10.2.1 Lobachevskii's first foundations of geometry 120
       10.2.2 Astronomical evidence 123

10.3 Lobachevskii's booklet of 1840 123
    10.3.1 Opening remarks 123
    10.3.2 Concluding remarks 126

11. To 1855 129
  11.1 Minding's surface 129
  11.2 The Bolyais read Lobachevskii 130
  11.3 Final years of János Bolyai 131
  11.4 Final years of Lobachevskii 132
  11.5 Gauss's death, Gauss's *Nachlass* 134

12. Writing 137
  12.1 Assessment and advice 137
    12.1.1 Reading and writing the history of mathematics 137
    12.1.2 Practice questions 141
  12.2 References and footnotes 145
    12.2.1 Appendices 146
    12.2.2 Names 147
    12.2.3 Your essays 147
  12.3 Assessment question 147
    12.3.1 Advice 147

13. Möbius 149
  13.1 Möbius's *Barycentric calculus* 149
    13.1.1 Barycentric coordinates 150
    13.1.2 Projective transformations 153
    13.1.3 Duality 155
    13.1.4 Central projection from one plane to another 157
  13.2 A note on duality 157
  13.3 Möbius's coordinates 159

14. The Duality Paradox 161
  14.1 Higher plane curves 161
    14.1.1 Cubic curves 163
  14.2 Plücker's resolution of the duality paradox 163
  14.3 Confirmation by others 165
  14.4 Plücker 166
  14.5 Hesse 168

15. The Plücker Formulae 173
   15.1 Singular points 173
      15.1.1 The non-singular cubic curve in the plane 174
      15.1.2 The non-singular quartic curve in the plane 174
      15.1.3 28 real bitangents 177

16. Higher Plane Curves 179
   16.1 Non-singular points and tangents 179
   16.2 Double points 180
   16.3 Homogeneous coordinates 182
   16.4 First and subsequent polars 183
      16.4.1 The first polars of a circle 183
      16.4.2 Inflection points 184
   16.5 Hessians 185
   16.6 Addendum 188
   16.7 References 190

17. Complex Curves 191
   17.1 Complex by necessity 191
      17.1.1 Complex numbers in geometry 193
      17.1.2 The introduction of complex curves 193
   17.2 Elliptic functions 194

18. Riemann 195
   18.1 Riemann 195
   18.2 Riemann's publications 197
   18.3 Riemann on geometry 198
      18.3.1 Surfaces 199
   18.4 Riemannian geometry 201
   18.5 From Riemann's *Habilitationsvortrag* 202

19. Differential Geometry of Surfaces 211
   19.1 Basic techniques 211
      19.1.1 Geodetic projection 212
   19.2 Introducing Beltrami's *Saggio* 215
      19.2.1 Beltrami's *Teoria* of 1868 217
   19.3 The *Saggio* 219
      19.3.1 From Beltrami's *Saggio* of 1868 219
   19.4 Legendre's error 225
   19.5 References 225

20. Non-Euclidean Geometry Accepted 227
20.1 Beltrami's version 227
20.2 Gauss's posthumous contribution 228
    20.2.1 Kant? 229
20.3 Felix Klein 230
    20.3.1 Klein at Erlangen 232
    20.3.2 . . . and beyond 232
20.4 Klein's Cayley metric 233
20.5 Klein's unification of geometry 234
20.6 The Erlangen Program in the 1890s 236
20.7 Weierstrass and Killing 237

21. Writing 241
21.1 Assessment questions 241
21.2 Advice 242
    21.2.1 Cremona 242
    21.2.2 Salmon 244
    21.2.3 Lobachevskii's account in 1840 245

22. Fundamental Geometry 247
22.1 The rise of projective geometry 247
22.2 Cremona 249
    22.2.1 Cremona's projective geometry 251
22.3 Salmon 254
22.4 Anxiety – Pasch 255
22.5 Helmholtz 256
    22.5.1 Free mobility 258

23. Hilbert 259
23.1 Hilbert 259
23.2 Hilbert and geometry 261
    23.2.1 The *Grundlagen der Geometrie* 262
    23.2.2 Desargues' theorem 263
23.3 Impact 266
23.4 References 267

24. Italian Foundations 269
24.1 Peano and Segre 270
24.2 Enriques 273
24.3 Pieri 276
24.4 Conclusions 277
24.5 Veronese 277

25. The Disc Model 281
  25.1 Poincaré 281
    25.1.1 A prize competition 282
    25.1.2 Poincaré's discovery of non-Euclidean geometry 283
    25.1.3 The Poincaré and Beltrami discs 285
  25.2 Poincaré and Klein 287
  25.3 Circumcircles 288
  25.4 Inversion and the Poincaré disc......................... 289
    25.4.1 Inversion 290
  25.5 References 297

26. The Geometry of Space..................................... 299
  26.1 How to decide?.. 299
  26.2 Poincaré's conventionalism.............................. 300
    26.2.1 Enriques disputes 301
  26.3 "Space and geometry" 304
  26.4 Poincaré's arguments 305

27. Summary: Geometry to 1900 309
  27.1 References 311

28. The Formal Side 313
  28.1 Nagel's thesis 313
  28.2 From Hilbert's *Grundlagen der Geometrie* 315

29. The Physical Side 321
  29.1 Geometry and physics.................................... 321
  29.2 Einstein 322
    29.2.1 The special theory of relativity 322
    29.2.2 The paradoxes of special relativity.................. 325
  29.3 Minkowski.. 326
  29.4 Einstein, gravity and the rotating disc 326
  29.5 Einstein's *Relativity* 328

30. Is Geometry True? 333
  30.1 Truth 333
    30.1.1 Mathematical truths 334
  30.2 Proof 335
    30.2.1 Frege versus Hilbert 336
  30.3 Relative consistency 337
  30.4 Poincaré on "Non-Euclidean geometries" 338

31. Writing 341

    31.1 Assessment questions 341

    31.2 Advice on writing such essays 342

    31.3 How the essays will be graded 343

A. Von Staudt and his Influence 345

    A.1 Von Staudt 345

        A.1.1 Von Staudt's *Geometrie der Lage* 346

        A.1.2 Klein's response to von Staudt 349

    A.2 Non-orientability 351

    A.3 Axiomatics – independence 354

    A.4 References 357

Bibliography 359

Some Geometers 377

Index 379

List of Figures

1.1 Monge 6
1.2 A line segment in space 9
1.3 Lengths in descriptive geometry 10

2.1 Poncelet 11

3.1 Pappus's theorem 26
3.2 Desargues' theorem 27
3.3 Pascal's theorem 27
3.4 "Seeing" a point at infinity 30
3.5 A special case of Pappus's theorem 31
3.6 A special case of Desargues' theorem 32
3.7 A very special case of Pascal's theorem 34
3.8 The projection from O of AB onto $A'B$ 35
3.9 The ratios AC/CB and $A'C'/C'B$ are not equal 36
3.10 The angles $\angle BAD$ and $\angle BA'D$ are not equal 36
3.11 Cross-ratio is preserved under projection 37
3.12 Uniqueness of the fourth harmonic point 38
3.13 Cross-ratio and the fourth harmonic point 40

4.1 Pole and polar 45
4.2 Poncelet on poles E, F, G and polars FG, EG, EF 47

5.1 Poles and polars 54
5.2 Desargues' dual 54
5.3 Brianchon's theorem 55

6.1 Chasles 67
6.2 Cross-ratio in terms of angles 70

7.1 The parallel postulate 80
7.2 Lambert 85
7.3 Legendre's figure 87

8.1 Gauss 91
8.2 Schweikart's square 94
8.3 An impression of the Gauss map 100

9.1 A parallel line as a limiting case 102
9.2 Parallels in Bolyai's geometry 104
9.3 Parallels and equal angles in Bolyai's geometry 105
9.4 Bolyai's L-curve 106
9.5 Bolyai's F-surface 106
9.6 Euclidean and non-Euclidean triangles in space 108
9.7 The angle of parallelism 109
9.8 The curve L' 109
9.9 The angles u and v 110
9.10 The ratio $X = AB/CD$ 111
9.11 The fundamental figure in Bolyai's geometry 113

10.1 Lobachevskii 115

11.1 A pseudosphere 130

13.1 Möbius 149

14.1 Plücker 162
14.2 Hesse 168

15.1 Plücker's quartic 177

16.1 Inflection points of a cubic picked out by its Hessian 186
16.2 The folium of Descartes and its Hessian 187
16.3 Klein's quartic curve and its Hessian 187

18.1 Riemann 195

19.1 Geodetic projection (1) 213
19.2 Geodetic projection (2) 215
19.3 Beltrami 216
19.4 The Beltrami disc 222

20.1 Felix Klein 230
20.2 Klein's disc 233

21.1 Six points in involution....................................... 243

22.1 Cremona 250
22.2 Cremona presents a theorem and its dual 252

23.1 Hilbert 259
23.2 Moulton's lines 264
23.3 The failure of Desargues' theorem 265
23.4 Veblen 266

24.1 Enriques 274

25.1 Poincaré 281
25.2 Poincaré's projections 286
25.3 The tessellation (2, 3, 7)...................................... 288

29.1 Changing coordinates 324

1

Mathematics in the French Revolution

1.1 The French Revolution

The French Revolution is usually said to have begun in 1789, when the National Constituent Assembly, a parliament-like body of aristocrats, churchmen and commoners, alarmed by the violent behaviour of the peasantry – a Parisian rabble had already destroyed the symbolic but by then almost empty prison of the Bastille on 14 July 1789 – hoped to appease them by abolishing the feudal regime (4 August 1789) and passing the Declaration of the Rights of Man and of the Citizen (26 August). This famous declaration begins:

Article 1 Men are born and remain free and equal in rights. Social distinctions may be based only on considerations of the common good.

Article 2 The aim of every political association is the preservation of the natural and imprescriptible rights of man. These rights are Liberty, Property, Safety and Resistance to Oppression.

I cannot see that it specifically addresses the question of rights of women, or if the declaration is gender blind and women were genuinely offered the rights of men.

The situation spiralled out of control. The National Constituent Assembly announced that a people had the right of self-determination, and that this justified a French invasion of the Papal territory of Avignon (13 September 1791). Tension between France and her neighbours rose until France eventually declared war on Austria and Prussia (20 April 1792). The first phase of the war (April–September 1792) went badly for France. The Austro-Prussian army

J. Gray, *Worlds Out of Nothing*,
Springer Undergraduate Mathematics Series,
DOI 10.1007/978-0-85729-060-1_1, © Springer-Verlag London Limited 2011

advanced rapidly toward Paris. Believing, not unreasonably, that France had been betrayed by the king, French revolutionaries imprisoned the royal family in the Temple (10 August 1792). In September there were violent riots, and aristocrats and churchmen already imprisoned were murdered in considerable numbers. But the revolution, with its real prospect of dramatic social change, and the dire national situation, brought many volunteers to the French army, and the invaders were repelled at the battle of Valmy, some 120 km east of Paris (20 September 1792). The next day a new governing body, the National Convention, abolished the monarchy and established the First French Republic.

Thus inspired, French troops now invaded nearby regions of other countries (Belgium, the Rhineland, Savoy and Nice – not then parts of France). Within the National Convention, opinion was split between those who wanted to create a bourgeois republic (better for the middle classes and men of property) and take that revolution to all of Europe, and those who wanted to spread the revolution to the lower classes (workers, craftsmen, shopkeepers, smallholders and agricultural labourers), but were not so keen on foreign adventures. This latter group prevailed for a time; it is they who executed the king, Louis XVI (21 January 1793).

The war now entered a third phase, in which France again initially did badly. Belgium and the Rhineland were recaptured, and Paris again came under threat. This played to the advantage of the radicals in the National Convention, and they pushed for taxation of the rich, national assistance for the poor and the disabled, declared that education should be free and compulsory, and ordered the confiscation and sale of the property of those who had fled the country. The violent reaction this provoked in some parts of the country, notably those with powerful branches of the church, was in turn countered by the Reign of Terror: at least 300,000 suspects were arrested, of whom 17,000 were sentenced to death and executed, while more died in prisons or were killed without any form of trial. At the same time the revolutionary government raised an army of more than 1,000,000 men.

Again the war then went France's way, with a decisive defeat of the Austrians and the reconquest of Belgium. And again, once the national situation seemed secure the populace lost its appetite for extreme measures. Robespierre, the radical leader, known ominously as "the Incorruptible" was overthrown in the National Convention (27 July 1794) and executed the very next day. Almost at once the radical social measures were withdrawn, and a year later there was even a Royalist attempt at a coup in Paris, but it was defeated by a young general of whom we shall hear more: Napoleon Bonaparte (5 October 1795). One of his first acts was to end the rule of the National Convention.

The war continued, and with it the endless opportunity for the use of force at home in response to national need. Napoleon took the war west, all the way to Portugal, south through Italy, and east across Prussia. Typically, as in Venice, the French armies installed the French constitution where ever they went, thus opening up the ghetto and integrating Jews into contemporary society. In the north, after failing to organise a cross-channel invasion of England, Napoleon instead decided to threaten the British in India by occupying Egypt. The invasion of Egypt went smoothly, but was utterly reversed when Horatio Nelson's fleet destroyed the French fleet at the Battle of the Nile (1 August 1798). The next year, on 9 November 1799, Napoleon became dictator (an event known as the 18th Brumaire of Napoleon Bonaparte because that was its date in the revolutionary calendar). He continued to pursue military plans with great genius, until he invaded Russia, meeting his defeat outside Moscow in 1812. This defeat cost him half a million men, and marked his end, which came at the Battle of Waterloo in Belgium in 1815.

What do we learn from this? It's a strange start to the idea that one has a right to liberty, property, safety and resistance to oppression. One might ask: who enjoyed those rights, who did not? Who wished to extend them, and who to restrict them? A reign of terror is not the obvious way to install these rights.

Next, in those 26 years from 1789 to 1815 (and the roller-coaster ride was not over even then) the French Revolution, and the French state, stood for a wholesale transformation of society. Almost every regime outside France felt threatened. At least in the early days, almost everyone who wanted a better world could echo Wordsworth's famous line "Bliss was it in that dawn to be alive". That legacy survived. Could the goals of the revolution be brought about if people tried again? If so, by persuasion, or violence? Or was the whole ideal hopelessly flawed?

Those 26 years mark a long time. An impressionable 15-year-old in 1789 would be, if he or she survived, a solid citizen of 41 in 1815. And in those 26 years, mostly the French state worked. It was, with the exception of Britain, briefly in control of the whole of Europe, the first time since the Romans that anyone could have made such a claim. It is usual to acclaim Napoleon's military genius, and he was a gifted general, but military success requires a trained army, with good engineers, properly equipped and with motivated soldiers. Whatever they were fighting for, a huge number of French citizens joined the French army.

From the other side, those 26 years were also a long time. Should Italian liberals be happy that a modern constitution based on the Declaration of Rights was installed where they lived? Or should they chafe under the foreign yoke? They were divided on the issue. Did those military defeats mean that something was wrong with the domestic order? And if so, what of the French system should be copied, and how? Could you adapt to the French model without bringing in all that talk of rights, and possibly all the social disruption it seemed to involve?

In all of that talk, one word stands out: "citizen". It goes with another: "state". Citizens are not subjects, it is citizens who have the ultimate power. To quote again from the Declaration of Rights:

Article 3 The source of all sovereignty lies essentially in the Nation. No corporate body, no individual may exercise any authority that does not expressly emanate from it.

Article 4 Liberty consists in being able to do anything that does not harm others: thus, the exercise of the natural rights of every man has no bounds other than those that ensure to the other members of society the enjoyment of these same rights. These bounds may be determined only by Law.

Article 5 The Law has the right to forbid only those actions that are injurious to society. Nothing that is not forbidden by Law may be hindered, and no one may be compelled to do what the Law does not ordain.

Article 6 The Law is the expression of the general will. All citizens have the right to take part, personally or through their representatives, in its making. It must be the same for all, whether it protects or punishes. All citizens, being equal in its eyes, shall be equally eligible to all high offices, public positions and employments, according to their ability, and without other distinction than that of their virtues and talents.

Whoever holds the real power, whoever actually, at any moment, exercises the general will, ultimate power is invested in the people, the citizens. The old order, the division of society into estates (nobles, churchmen, others) is abolished, along with the inherent rights of those groups. Everyone is plunged equally into a new order, the French Republic. But this is not anarchy, it is a structured society. It is a fully functioning state, on a war footing. People have to be educated for the new state. Enter education. Enter, remarkably, mathematics (and so, for us, the history of mathematics).[1]

1.2 Some mathematicians

Gaspard Monge was born in 1746 in the old French town of Beaune, and as a young man in 1764 made a plan of his native town which he took to the École

---

[1] By far the richest source on mathematics in France in the immediate post-revolutionary period, full of detail about the personal relationships and institutional settings of a wide range of mathematicians and physicists, and one representing years of research in the archives, is Grattan-Guinness [97].

Royale du Génie at Mézières.[2] This is the Royal School of Engineers, and was a prestigious military academy, offering scientific and practical training. The plan so impressed the people there that Monge was enrolled as a draughtsman. In career terms, this was not much of a success, but within a year Monge was given an exercise in defilading – the design of a fort that will protect the lines from an enemy's guns firing frontally (enfilading fire) and the interior from plunging or reverse fire. Monge solved the problem by a new method, one much more rapid than the established one. He was thereafter rapidly promoted, and became the Royal Professor of Mathematics and Physics in 1775. During those years he worked on improving his new method, which had been made a state secret, and his method became known eventually as descriptive geometry. He also began work as a research mathematician: there is a method in the theory of first-order partial differential equations that bears his name today, and he worked on other topics besides. In 1784 he left Mézières and moved to a position where he oversaw the education of naval cadets, and became very interested in the experimental study of chemistry and physics. He was therefore able to speak from a position of real knowledge on questions of military technology.

So, in career terms, where was Monge in 1789? A proven success in the education of military engineers and naval cadets, well in with the educational establishment, and an expert in chemistry and physics as well as mathematics. He had a lifelong commitment to the useful aspects of mathematics, not exclusively, but predominantly. Among his rivals, contemporaries and colleagues are three of the most famous mathematicians: Pierre Simon Laplace, the greatest applied mathematician of his time, the most profound student of celestial mechanics since Newton and active behind the scenes as well as in front; Joseph Louis Lagrange, an Italian from Turin, perhaps the leading pure mathematician of his day, a remarkable algebraist but not a strong personal presence; and Adrien Marie Legendre, the leading analyst, not quite in the class of the previous two, but very important nonetheless, who also worked on number theory and geometry. Two others are not so well known: Sylvestre François Lacroix, the prolific author of state-of-the-art calculus textbooks; and the Marquis de Condorcet, for a time the expert on the burgeoning theory of probability. These six men were forced into the torrents that constituted French public life for the next 26 years. How did they fare?

Laplace did well.[3] The consummate academic politician, even briefly brought into Napoleon's cabinet as Minister of the Interior, he was 66 in 1815. In the first years of the new Royalist regime, Louis XVIII made him a marquis, and he was elected to the Académie Française, the highest intellectual honour in France.

---

[2] See the biographies of him by Dupin [56], who knew and admired him, and Taton [235].

[3] See the biography by Gillispie [91].

Lagrange also prospered under Napoleon. Rather than mention his honours, it is easier to say that when he died in 1813 at the age of 77 he was buried in the Panthéon and the funeral oration was given by Laplace. Legendre did well, serving like the others on various committees, and succeeding Lagrange at the important Bureau des Longitudes in 1813. Lacroix, the youngest of them all, also managed his way to safety, and by 1815, when he turned 50, he was at the Collège de France, the plum among teaching positions, and by no means his only job. The Marquis de Condorcet, however, was drawn into politics and backed the wrong side in 1793. When a warrant was issued for his arrest he went into hiding, and wrote his most famous book, a work of philosophy called *Esquisse d'un tableau des progrès de l'esprit humain*. But when he attempted to move to another safe house something went wrong and he was arrested and died in prison, whether as a result of a suicide attempt or not has never been conclusively established.

1.2.1 Monge

Figure 1.1 Monge

And Monge? After the fall of the monarchy, he became Minister for the Navy, a post he held for eight months until he resigned in April 1793 under criticism for being too moderate. In the more and more heavily politicised climate he preferred to be an ardent patriot than associated with any particular faction, and in that capacity he served on numerous commissions. In March 1794 he was appointed to the commission responsible for creating a new École Centrale des Travaux Publics (Central School for Public Works). With his background it was natural he should throw himself energetically into this task, and he did.

He became the instructor for descriptive geometry there in November of that year, teaching so-called "revolutionary courses". Politics intervened to create a delay, but by June 1795 the school, soon to be called the École Polytechnique, was running smoothly, and Monge was teaching differential geometry. He was also involved in the short-lived École Normale de l'An III (the revolution at its height restarted the calendar and renamed all the months). This was intended to be the main college for the training of teachers. It enrolled 1,200 students, and Monge taught there with Lacroix, but it soon folded and was restarted in 1810 as the École Normale Supérieure. Monge was also active in re-establishing the scientific societies that the revolution had regarded as hopelessly corrupt, such as the Académie des Sciences and the Institut de France (of which he became the first president).

Just when his career might have been settling down (Monge turned 50 in 1796) he was appointed to the commission that selected artworks from Italy to be brought to France (the war booty that stocks the Louvre to this day). This brought him into contact with Napoleon, and the two got on. On his return to Paris, Monge was appointed director of the École Polytechnique, but he was soon drawn into the preparations for the expedition to Egypt. He arrived in Cairo in July 1798 and ran the scientific side of the expedition, and returned to France only in October 1799. He briefly became director of the École Polytechnique again, but only for two months because Napoleon, after staging his successful coup d'état, now appointed him a senator for life. Monge then led the life of the successful politician, doing some teaching at the École Polytechnique in differential geometry, publishing his descriptive geometry and editing the monumental account of the expedition to Egypt. Arthritis brought his teaching career to an end in 1809, and he appointed Arago to succeed him. During this period, to be precise in 1804, the École Polytechnique became a military school, and the students military cadets. The success of Napoleon's armies, and a residual affection for the French among Americans, who remembered the French support for their revolution in 1776, made the military École Polytechnique the model for West Point, America's premier military academy.

The defeat of Napoleon in 1812 seems to have gravely undermined Monge's health, but in 1815 he rallied to Napoleon's side one last time in the regime of the Hundred Days, which ended with Napoleon's final exile to Elba. Politics is a cruel business, and now, in old age, Monge was only too prominently associated with the defeated dictator, and he was harassed politically by the new regime, who expelled him from the Institut de France. When he died in July 1818, the newly re-established Bourbon monarchy even opposed his funeral being made a major occasion, but many current and former students attended and paid tribute, alongside former officers of all the public services, and, among the mathematicians, Laplace and Legendre. Appearances can be deceptive: Laplace had not originally been a supporter of the École Polytechnique, only coming round to it when he was briefly Minister of the Interior, and had kept out of the way during most of the French Revolution. The first director of the École Polytechnique spoke of Laplace's "hateful jealousy of citizen Monge". He was instrumental in replacing Monge with Arago in 1815, and soon downgraded the status of descriptive geometry in favour of more analysis.[4]

Balzac, in his novel *Eugénie Grandet*, gives a vivid impression of the high-wire nature of Parisian life in the period when he makes one of his characters say: "Children, as long as a man is a minister, adore him; if he falls, help to drag him to the refuse-dump. In power, he is a kind of God; out of office, he is below Marat in his sewer, because he is alive and Marat is dead. Life is a series of combinations, and you have to study and adapt to them if you are to succeed in maintaining a good position."[5] Marat had been one of the leaders of the revolution. On his death in 1793 – he was murdered in his bath – he was buried with great pomp in the Panthéon, but in 1795, with the change of regime, he was dug up and reburied elsewhere, while his heart was paraded through the streets of Paris, insulted, and thrown into a sewer in rue Montmartre.

1.3 Descriptive geometry

Monge's descriptive geometry is the one accomplishment of his we shall be most concerned with. The first problem he set himself was to depict in some way the position of a point in space, and then to depict lines and curves and therefore the intersections of two surfaces in space. The aim is to do so by a graphical method, so that actual lengths can be read off very simply. If you want to get a sense of this, try to imagine how you would tell a non-mathematical friend how to cut an octahedron out of a piece of foam rubber, or how to cut a cube by a plane at right angles to a body diagonal and one-third, say, of the way up.

Monge considered two planes at right angles to one another, which we can take to be a horizontal plane and a vertical plane and call them the (x, y)-plane and the (y, z)-plane respectively. A point in space can be projected vertically onto the horizontal plane and horizontally onto the vertical plane. A straight line can likewise be projected in these ways. Its images are two straight lines. Now suppose A and B are two points in space and you want to know the length of the segment AB. For example, it might be the edge of two parts of a roof.

---

[4] See Belhoste [11, pp. 89, 91].
[5] English translation [8, p. 111].

Notice first of all that the line is known once its intersections with the two (distinct) planes is known; Monge called these points the traces of the line. (See Figure 1.2.)

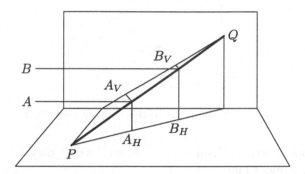

Figure 1.2 A line segment in space

In Figure 1.2, the traces of the (thick) line are called P and Q. The projections of the points A and B on the two planes are shown as A_H, A_V, B_H and B_V; see how everything lines up.

Monge introduced the trick of folding out the two planes to get the next figure (see Figure 1.3). The difference in the height of A and B is exactly the difference in the height of A_V and B_V. Draw that in as CB_V. Draw the right-angled triangle $A_H B_H C'$, where $B_H C'$ is equal to CB_V, then $A_H C'$ is the length you want.

The method does not work for all (or even many) surfaces, because one loses sight of what point in one plane corresponds to what point in the other. But some surfaces can be dealt with: for example, a plane in space is determined by the lines of intersection it has with the horizontal and vertical planes (its "traces"). Spheres and cylinders likewise have traces that are easy to handle, and because the outline of these surfaces is lines and circles, their shadows are also easy to find. Whence the standard question at the École Polytechnique until the First World War: draw a specified collection of planes and cylindrical (but perhaps oblique) columns and their shadows under certain lighting.

It is easy to see that students at the École Polytechnique could learn how to do such non-trivial things in a term, given an enthusiasm for the task. The historian's question is: why would people have cared in and around 1800? One answer is that the uses of descriptive geometry in engineering were many and varied. It was an essential tool for anyone seeking a career in engineering, and who presumably had already worked hard. Another answer is that Monge had a particular gift: the remarkable ability not only to work in three dimensions in his mind's eye, but to convey such pictures to his students. This was a

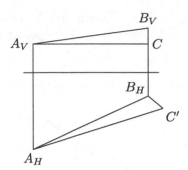

Figure 1.3 Lengths in descriptive geometry

tremendous motivation for them to do geometry, which they often mentioned in their writings about him.

As an example of this, consider the theorem called the radical axis theorem. Suppose you have two circles, with equations $S = 0$ and $S' = 0$. Then the circles with equations $\lambda S + \lambda' S' = 0$ form what is called a coaxial family, and the one with $\lambda + \lambda' = 0$ is the only line in the family. It is called the radical axis of the family. Now suppose that you have three circles, and so three pairs of circles and so three coaxial families and three radical axes. The radical axis theorem says that these three radical axes meet in a point. Here is the insight that Monge's proof rests upon. We can take two of the circles and imagine spheres passing through them. Suppose these spheres intersect. Then they do so in a plane, and this plane meets the plane of the original two circles in a line: the radical axis of the circles. Now imagine spheres through each of the circles, and choose them so that each pair of spheres meets in a plane. We have three radical axes, where the plane of the three circles meets each of these new planes. But three planes meet in a point, which must be common to the three radical axes, so they all meet in a point. Think about it!

2

Poncelet (and Pole and Polar)

Figure 2.1 Poncelet

2.1 Poncelet reminisces

Listen to his words:

Following the example of a celebrated contemporary novelist, whose statue stands at the entrance to the room where our academicians hold their private meetings to glorify, without doubt, a political and religious system from the day before yesterday and which is still fashionable today, I could have entitled this work, which is purely mathematical, *Memoirs from beyond the tomb*. It is, in fact, the fruit of the meditations of a young lieutenant of the engineers, left for dead on the fatal battlefield of Krasnoy, not far from Smolensk, and for a long time strewn with the bodies of the French army. There, in that terrible retreat from Moscow, seven thousand Frenchmen, exhausted by hunger, cold and fatigue, under the orders of the unfortunate Marshal Ney, came, deprived of all artillery, on the 18th of November 1812, the an-

J. Gray, *Worlds Out of Nothing*,
Springer Undergraduate Mathematics Series,
DOI 10.1007/978-0-85729-060-1_2, © Springer-Verlag London Limited 2011

niversary of the Russian Saint Michael, to fight a furious, bloody and final combat with twenty-five thousand soldiers, fresh and equipped with forty cannons of Field-Marshal Prince Miliradowitch, who himself would soon become the victim of a military conspiracy hatched in the bosom of the modern capital of the Muscovite Tsars. But the adoption of such an ostentatious title, however justifiable it might seem, would seem with good reason to be a ridiculous plagiarism, an overweening imitation with perhaps a permitted licence, of the avowed leader of the romantic novel in our France, at a time of moral perturbation as much political as literary. A similar title, besides, would suggest of this modest book neither the serious and reserved habits of the author, still less the character, the aptitudes, the tastes which presume a sincere love of the truths of geometry, whose profound culture calls for a spirit disengaged from all foreign passion and, one might say, of any earthly interest.

Now, such is precisely, and in some way inevitably, the moral and mathematical position of the author of this work in the distant prisons of Russia. Much later, when he appeared to neglect the study of this geometry in favour of teaching the mechanical and industrial sciences, he had in reality no other purpose but to make it useful to the working class and the youth of our schools; he wished to inspire them with a love of the eternal truths of science, a hatred of the intrigue and the sophisticated subtleties of a greedy charlatanism, which signals an epoch where, among the conquests of the modern spirit, one deplores with sorrow the aberrations, the passion for money which dishonours our character, our customs, and even our national literature. Finally, if in the honourable steps of Vauban and Belidor, of Bézout, of Borda and Coulomb, of Daniel Bernoulli, of Euler and so many other illustrious benefactors of humanity, he has attempted to make useful to the class of artists or engineers, in writing for the general public in such a way as to avoid the reproaches too often and rightly addressed to the members of the profession.

The novelist referred to is François-Auguste-René, Vicompte de Chateaubriand (1768–1848), one of France's first Romantic authors. Chateaubriand had initially refused to side with the Royalists in the French Revolution, but eventually did so (after the flight of Louis XVI in June 1791) and was wounded in action. He left for England in May 1793, where he wrote his *Essais sur les révolutions* (*Essays on revolutions*, 1797) [34]. In 1800 he returned to Paris, and in 1802 to the traditional Christianity he had once disclaimed. *Encyclopaedia Britannica* [63] comments that "His apologetic treatise extolling Christianity, *Le génie du christianisme* (*The genius of Christianity*, 1802) [35], won favour both with the Royalists and with Napoleon Bonaparte, who was just then concluding a concordat with the papacy and restoring Roman Catholicism as the state religion in France." Napoleon made him first secretary to the embassy to Rome on the strength of it, but Chateaubriand resigned in 1804 in protest at Napoleon's execution of a supposed conspirator, and threw himself into the literary life with many love affairs. The Bourbon monarchy that reigned from 1814 to 1830 favoured him with many appointments, but after then he lived a private life. His *Mémoires d'outre-tombe* (*Memoirs from beyond the tomb*, 1849–50) [36], was written for posthumous publication. It is an account of his emotional life, mixed up with contemporary French history, his unstinting appreciation of women and his sensitivity to nature; it succeeds (or succeeded) in evoking vividly the spirit of the Romantic epoch, and became his best-remembered work.

The long quotation is from Jean-Victor Poncelet in 1862, in the preface of a book (*Applications d'analyse et de géométrie* [205]), which was an annotated set of the notes he made while a prisoner in Russia in 1813 and 1814. He fought in the terrible battle at Krasnoy but was taken prisoner, and survived the years that followed only by luck. In the winter of 1812 it became so cold that even the mercury in the thermometers froze (which occurs at a temperature of $-39°C$). He managed to get to the hospital at Saratov, where he remained a prisoner until the defeat of Napoleon and the Treaty of Paris was signed on 30 May 1814; his journey home took two and a half months, and he arrived in Metz on 7 September. In prison there was nothing to restore him to health but the April sun, and there, to distract his spirits, he resumed his study of mathematics, even though there was not even the distant echo of the profound analytical works of Euler, Bernoulli, Huygens, Newton and d'Alembert, not to mention the more recent and no less admirable work of Lagrange, Legendre, Laplace, Monge and their disciples.[1]

He recalled that he had graduated from the École Polytechnique in November 1810, and left the Applied Engineering School in Metz in March 1812 to work on the fortifications of the Dutch island of Walcheren. At the École Polytechnique he had acquired a taste for the work of Monge, Carnot and Brianchon, but, completely cut off as he was in Saratov, he knew nothing of their recent work published before his return to France in 1814. This is why he occupied himself summarising all he knew of the mathematical sciences in notebooks that he then distributed to his fellow prisoners who wanted to finish an education disrupted by the incessant military campaigns.[2]

---

[1] For a biography of Poncelet, see Tribout [239], who points out that much original documentation about Poncelet was destroyed in the First World War.

[2] Carnot's study of the properties of pairs of intersecting curves had been published in 1806 as the third of his books aimed at revising and extending the science of geometry. Lazare Carnot (1753–1823) was another mathematician and scientist ac-

Poncelet was struck by the observation that the elementary parts of the differential and integral calculus and of algebra had left the most vivid impression on his mind, and he could rediscover the basic results about areas and volumes, even though he had forgotten them. It seemed that those ideas would not be forgotten at any stage in life. But the complicated and laborious methods, whatever their interest or scientific merit, the abstract and spiny proofs which have been introduced into mathematics, and which would never have been recommended by Lagrange, Laplace, or Monge in their admirable lectures in the early École Polytechnique and École Normale – they vanished entirely. As for mechanics, Poncelet confessed that apart from purely geometric theorems on the composition of forces he remembered nothing. Galileo's laws left absolutely no trace in his mind. It was in vain that he tried to write the differential equations of motion with respect to the coordinate axes, which is why, when in charge of creating a course in mechanics at the Applied Engineering School in Metz he became an innovator by conviction and a reformer by necessity.

Poncelet's purpose in publishing these notes in 1862, the 50th anniversary of his capture, was not just to cheat death by writing a book that would survive. Poncelet, like so many old men, was fighting his old battles one more time. These notebooks were the proof that he had had the priority for several discoveries over others who had not been deprived of their liberty in the service of France. He had argued his case before, when these theorems had been published; with this book he would argue them from beyond the grave.

Poncelet had published several theorems in the years 1817 to 1832, and one remarkable book. Not all of these results were contested by Gergonne, one of his rivals, and most were published in Gergonne's *Annales de Mathématiques Pures et Appliquées*, the only journal at the time entirely devoted to mathematics. (The journals of the learned societies at the time covered all of science, loosely divided into topics in some journals but not others.) The first of these results goes to this day by the name of Poncelet's porism; what it is will be made clear in due course. Another, published in 1818 (no. 2 in Poncelet's list[3]), established that the number of tangents common to two curves of degrees m and k is in general, and at most, $mk(m-1)(k-1)$.

The third was a novel solution to a problem first raised by the ancient Greek geometer Apollonius: find a circle tangent to three given circles. Yet another was a long article, published in 1820, covering some of the same material later

---

tively involved in politics. His defence of Paris, in 1794, when he was in charge of the revolutionary army, had earned him the popular title of "Organiser of the Victory". For a thorough account of his life and work, see Jean and Nicole Dhombres, *Lazare Carnot* [52].

[3] This list is Poncelet's report on his own work, the *Notice analytique sur les travaux de M. Poncelet*, Paris 1834 [204].

treated in the book of 1822 on conic sections and quadric surfaces. We shall see that this book is his most remarkable and lasting claim to originality. In 1829 he wrote an important paper extending the theory of pole and polar to curves of degree greater than 2.

With all this in mind, let's go back over the long passage above that opened this chapter. Unexpected, isn't it, in a mathematical book? Personally, I love these long French sentences. You can easily imagine the old man grabbing you by the arm, telling you of the lofty mission of his life that began so near to death so many years ago, on the field that marked the final defeat of the country's greatest military leader. Don't you feel, just as certainly, the literary power of this? Who is this strange, passionate, eccentric old man, straight out of fiction?

Listen to the text:

> Memoirs from beyond the tomb . . . the meditations of a young man left for dead . . . seven thousand exhausted men, ill equipped, facing twenty-five thousand well-equipped soldiers. . . . And yet, you know, the Russian Field-Marshal dies, and I, young Poncelet survived. . . . And I (if you can believe it) am "the author of a modest book, and a person of serious and reserved habits".

Did Poncelet himself really believe it? Or is that his deluded self-image? Or is it a permissible view of himself, and yet false? People are different people at different times. What did it mean to escape death by the merest chance, when so many you know died, and then to endure two years in limbo, from the age of 24 to 26, not knowing for most of that time what your future would be? Denied, in any case, the pleasures of the prime of one's life. To find survival in mathematics, and perhaps to dedicate oneself to it on one's return.

> A love of the eternal truths of science, a hatred of the intrigue and the sophisticated subtleties of a greedy charlatanism.

Some things matter; some things don't. The eternal truths of science . . . this modest book. False modesty? An honest recognition that, amid all the desolation of the Russian prison, there came no great mark of redemption? Or in the end did Poncelet make eternal mathematics?

Some things matter; some things don't: to have seen mercury freeze, and to survive even that. What does that count for, when one is old and thinks over one's life?

Poncelet's antithesis was between the eternal truths of geometry and the grubby world of intrigue and charlatanism. Real enough, perhaps. But we have to live in it. Did Poncelet pass beyond any earthly interest? His mathematics, shortly to be described, was not to everyone's taste. The mathematician may like it, the scientist may not. But Poncelet eschewed the world of recondite learning in favour of popular instruction. Even when he moved over to the study of machines, this was, he now said, "to make [geometry] useful to the working class and the youth of our schools; and to inspire them with a love of the eternal truths of science" [202]. They go together, it seems, utility and eternal truth.

And who will be helped? The working class, and the young in general, of France. They will be rescued, Poncelet evidently hoped, from "the passion for money which dishonours our character, our customs, and even our national literature". "Ours" here means that of France. The whole peroration at the start is about the French defeat. It is Frenchmen who die, bravely we must presume, and glory with it. French glory. It is France that stands in a moral swamp in 1862. It is French dignity that Poncelet has spent his life, in his way, trying to restore. There are some typically French notes struck, here and later. The pantheon of great names: Lagrange, Legendre, Laplace and Monge, the great days of the École Polytechnique and the École Normale; Metz, where the Applied Engineering School was.

Finally, the mathematics. The honest recognition that only the elementary bits survived, and all that subtle stuff went clean out of his head. Poncelet, at least, was not one of those mathematicians who only wake up when the rest of us find it too difficult. He was one of those mathematicians who remade things, made them new, made them according to his rules. The two years as a prisoner in Russia made him an original, someone who did things his way. I think that fits with the rest of the person he made himself become between 1814 and 1867, when he died. But he also had a dislike of clever, tricky mathematics. He won't teach that kind of thing, if he can avoid it. He wants to be understood by the common man, if you like. Is this part of his dislike of "sophisticated subtleties"?

2.2 Poncelet's mathematics

What, then, was Poncelet's original mathematics? I shan't take us into his study of machines. Poncelet's porism will be discussed briefly below; a porism – the term is Greek – is a striking thing: a problem that either cannot be solved, or has infinitely many solutions. The paper of 1818 (no. 2 in Poncelet's list) is on the number of tangents common to two curves, and the study of curves other than conics was very little understood at that time. Since conic sections were Greek in origin, it's clear that any exploration of this topic was like entering a new continent. Then we go back to a problem posed by the ancient Greek geometer Apollonius: find a circle tangent to three given circles. This won't detain us. Suffice it to say that the classical solutions are long and difficult.

I have already said that the long article, published in 1820, and the book of 1822 on conic sections and quadric surfaces are his most remarkable. The paper of 1829 on the theory of pole and polars for curves of degree greater than 2 is, it turns out, another major paper opening up the study of curves other than conics.

Let us open the book of 1822 [202]. It too has an account of the circumstances in which he was led to his discovery of projective geometry, during the months following his capture by enemy troops during Napoleon's ill-fated invasion of Russia.

> This book is the result of researches which I undertook in the spring of 1813 in the prisons of Russia: deprived of every kind of book and help, and the proper facilities, above all distracted by the misfortunes of my country, I was unable to give it all the perfection desirable. However, I had at the time found the fundamental theorems in my work: that is to say the principles of central projection of figures in general and conic sections in particular, the principles of secants and tangents common to those curves, those of polygons which are circumscribed or inscribed to them.

And he gave this account of the aim of his work:

> The point of this book, voluminous as it may appear, is less to increase the number of properties [of figures] than to indicate the route by which they are found. In a word, I have sought above all to perfect the method of proof and discovery in elementary geometry.

In his *Treatise on the projective properties of figures* [202] (see the extract below), Poncelet contrasted what he called the geometry of particulars with analytic geometry, which he also called algebraic analysis. He found algebraic analysis to be well developed, but the geometry of particulars ("individual curves and surfaces") to be lacking in some respects. He regretted that the arguments used in synthetic geometry lacked generality. By contrast with algebra, which can handle negatives and even imaginary magnitudes, synthetic geometry "is more timid or more severe". For instance, if three points A, B, C lie on a line and C is between A and B then we write $AC + CB = AB$, but if it lies outside it we write $AC - CB = AB$ or even $CB - CA = AB$. So if in the course of a proof a perpendicular DC from D to AB falls inside or outside AB the whole argument must be carried on in two variant forms.

Poncelet gave examples of what he thought was an acceptable general rule for applying the same argument to different figures. If one figure could be obtained from another by changing it by insensible degrees, say by an arbitrary continuous movement, then he considered it obvious that some properties of the first figure would persist through these changes to the final figure, provided, of course, that one took note of the fact that some quantities (which could be specified in advance) change in size, vanish, or become negative. He called this proposal the principle or law of continuity. It is, to be frank, somewhat vague. It was never made rigorous by Poncelet, and it was strongly attacked as soon as it was published, as we shall see. He admitted it led to paradoxes; for example, where are the common points – which he called ideal points – to two seemingly non-intersecting circles? According to his principle of continuity, these points should be obtained by a continuous movement of a pair of intersecting circles. But, he said, the paradoxes do not go away if you use algebraic, rather than geometric, analysis. So the problem was to explain them directly and not to let them halt progress.

It is possible to feel here that Poncelet has worked something long and difficult out for himself, building on the insights obtained in Saratov. He is in possession of not just a theorem, but a theory – a whole way of thinking about projective geometry. Naturally he believes that it speaks clearly, directly and accessibly to everyone. It is not the spiny stuff that does not stay in the mind, but clear basic principles. Nothing sophisticated, only limpid truth. If we find it obscure, well, give it time seems to have been his view.

The terms Poncelet applied to geometry ("analytic", "synthetic", "algebraic") are not entirely easy to use; they shift their meanings a little from user to user and date to date. The principal distinction being made is between analytic or algebraic geometry, which can even be called coordinate geometry, on the one hand, and synthetic geometry on the other. Synthetic geometry then means geometry (loosely) in the style of Euclid's *Elements* in which what is discussed are curves, lines, angles and areas, and algebra is avoided.

2.3 Poncelet, *Traité des propriétés projectives des figures*, 1822 [202, pp. xix–xxvii]

In ordinary geometry, which one often calls synthetic, the principles are quite otherwise, the development is more timid or more severe. The figure is described, one never loses sight of it, one always reasons with quantities and forms that are real and existing, and one never draws consequences which cannot be depicted in the imagination or before one's eyes by sensible objects. One stops when those objects cease to have a positive, absolute existence, a physical existence. Rigour is even pushed to the point of not admitting the consequences of an argument, established for a certain general disposition of the objects of a figure, for another equally general disposition of those objects which has every possible analogy with the first. In a word, in this restrained geometry one is forced to reproduce the entire series of primitive arguments from the moment where a line and a point have passed from the right to the left of one another, etc.

Now here precisely is in fact the weakness; here is what so strongly puts it below the new geometry, especially analytic geometry. If it was possible to apply implicit reasoning having abstracted from the figure, if only it was possible to apply the consequences of that kind of reasoning, this state of things would not exist, and ordinary geometry, without needing to employ the calculus and the signs of algebra, would rise to become in all respects the rival of analytic geometry, even if, as we have said already, it was not possible to conserve the explicit form of the reasoning.

Let us consider an arbitrary figure in a general position and indeterminate in some way, taken from all those that one can consider without breaking the laws, the conditions, the relationships which exist between the diverse parts of the system. Let us suppose, having been given this, that one finds one or more relations or properties, be they metric or descriptive, belong to the figure by drawing on ordinary explicit reasoning, that is to say by the development of an argument that in certain cases is the only one regards as rigorous. Is it not evident that if, keeping the same given things, one can vary the primitive figure by insensible degrees by imposing on certain parts of the figure a continuous but otherwise arbitrary movement, is it not evident that the properties and relations found for the first system, remain applicable to successive states of the system, provided always that one has regard for certain particular modifications that may intervene, as when certain quantities vanish or change their sense or sign, etc., modifications which it will always be easy to recognise a priori and by infallible rules?
...

Now this principle, regarded as an axiom by the wisest mathematicians, one can call the principle or law of continuity for mathematical relationships involving abstract and depicted magnitudes.

In the last analysis the principle of continuity has been admitted in its full extent and without any restriction by different geometers, who have employed it either overtly or tacitly, because without it they would be plunged into all the metaphysical considerations of imaginaries which have always been driven back from the narrow sanctuary of rational geometry. Its explicit use in this science is almost always limited to real states of a system which is transformed by insensible

degrees. And even there it gives rise to the infinitely little and the infinitely great which geometers still seek, in our day, to banish from the
domain of the exact sciences.

[...]

However, it will still not be difficult to establish this principle in
an entirely direct and rigorous manner, with the aid of a calculus just
like algebra the certainty of which is not the least to be doubted in our
time, thanks to two centuries of efforts and success!

In any case will it be necessary, and will one not immediately admit
the principle of continuity in its full extent into rational geometry, as
one does at once in algebraic calculus and then in the application of
calculus to geometry, if it is not a means of proof but rather as a means
of discovery or invention? Is it not at least as necessary to point out the
resources employed at various times by men of genius for discovering
the truth, as the feeble efforts they have then been obliged to use to
prove them according to intellectual taste, either timid or less capable
of bringing them home?

Finally, what harm can result, above all if one is restrained in one's
conclusions, if one never uses half-truths, if one never admits analogy
or induction, which are often deceptive, and which it is not necessary
to confound with the principle of continuity? In fact, analogy and induction conclude from the particular to the general, from a series of
isolated facts not necessarily related, in a word discontinuous, to a general and constant fact. The law of continuity, on the contrary, starts
from a general state and some sort of indeterminacy of the system (that
is to say that the conditions which govern it are never replaced by still
more general ones) and they remain in a series of similar states going
from one to the other by insensible gradations. One insists, besides,
that the objects to which it is applied are by their nature continuous
or submit to laws which can be regarded as such. Certain objects can
even change their position by a series of variations undergone in the
system, others can move away to infinity or approach to insensible distances, etc.; the general relations survive all the modifications without
ceasing to apply to the system.

The only difficulty consists, as we have seen, in understanding fully
what one wants to convey with the word general or indeterminate or
particular state of a system. Now in each case the distinction is easy.
For example, a line which meets another in a plane, is in a general
state by comparison with the case where it becomes perpendicular or
parallel to that line. Similarly a line (straight or curved) which meets
another in a plane, is in a general or indeterminate state with regard to

that other and the same thing takes place even when it ceases to meet it, provided that the two states do not suppose any particular relation of size or position between these lines. The contrary will evidently hold when they become tangents, or asymptotic, or parallel etc.; they will then be in a particular state with regard to the primitive state.

2.3.1 Commentary

Poncelet's account of these mysterious points of intersection was obscure. It was not equivalent to what a geometer relying on algebraic methods would say – namely, that such circles meet in complex points – we note only that in this respect Poncelet's presentation of his ideas, however "geometric", was not likely to replace the algebraic formulations of the previous hundred years. Nor indeed did it. One interpretation of what it means is given below.

So far in this account, Poncelet could appear as something of a crank, an oddball who did well in the French academic system of the early 19th century. It is time to begin to explain why he merits the attention of historians of mathematics, and the attention of mathematicians in his day and subsequently.

Let us start with the idea of the polar line of a point with respect to a conic. This is a topic worth following in more detail, and Poncelet made considerable use of it. It is not original to him, indeed the idea goes back to Apollonius, and was taken up in an original way by Brianchon before Poncelet, as he happily admitted. If a line ℓ meets a conic at Q_1 and Q_2 then the tangents to the conic at Q_1 and Q_2 meet at point P, say, called the pole of ℓ, which lies, of course, outside the conic. He deduced that to each line ℓ one can associate a point P, and the converse is also true; given a point P outside the conic, one can draw through P two tangents to the conic. Label the points where they touch the conic Q_1 and Q_2; then to P one can associate the line Q_1Q_2, the polar of P.

What happens if you start with a line that does not meet the conic? Well, the line is made up of points. Let P be one of them, and construct the polar line Q_1Q_2. Let P' be another, and construct its polar line $Q_1'Q_2'$. Let P'' be a third, and construct the polar line $Q_1''Q_2''$. Then rather wonderfully, it turns out that these three polar lines meet in a point, let us call it Q (from which it follows that the polar line of each point on the line ℓ passes through the point Q). So we have a natural candidate for the pole of the line: the point Q.

This is the occasion to state my policy on old proofs in this book on the history of mathematics. In principle, every fragment we have of the past is part of the evidence. In practice, for the 19th century, there's too much evidence, and a selection has to be made. Sometimes I shall select a proof as evidence.

I shall then make it clear what I take it to be evidence of. Sometimes, and this business of poles and polars is a case in point, I just want you to know what the words mean and that the statements are true. Generally then I shall give a modern proof, which may be quite different from the ones used at the time. Such proofs are presented in order to make the story easier to understand. In this case, the proof displays a profound insight into the geometry, which is part of our historical case for the importance of Poncelet.

2.4 Pole, polar and duality

Result 1

With respect to the unit circle, $x^2 + y^2 = 1$, the point $P = (u, v)$, the pole, has the polar line $\ell : ux + vy = 1$.

Suppose that the point lies outside the circle, and the two tangents, call them t and t', from it to the circle meet the circle at the points $T = (a, b)$ and $T' = (a', b')$ respectively. The equations of these tangents are $ax + by = 1$ and $a'x + b'y = 1$. Now, (u, v) lies on t and t' if and only if $au + bv = 1$ and $a'u + b'v = 1$, and (u, v) is the unique point satisfying those two equations. However, $(a, b), (a', b')$ lie on the line $px + qy = 1$ if and only if (p, q) satisfies $ap + bq = 1$ and $a'p + b'q = 1$, so, by uniqueness, $(p, q) = (u, v)$ and the line through T and T' has the equation $ux + vy = 1$. This line is called the polar line (or polar) of the point (u, v), and that point is called the pole of the line.

Now consider the line $xu + yv = 1$. A typical point on it has coordinates $\left(t, \frac{1-tu}{v} \right)$.

The polar line of this point is $xt + y \left(\frac{1-tu}{v} \right) = 1$.

Consider the points for which $t = 0$ and $t = 1$: $\left(0, \frac{1}{v} \right)$ and $\left(1, \frac{1-u}{v} \right)$ respectively. The corresponding polar lines are $y \left(\frac{1}{v} \right) = 1$ and $x + y \left(\frac{1-u}{v} \right) = 1$ respectively, which clearly meet at (u, v) – solve for y, to obtain $y = v$, and substitute that in the equation for x.

This implies that the polar lines of points on the line $xu + yv = 1$ all pass through the point (u, v), as you can check directly.

Result 2

The polar lines of points on the line $xu + yv = 1$ all pass through the point (u, v).

So (by Result 1) we can start with a point P outside the circle and obtain its polar line. We can then take points on that line (most of which, after all, lie outside the circle) and take their polar lines: they all meet at P.

This is what is called a duality: from a point obtain a line, from a line obtain a point. Do it twice and the original point is returned.

We extend this to points inside the circle entirely formally, so to each point we have a line, and to each line we have a point, and duality applies. We can even start with the line, pass to its pole, and obtain the polar line of that point: it will be the line we began with.

Some consequences of, and observations about, this result: for example, if the point (u, v) lies on the circle, its polar line is the tangent to the circle at that point. You can see by drawing a diagram that as the point P gets closer to the circle, the corresponding points Q and Q', also get closer to P, and the line joining them gets closer to being a tangent.

On the other hand, not every line has an equation of the form $xu + yv = 1$. Those that don't are those of the form $xu + yv = 0$, the lines through the origin. And indeed, what is the polar line of the origin? The point $(0, 0)$ would seem to have the line $0x + 0y = 1$, which is $0 = 1$, as its polar, but that's nonsense. So there's a problem with such lines.

Can you prove both these results without algebra? Yes. Much of it goes back to Apollonius, and everything that was needed was in place in the 17th century. Poncelet knew that much of it was in a fine book by La Hire, his *Sectiones conicae* of 1685 [144], and Brianchon had published some ideas about poles and polars before his journey to Russia. He knew of Desargues, however, only from secondary accounts, because at that time no copy of Desargues' major work had apparently survived, and he praised him handsomely.[4]

Does this just work for circles? Indeed not. It certainly works for any conic section. You can see that from the algebra. It would become more complicated, but the main result, the passage from pole to polar and back, all of duality, would survive. If you don't believe that, you have three options:

1. slog through the algebra for a general conic;

2. find algebraically a transformation that maps the circle to the conic, and use it to find the formulae for pole and polar;

3. work out a way of seeing it. That is for the next chapter.

One final remark, why the names pole and polar? They are faintly reminiscent of geography – and so they should be. Their origins are in spherical trigonometry,

---

[4] Desargues' main work on geometry, his *Brouillon Projet* was published in an edition of 50 copies in 1639 which were as good as lost by Poncelet's time. Desargues' theorem was described in a work by Abraham Bosse in 1648, Pascal and La Hire also mentioned Desargues by name in their work on what we would call projective geometry, and in these tenuous ways his reputation was kept alive. But it was only in the mid-19th century that Michel Chasles found a copy of the *Brouillon Projet* made by La Hire in 1679, and this was published by Poudra in his edition of Desargues' works in 1864 [51]. The only surviving original copy of the *Brouillon Projet* was found by P. Moisy and communicated to R. Taton in 1951; it forms the basis of the edition of 1951 [234]. For an English translation, see [75].

as Chemla has described [37]. To each point on a sphere there is a natural great circle that comes with it: join the point to its diametrically opposite point, there is exactly one great circle perpendicular to that diameter. In the case of the north pole, that great circle is the equator. Conversely, to every great circle there are two points naturally associated with it, and they are antipodal – at opposite ends of a diameter of the sphere. That is where the word "pole" comes from in "pole and polar". "Polar" is a natural word to choose, given that "equator" is too strong.

Theorems in Projective Geometry

3.1 The theorems of Pappus, Desargues and Pascal

The flavour of this chapter will be very different from the previous two. It is chiefly devoted to giving an account of some theorems which establish that there is a subject worthy of investigation, and which Poncelet was rediscovering. I shall state what they say, and indicate how they might be proved. Then I shall indicate a way of proving them by the tactic of establishing them in a special case (when the argument is easy) and then showing that the general case reduces to this special one. I shall prove them in the special case, and indicate how the reduction from general to special can be carried out. This method of reduction is the key idea in projective geometry, and in that way we shall begin our study of the subject. Towards the end of the section we shall work our way back to Poncelet and see what he required of projective geometry.

The theorem known as Pappus's theorem, after the Hellenistic writer Pappus, is an example of a theorem in projective geometry. Be careful, this doesn't mean there was a subject called projective geometry in Pappus's day. The subject came along later. And indeed the theorem we call Pappus's theorem is itself a reworking of what Pappus wrote, which suggests that he himself might have had something different in mind.

J. Gray, *Worlds Out of Nothing*,
Springer Undergraduate Mathematics Series,
DOI 10.1007/978-0-85729-060-1_3, © Springer-Verlag London Limited 2011

Theorem 3.1 (Pappus)

If A, B, C are three points on a line ℓ, and A', B', C' are three points on a line ℓ', and the lines AB' and $A'B$ meet at R, the lines BC' and $B'C$ meet at P, and the lines CA' and $C'A$ meet at Q, then the points P, Q and R lie on a line m. (See Figure 3.1.)

(It is not true in general that the line m passes through the intersection point of the lines ℓ and ℓ'.)

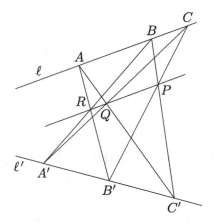

Figure 3.1 Pappus's theorem

Desargues' theorem, discovered in the first third of the 17th century by the French mathematician and architect Girard Desargues, says the following.

Theorem 3.2 (Desargues)

If ABC and $A'B'C'$ are two triangles in perspective from O (so O, A, A' lie on a line, as do O, B, B' and as do O, C, C') and if L is the point common to BC and $B'C'$, M is the point common to $C'A'$ and CA, and N is the point common to $A'B'$ and AB, then the points L, M, N lie on a line. (See Figure 3.2.)

Theorem 3.3 (Uniqueness of the fourth harmonic point)

Let A, B and C be any three distinct points on a line, and let an arbitrary point P not on this line be chosen. Draw the lines PA and PB, and draw an arbitrary line through C (other than AB) meeting AP at Q and BP at R.

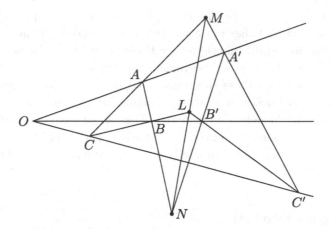

Figure 3.2 Desargues' theorem

Draw the lines BQ and AR and let them meet at S. Draw the line PS meeting AB at C'. Then the position of C' is independent of all the choices made (it depends only on A, B and C).

The name "fourth harmonic point" will be explained below.

Theorem 3.4 (Pascal)

If ABC and $A'B'C'$ are six points on a conic section, and if P is the point common to BC' and $B'C$, Q is the point common to CA' and $C'A$, and R is the point common to AB' and $A'B$, then the points P, Q, R lie on a line. (See Figure 3.3.) (Draw them in the order A, B, C, C', B', A'.)

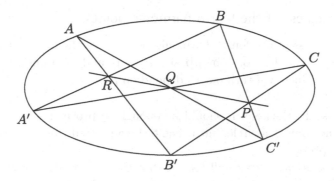

Figure 3.3 Pascal's theorem

We shall see in due course why these theorems are called projective. Just note for the moment that they make a rather unusual claim: all but one of them says that if you do certain things you get three points and (theorem) they lie on a line (and not a triangle, as you might suspect).

How might we prove these theorems? Originally they were proved as ingenious exercises in Euclidean geometry, making use of theorems about ratios (such as Ceva's and Menelaus's theorems). They are doable by coordinate geometry, but tough going, because the expressions get complicated. Another, even later, way of proceeding, but one that is less complicated, is to use vector methods, as follows.

Proof (Pappus's theorem)

Let the lines ℓ and ℓ' meet at O and take vectors in the plane based at O. Let A be represented by the vector $\mathbf{a}$ and A' by $\mathbf{a}'$. Then B is represented by $\lambda\mathbf{a}$ and C by $\mu\mathbf{a}$, for some λ and μ, with similar expressions for B' and C'. The point P is represented by a vector of the form $s\lambda\mathbf{a} + (1-s)\,\mu'\mathbf{a}'$, which says that it is on BC', and also by a vector of the form $t\mu\mathbf{a} + (1-t)\,\lambda'\mathbf{a}'$, which says that it is on $B'C$. This gives you two equations for s and t, and so the position of P is determined. Points Q and R can be found the same way, and then you have to show that P, Q and R lie on a line.

Proof (Desargues' theorem)

Desargues' theorem comes out somewhat the same way. Let the point O be the origin, so A is represented by $\mathbf{a}$ and A' by $\mathbf{a}'$, and so on. As before you get expressions for L, M and N and then you have to show that they lie on a line.

Proof (Uniqueness of the fourth harmonic point)

For the uniqueness of the fourth harmonic point, take the origin at A, so the point B is represented by $\mathbf{b}$, C by $\mu\mathbf{b}$ and P by $\mathbf{p}$, and then find Q, R and S, and finally C', which will turn out to depend on μ alone.

As for Pascal's theorem, it should be enough to prove this for the special case where the conic is a circle, but it has to be admitted that even then there isn't an easy proof.

By Poncelet's day it was well known that there were also easier proofs of these theorems. The relevant techniques form the subject of projective geometry. Let us take Desargues' theorem. As Desargues had observed when first presenting the theorem, it is easy when you allow yourself to draw the figure in three dimensions.[1] Let triangle ABC lie in one plane, and triangle $A'B'C'$ in another. Look at the lines OAA' and OBB', they define a plane, and the lines AB and $A'B'$ lie in this plane. So they meet, at N, a point common to the two planes containing the triangles. Similarly, the points M and L lie in the two planes containing the triangles. But two planes meet in a line, so the points L, M, N lie in a line.

How do we feel about this proof? We've changed the subject, of course, from two dimensions to three. We need to convince ourselves that any two-dimensional figure can be drawn in three dimensions. That's easy enough if the triangles don't cross, but what if they do? Still, this ability to see the figure and see the truth of the theorem is a very powerful guide to understanding it. It conveys what a lengthy calculation may not always manage, a sense of the inevitability of the result.

In fact, any plane figure for Desargues' theorem lying in a space of three dimensions can be turned into a three-dimensional one. This is easier to describe in words than to draw. Take the plane figure, and let O' be a point not in the plane of the figure. Draw the lines $O'B$ and $O'B'$, and a line through O crossing $O'B$ at the point D and $O'B'$ at the point D'. You now have two triangles in space, ADC and $A'D'C'$ that are in perspective from O, so there is a line through the points N' (where AD and $A'D'$ meet), L' (where DC and $D'C'$ meet) and M (where CA and $C'A'$ meet). Now you project the triangles ADC and $A'D'C'$ from O' back onto the triangles ABC and $A'B'C'$ respectively, and deduce that the line $L'MN'$ projects onto the sought-for line.

There is, however, a difficulty that we can avoid no longer, whether we wish to prove Desargues' theorem as a result in plane geometry or in the geometry of three dimensions: what if some of the lines in the figure are parallel, and some of the points simply disappear?

One answer, with much to recommend it, is that you sit down with each and every case of this, and write down what the theorem says. When you do, you find yourself writing this sort of thing: either "let lines l and m meet at the point P'", or "the lines l and m are parallel".

Another answer is to invent some points, said to be "at infinity", and to say either "let lines l and m meet at the point P'", or "the lines l and m meet at infinity". Why is this tempting? Because every one who has seen classical painting knows that parallel lines receding to the far horizon appear to meet there. In Figure 3.4 an observer with a single eye at the point O sees the intersecting lines in the vertical plane exactly line up with the parallel lines in the horizontal plane. The point X in the ver-

---

[1] In the *Three geometrical propositions of 1648*, published in *La perspective de Mr Desargues* by A. Bosse; see Field and Gray [75, p. 163].

tical plane, where the lines intersect, has no companion in the horizontal plane, but it is very tempting to say that it does – "at infinity".

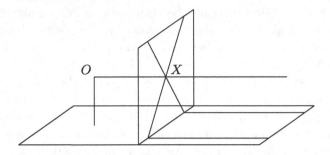

Figure 3.4 "Seeing" a point at infinity

If we define a map from the vertical plane, V, to the horizontal plane, H, by sending a point P in V to the point in H where the line OP meets H, we can say some things very quickly about this map.

1. It sends points to points.

2. It sends lines to lines. Unless ... it sends a point to infinity. Or, indeed, a whole line to infinity.

The reverse map, from the horizontal plane to the vertical plane, has the same features: the map is one-to-one where it is defined, and it maps onto not all of the image plane, but the image plane with a line removed. To see that, consider the map from V to H and the plane through O parallel to V. It meets H in a line, I_V, say, but since it is parallel to V and therefore never meets it, there are no points on V that are mapped to this line.

The mathematician attracted to the theorems of projective geometry has a choice: stick with Euclidean rigour and numerous special cases, or make the language of points and lines at infinity work. Now we come to an interesting historical point. It's not our job to rescue old mathematicians from their mistakes. We don't have to make their arguments rigorous by replacing them with modern arguments, when they exist. Our interest is in how original mathematics got done, and in what these people actually did. If for other reasons, say because the mathematics is lovely, other mathematicians come along and do it differently, that's interesting too. If simpler methods come along, or more rigorous ones, that's fine. But we don't want to attribute later ideas to earlier people, and credit early people with what somebody later might regard as a legitimate improvement, even a breakthrough. So it's not the historians' job to spell out how talk of points at infinity can be made rigorous – although it can be, as you will see in due course. We do observe, however, that mathematicians came to feel comfortable with them, and to use them. In particular, Poncelet did, and so did those who came after him in this line of work.

We may, however, want to orient ourselves by knowing that we are looking at true or false statements, but that's another matter. So let's look back at these theorems, and satisfy ourselves that we can prove them if we buy the idea of a line of points at infnity.

Let's take Pappus's theorem. If we want the points P, Q, R to lie on a line, let us send the line QR to infinity and see if the point P goes to infinity as well. It will if and only if it lies on the line QR, so it will if and only if Pappus's theorem is true. So we have the situation where A, B, C are three points on a line l, and A', B', C' are three points on a line l', and the lines AB' and $A'B$ are parallel, the lines BC' and $B'C$ are parallel, and we want to prove that the lines CA' and $C'A$ are parallel. In Figure 3.5, one pair of parallels is marked with single arrows, the other with double arrows, and the ones we want to show to be parallel are marked with question marks.

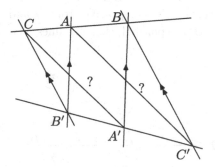

Figure 3.5 A special case of Pappus's theorem

We are going to use what was once called the intercept theorem, which says: if the lines XX' and YY' are parallel, then $\frac{OX}{OY} = \frac{OX'}{OY'}$ and conversely, if those ratios are equal, then the lines are parallel. This is a straightforward consequence of the fact that the triangles OXY and $OX'Y'$ are similar.

In the Pappus figure with parallel lines (Figure 3.5), let the lines l and l' meet at the point O. (If those lines are parallel a separate argument is needed.)

Because AB' and $A'B$ are parallel, $\frac{OB'}{OA'} = \frac{OA}{OB}$.

Because BC' and $B'C$ are parallel, $\frac{OC'}{OB'} = \frac{OB}{OC}$.

Multiply these ratios together, and you find that $\frac{OC'}{OA'} = \frac{OA}{OC}$.

So, by the converse part of the intercept theorem, the lines AC' and $A'C$ are parallel, as was to be proved.

What is satisfying about this proof is that once you have the talk in place about points and lines at infinity, the argument is very short, and you can almost literally see what you have to prove. The special case is the case where various lines are parallel. The reduction from the general to the special case is the proof that we can always, by means of a projection, arrange for the lines to be parallel.

Let's push our luck, and try to prove Desargues' theorem this way.

Proof (Desargues' theorem – special case)

We want to prove that if ABC and $A'B'C'$ are two triangles in perspective from O (so O, A, A' lie on a line, as do O, B, B' and as do O, C, C' – see Figure 3.2) and if L is the point common to BC and $B'C'$, M is the point common to AC and $A'C'$, and N is the point common to AB and $A'B'$, then the points L, M, N lie on a line. As with Pappus's theorem, we send the points M and N to infinity, and see if the point L goes with them, which it will if and only if the points L, M, N lie on a line. So we want to prove that if ABC and $A'B'C'$ are two triangles in perspective from O and if AC and $A'C'$ are parallel, and AB and $A'B'$ are parallel, then the lines BC and $B'C'$ are parallel (see Figure 3.6).

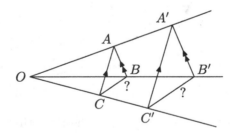

Figure 3.6 A special case of Desargues' theorem

In this case, using the first pair of parallels and the intercept theorem, we deduce that $\frac{OC}{OC'} = \frac{OA}{OA'}$.

Using the second pair of parallels and the intercept theorem, we deduce that $\frac{OB}{OB'} = \frac{OA}{OA'}$.

Therefore $\frac{OB}{OB'} = \frac{OC}{OC'}$, and so, by the converse part of the intercept theorem, the lines BC and $B'C'$ are parallel, and Desargues' theorem is proved.

Again, the special case is the case where various lines are parallel. The reduction from the general to the special case is the proof that we can always, by means of a projection, arrange for the lines to be parallel.

Finally we come to the uniqueness of the fourth harmonic point. Recall that this says that if ABC is any line, and an arbitrary point P is chosen not on this line, and an arbitrary line is drawn through C meeting PA at Q and PB at R, and AR and BQ are drawn meeting at S, and PS is drawn meeting AB at C', then the position of C' is independent of the position of P, Q, R and S and depends only on A, B and C. Here the cunning choice is to send the points P and C to infinity. We then have to show that if AQ and BR are parallel lines, and AB and QR are parallel lines, and AR and BQ are drawn meeting at S, and a line is drawn through S parallel to AQ and BR meeting AB at C', then the position of C' is independent of the choice of line QR. But this is almost immediate. The figure reduces to a parallelogram $ABRQ$, and the point S is its centre, where ever the parallel QR is drawn. As before, the special case is the case where various lines are parallel. The reduction from the general to the special case is the proof that we can always, by means of a projection, arrange for the lines to be parallel.

The name of the theorem derives from the fact that the three points A, B and C determine a fourth point uniquely (the point C'). The term "harmonic" is an old one which in a different language picks out the fact that the cross-ratio (a term to be defined below) of the four points A, C', B, C is -1.

It is an interesting and non-trivial exercise to show that if you start with the points A, B and C on a line and use this construction to obtain the point C', and then repeat the construction but with the points A, B and C' to obtain the point C'', that in fact C'' coincides with C.

The point of this excursion into mathematics is to show you that the idea of points on a line at infinity in the plane enables mathematicians to prove some striking theorems directly. This is the sort of thing that makes mathematicians excited. Their instinct is to go with the idea and to come back and make it more rigorous later, hoping, of course, that it doesn't evaporate once the euphoria is over. Probably most creative mathematicians know that they got all excited about an idea at one time and found out sooner or later that it didn't work. This is raw mathematics, before it's been polished, before one can be quite sure that it isn't just going to fall apart.

There is, of course, more. A conic section (an ellipse, parabola, or hyperbola) is a section of a suitable cone. A cone is a figure obtained from a circle by joining each point of the circle to a point in space (the vertex of the cone) that does not lie in the plane of the cone. So if you start with a conic, put it in the cone and put your eye at the vertex of the cone, the conic and the circle line up exactly. It follows that some theorems about conic sections are reducible to theorems about circles. What theorems? Clearly, if we can turn circles into hyperbolas, the theorems can't be about some length being equal to some other length. It turns out that they can't even be about some ratio equalling some other ratio.

But they can be about three points lying on a line, or about three lines meeting in a point. So, for example, Pascal's theorem.

Pascal's theorem for a circle says that if ABC and $A'B'C'$ are six points on a circle, and if P is the point common to BC' and $B'C$, Q is the point common to CA' and $C'A$, and R is the point common to AB' and $A'B$, then the points P, Q, R lie on a line. (Draw them in the order A, B, C, A', B', C'.) Let us take the special case where the line QR is projected to infinity but the circle remains a circle. Then we have to prove that if ABC and $A'B'C'$ are six points on a circle, and if the lines CA' and $C'A$ are parallel, and the lines AB' and $A'B$ are parallel, then the lines BC' and $B'C$ are parallel. (See Figure 3.7.)

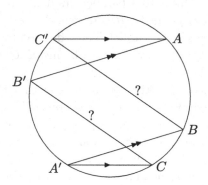

Figure 3.7   A very special case of Pascal's theorem

Proof (Pascal's theorem – very special case)

To prove this, draw BB'. We have $\angle B'AC' = \angle BA'C$, because the lines are parallel in pairs. We have $\angle BA'C = \angle BB'C$ (angles on the same segment) and $\angle B'AC' = \angle B'BC'$ (for the same reason). So $\angle CB'B = \angle C'BB'$, but they are alternate angles on BB' and so BC' is parallel to $B'C$, as required.

But now you may legitimately say that this has been proved only in a special case that seems very unlikely to be general, in which the pairs of lines are parallel and the conic is a circle. This will occupy us later on.

There is another class of results that are also highly relevant. In the previous chapter we saw that there is a theory of poles and polars, and we proved that it made sense for the circle. You now know that it makes sense for any conic section. Let us check the details.

We started with a point outside a circle, and we drew tangents from it to the circle. That becomes a point, a conic and two tangents from the point to the conic (if you can do this with a hyperbola, you're outside it). We joined the points of tangency by a line. Repeating this construction for points on the line that lie outside the conic, we found that their polar lines met in a point. All of that is preserved under projections, so we find that every line has a pole, and indeed that duality holds for any conic. But now we have the language of a line at infinity, we can say what the polar line of the centre of the circle is: it is the line at infinity. And any line through the centre of the circle has a pole that lies on the line at infinity.

To sum up, projective geometry is theorems about points lying on lines (collinearity) and lines meeting in points (concurrence). It is about conic sections being equivalent to circles. It is about properties of figures, in short, that are true of a plane figure and any of its projections onto other planes.

Projective geometry requires that the usual plane is enriched with a line of points at infinity. That any line can be sent to infinity (and brought back again).

With that mathematics granted, we can prove Pappus's theorem, Desargues' theorem, and the theorem on the uniqueness of the fourth harmonic point, but our proof of Pascal's theorem would only be general if we could show that a conic and a line can be mapped to a circle and a line at infinity. This is a modest request: Poncelet's proof of Poncelet's porism requires that any two conics can be mapped to two circles. But in each of these cases there would seem to be a difficulty: what if the conic and the line cross? What if the conics meet in four points (distinct circles can meet in only two)?

3.2 Some properties of some transformations

It is clear that a projection does not preserve length: you are not the same height as your shadow. (See Figure 3.8.)

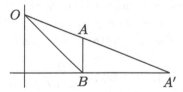

Figure 3.8 The projection from O of AB onto $A'B$

Nor do projections preserve ratios, as Figure 3.9 shows.

If the ratios AC/CB and $A'C'/C'B$ were equal, the lines AA' and CC' would be parallel, but they meet at O.

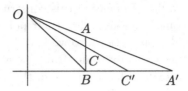

Figure 3.9 The ratios AC/CB and $A'C'/C'B$ are not equal

Projection does not preserve angles, as shown in Figure 3.10.

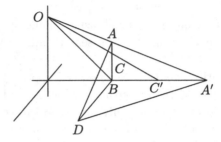

Figure 3.10 The angles $\angle BAD$ and $\angle BA'D$ are not equal

The angles $\angle BAD$ and $\angle BA'D$ are not equal (although the angles $\angle DBA$ and $\angle DBA'$ are equal!). In fact (do this as an exercise – it's easier in two dimensions than three) any two triangles are in perspective, that is, given a triangle ABC and a triangle $A'B'C'$ and a point O it is possible to arrange the triangles so that the points O, A, A' lie on a line, as do the points O, B, B' and the points O, C, C'.

All this might suggest that there is no property at all that projective transformations preserve, other than sending points to points and lines to lines, but there is. It is a curious property called cross-ratio. I shall introduce it here because it belongs in a mathematical chapter at this point, but the history of its introduction is another story, largely omitted from this book for reasons of space. First, to define it. The cross-ratio of four points A, B, C, D on a line is defined to be

$$(A, B; C, D) := \frac{AB \cdot CD}{AD \cdot CB}.$$

There are 24 permutations of the symbols A, B, C, D and you might think that that means there are 24 possible cross-ratios, but in fact there are only six. (This is another nice exercise, but left to the reader.) For example, if you apply the permutation $(AB)(CD)$ so the points are taken in the order B, A, D, C, the cross-ratio becomes $\frac{BA \cdot DC}{BC \cdot DA}$, which is the same as we had before. To show that cross-ratio is preserved by a projective transformation, consider Figure 3.11.

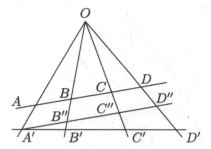

Figure 3.11 Cross-ratio is preserved under projection

We want to show that $(A, B; C, D) = (A', B'; C', D')$. The first thing to notice is that the figures $OABCD$ and $OA'B''C''D''$ are similar, so corresponding lengths are scaled by the same amount and so $(A, B; C, D) = (A', B''; C'', D'')$. So it is enough to show that $(A', B'; C', D') = (A', B''; C'', D'')$.

For that we use Menelaus's theorem twice. Applying it to triangle $A'C'C''$ with transversal $OD'D''$ we get

$$\frac{A'D'}{D'C'} \cdot \frac{C'O}{OC''} \cdot \frac{C''D''}{D''A'} = -1.$$

This is chosen because it has some of the line segments we like, but the terms involving O are not wanted. So we pick another transversal which will also introduce those symbols. Applying it to triangle $A'C'C''$ with transversal $OB'B''$ we get

$$\frac{A'B'}{B'C'} \cdot \frac{C'O}{OC''} \cdot \frac{C''B''}{B''A'} = -1.$$

From this we deduce that

$$\frac{A'D'}{D'C'} \cdot \frac{C'O}{OC''} \cdot \frac{C''D''}{D''A'} = \frac{A'B'}{B'C'} \cdot \frac{C'O}{OC''} \cdot \frac{C''B''}{B''A'}.$$

Cancelling the repeated terms, we find that

$$\frac{A'D'}{D'C'} \cdot \frac{C''D''}{D''A'} = \frac{A'B'}{B'C'} \cdot \frac{C''B''}{B''A'},$$

which rearranges to give

$$\frac{A'B'}{A'D'} \cdot \frac{C'D'}{C'B'} = \frac{A'B''}{A'D''} \cdot \frac{C''D''}{C''B''},$$

which says exactly what we want:

$$(A', B'; C', D') = (A', B''; C'', D'').$$

We can use cross-ratio to deduce the uniqueness of the fourth harmonic point.

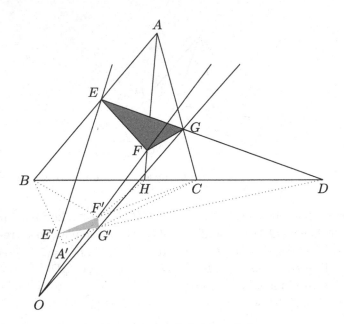

Figure 3.12 Uniqueness of the fourth harmonic point

In Figure 3.12, I have drawn the triangle ABC and a copy, $A'BC$. In the original, I draw the transversal DGE; in the copy, I have drawn a different transversal, $DG'E'$.

Look at the triangles EFG and $E'F'G'$. The lines EF and $E'F'$ meet at C, the lines FG and $F'G'$ meet at B, and lines GE and $G'E'$ meet at D, so (by the converse of Desargues' theorem) the triangles EFG and $E'F'G'$ are in perspective from some point, O. Now consider the lines BE and BE'; they are in perspective. So are the lines CG and CG'. It follows that their points of intersection, A and A' are in perspective. Therefore the lines AF and $A'F'$ are in perspective. We can finish the uniqueness theorem in two ways. We can argue that trivially the line BD is in perspective with itself, so the point where AF and BD meet is in perspective with the point where $A'F'$ and BD meet, so H must be the point where $A'F'$ and BD meet. Or we can argue that if where $A'F'$ and BD meet is called H', then these cross-ratios are equal: $(B, C; D, H) = (B, C; D, H')$, and so $H = H'$.

It is important to observe that we can even calculate the cross-ratio $(B, C; D, H)$. It is the same if we send the points A and D to infinity, and reduce to the figure of a parallelogram. But that has $BH = HC$, and $BD = CD = \infty$ (see the exercises for a defence of this symbol and its meaning) so the cross-ratio is $(B, C; D, H) = -1$. This shows that cross-ratio minus one is projectively equivalent to a midpoint and a point at infinity.

EXERCISES

3.1. Show that if coordinates are chosen on the line such that A is the origin, and B, C and D have coordinates x, c and 1 respectively, then $(A, B; C, D) = \frac{x(1-c)}{x-c}$.

3.2. Deduce that as $c \to \infty$ $(A, B; C, D) \to x$. Persuade yourself that it's okay to use the symbol ∞ and to speak of the cross-ratio with a point, C, at infinity.

3.3. Deduce that if C and D are fixed, then for each α there is a unique position for B such that $(A, B; C, D) = \alpha$. When do you need to use infinity?

3.4. Show that if A, B, C, D, E are five points in a line, then
$(A, B; C, D)\,(A, D; C, E) = (A, B; C, E)$.

This looks prettier if we write $(A, B; C, D)$ as $\langle BD \rangle$. It then says: $\langle BD \rangle \cdot \langle DE \rangle = \langle BE \rangle$. We shall see a good reason for doing this much later in the book (see §20.4 Klein's Cayley metric).

3.5. Choose coordinates as in Exercise 3.1 above, and let E have coordinate y. Confirm the calculation in Exercise 3.4 in the special case when C is at infinity.

3.3 Alternative treatment of cross-ratio and the fourth harmonic point

This treatment uses homogeneous coordinates, which we shall meet in Chapter 13. If you do not know about this system of coordinates, skip this subsection and return later if you wish, after consulting §6.7.

Given $A = [t, 0, 1]$, we want to find $B = [t', 0, 1]$. Choose P to be $[0, 0, 1]$, Q to be $[1, 0, 0]$ and $Y = [0, 1, 0]$ (see Figure 3.13).

Then line AY has equation $x = tz$.

Let $F = [t, s, 1]$. Line QE has equation $y = sz$, and meets PY (with equation $x = 0$) at $[0, s, 1]$.

Line PD has equation $ty = sx$, and meets QY (with equation $z = 0$) at $[t, s, 0]$.

Line ED has equation $sx - ty + stz = 0$, and meets PQ (with equation $y = 0$) at $[-t, 0, 1]$. So $t' = -t$ and $B = [-t, 0, 1]$.

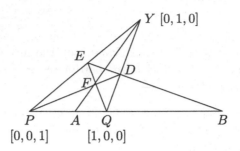

Figure 3.13   Cross-ratio and the fourth harmonic point

To find the cross-ratio of the four points P, A, Q, B with these coordinates, let A be $\lambda[1,0,0] + (1 - \lambda)[0,0,1] = [\lambda, 0, 1 - \lambda] = \left[\frac{\lambda}{1-\lambda}, 0, 1\right]$. So $t = \frac{\lambda}{1-\lambda}$.

Similarly, B has coordinates $\left[\frac{\lambda'}{1-\lambda'}, 0, 1\right]$, and

$$t' = -t = \frac{\lambda'}{1 - \lambda'}.$$

It follows that

$$CR(P, A; Q, B) = \frac{\lambda\left(\lambda' - 1\right)}{\lambda'\left(\lambda - 1\right)} = -1.$$

3.3.1 Porismata

Now for the long-delayed Poncelet's porism, also known as Poncelet's closure theorem.

Take two conics, one inside the other for simplicity. Pick a point P_1 on the outer one, and draw a tangent to the inner conic through it, meeting the outer conic again at P_2. Repeat the construction, starting again at P_2, to get P_3, and carry on. One of two things happens: this process goes on for ever, or at some stage you return to P_1 and for some n, $P_n = P_1$. Poncelet's porism asserts that if the second of these happens it will happen for the same n and whatever point is chosen as P_1. It depends, therefore, only on the conics, and not the starting point.

Let's think about this. Could it ever be true that the chain of points P_1, $P_2, \ldots$ closes up? Let us take a circle as the outer conic and draw a triangle in it. Every triangle has an incircle, as it is called, which is a circle having the three sides of the triangle as tangents. So we can certainly display the phenomena, and indeed it was known, but not well known before Poncelet that you could do this with circles. There is even a formula connecting the radii of the inner and outer circles, which was proved (and forgotten) quite a few times.

The formula is $a^2 = R^2 - 2rR$, where the radii are r and R, $r < R$, and a is the distance between their centres. This formula was probably known to Gergonne. It underpins a problem he set in his journal for 1810 and yields the closure theorem in this case, as Lhuillier proved in answering the problem.

Now that we know at least a special case of the theorem, we can see a few more. Draw a circle, and a square in it. Can you fit a circle inside the square touching each side of the square? Of course you can, and once you've done it, you can rotate the square too. The porism holds up, and it is clear that any regular n-gon will do.

What about the theorem in its full generality? We take two conics, one inside the other, and our previous analysis stops right there. We have not even seen a special case for an ellipse. One remedy, if it is available, is to look for a projection that sends the two conics to two circles – we cannot insist that the circles are also concentric. But suppose there is a projection that sends two conics to two circles. Then we can try to prove the closure theorem in this special case, knowing that it will then be true in general. I hope you will not be surprised to learn that we shall not be giving a proof here, but be content with the audacity of Poncelet's claims about it. A full account of Poncelet's approach and the rigorous mathematics that it has inspired will be found in the paper by Bos et al. [22].

4
Poncelet's Traité

4.1 Poncelet's singular claims

The singular novelty of Poncelet's geometry is that he was seldom interested in the metrical properties of figures. Everything is studied at the level of what properties a figure has in common with its shadows (its projections). This was not the first time such an idea had been presented, but this time the message caught on. Poncelet was quite explicit, and reasonably clear, that there was a class of geometric properties that deserved to be singled out, and these were the projective ones. For example, and this needs to be shown, the property of being pole and polar is a projective property.

The wholly original part of Poncelet's vision of geometry is also the one that caused (and causes) problems. We saw at the end of the last chapter that Poncelet wanted a very general way of reducing questions in the projective geometry of conics and pairs of conics to questions about circles and pairs of circles. It is true that any conic section is projectively equivalent to a circle. It is wholly bizarre that any two conics tangent to each other at two points can be regarded as projectively equivalent to a pair of concentric circles. A projective transformation sends tangents to tangents, but if one conic is transformed into a circle, the other cannot be (two circles cannot touch at two points, unless they coincide entirely). What is going on?

Let us take the middling case: Poncelet's claim that a conic and a line can be transformed projectively to a circle and a line at infinity [202, §109].[1] We

---

[1] That is, in §109 of Poncelet's *Traité* [202]. This style of referencing will be used throughout this book.

J. Gray, *Worlds Out of Nothing*,
Springer Undergraduate Mathematics Series,
DOI 10.1007/978-0-85729-060-1_4, © Springer-Verlag London Limited 2011

call the conic C and the line L, and we assume first of all that they do not intersect. Then it is reasonably clear (I skip the proof) that it can be done. But it is surely even clearer that if the circle C and the line L intersect, then there is no hope. The common intersection points must surely go to infinity, making the image of the conic a hyperbola.

Before we consider how Poncelet tried to talk his way out of this seemingly obvious objection, note that were it to be true that any conic and a line are equivalent to a circle and a line at infinity, there would be a simple general proof of Pascal's theorem.

Poncelet had a novel way of dealing with intersections. To understand it, we need a useful construction in the theory of conic sections that goes all the way back to Apollonius, the Greek geometer who was the first to write about them at length (and whose work has survived, almost everything written before him has been lost). It's called the theory of conjugate diameters. These are diameters that come, somehow, twinned together.

Given a conic section C and a line L, draw the two tangents that are parallel to L. These touch the conic at points P and P', say. Draw the line L' through P and P', it is a diameter of the conic. Now repeat the construction, starting with L', drawing the two tangents that are parallel to L', letting them touch the conic at the points Q and Q', and drawing the line L'' through Q and Q' (it is another diameter of the conic). Then (theorem) the line L'' is parallel to L. The diameters L and L' are said, accordingly, to be conjugate.

To see why the theorem is true, consider the conic and a point A, for convenience outside the conic. Let the polar line of A be denoted L_A. Pick a point A' on L_A outside the conic, and draw the polar line of A', $L_{A'}$. This polar line passes through A, by duality. (See Figure 4.1.) The construction described in the paragraph above is this construction with the points A and A' at infinity.

Conjugate diameters are familiar in a special case: the major and minor axes of a conic are mutually conjugate. Quite generally, when a circle is projected onto a conic, any pair of orthogonal diameters of the circle is projected onto a pair of conjugate diameters of the conic.

4.1.1 Meeting

Now to explain Poncelet's theory of "meeting".[2] Suppose one has an ellipse C and a line L that does not meet it. Poncelet wanted to find the points where the line "met" the conic. To this end, he considered the diameter of C that is parallel to L, and the corresponding conjugate diameter, which cuts C at the

---

[2] This account follows Bos et al. [22].

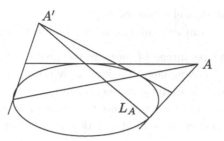

Figure 4.1 Pole and polar

points A and B, say. Let B be on the same side of the centre as the line L. Let me now replace Poncelet's elegant argument here with an algebraic one (that he would have rejected out of hand!) to the same effect. When a conic is given its equation with respect to a pair of conjugate diameters, the equation has the form $x^2 + py^2 = 1$. Among the infinitely many hyperbolas with the same conjugate diameters as the given ellipse, and which touch the ellipse at B, one has the equation $x^2 - py^2 = 1$. It genuinely meets the line L at the points R and R', say. Poncelet saw that the line segment RR' was the ideal chord of the line L with the ellipse. If one had started with a hyperbola, the constructed conic would be an ellipse; the two conics are said to be supplementaries of each other.

It is important to distinguish carefully here between what Poncelet called "imaginary" and what he called "ideal". The points R and R' and the chord they define are perfectly real, and not therefore called imaginary. The sense in which they may be said to be where the line L meets the conic C is strange, and that is what he called ideal. Points or lines that do not exist in the (real) plane but must be imagined to exist in order not to block the argument he called "imaginary". Typically, anything Poncelet did can be rescued at the price of working with complex numbers and complex coordinates, but that was not at all what Poncelet was advocating. He wanted a way of reasoning geometrically that was as general as algebra. As he put it:

> We can define by the adjective imaginary any object which, absolute and real as it was in a certain figure will become entirely impossible or inconstructible in the correlative figure, which it is counted as having been derived from by a progressive and continuous movement of some parts, without violating the primitives of the system. [202, p. 28]

On the same page he wrote that if mathematicians wish to persist in regarding a line as a secant when it no longer cuts a conic then the proper thing to do is to call it an ideal secant, and to say that its points of intersection with the curve are imaginary. Put that way, it doesn't sound quite so bad.

It seems clear that Poncelet himself knew he was extending the meaning of words such as "meet" from this remark in the preface:

> To extend the resources of elementary geometry by generalising the concepts and the language that is usually quite restricted, to bring them closer to those of analytic geometry, and above all to offer general and appropriate means to prove and to discover, in an easy manner, this class of properties that figures enjoy when they are considered in a purely abstract manner and independently of any absolute and determined quantity – this is the particular object of this work. [202, p. xxxiii]

Now we can see how Poncelet would deal with an ellipse and a line that crosses it. He would pass to the supplementary hyperbola, which does not meet the line, and project the hyperbola and the line to the circle with the line at infinity.

It helps at this point to recall where Poncelet was coming from, which was the École Polytechnique, with its heavy emphasis in geometry, and especially the descriptive geometry introduced by Monge. This taught him the importance of transformations; not the full-blooded projective transformations but the more restricted class of transformations used by Monge, and the idea of using particular cases and showing that they were, in fact, equivalent to the general case. He was also coming from Saratov: two terrifying years where he endeavoured to reconstruct what he knew, and to give it the simplicity he felt mathematics properly required.

In those two years he had ample time to calculate, and many of his theorems were discovered as a result of quite appalling calculations. It is worth remembering when reading mathematics that sometimes the original author had nothing else to do all day, day in, day out, but to calculate, to grind out example after example, case after case. There is a great deal of hard work hidden in most presentations of mathematics. The lesson Poncelet set about drawing from his own immersion in calculation when he returned to France was that there should be a better way of reasoning geometrically, one that did not pursue the argument down a maze of bifurcating cases: one when there are four points, another when there are two, a third when there are none, a fourth when two points coincide; one when this segment is less than that one, another when it is greater .... Poncelet argued that if an argument is true in infinitely many cases of a certain kind, it will always be true, subject at most to some trivial and obvious modifications. The case of a line and a conic is typical. A segment can be produced of the right length, with just the properties required to keep the theorems true, provided one stretches the meaning of the term "meet".

In this spirit he set to work, and in 1822 published his major work, the *Traité des propriétés projectives des figures* [202]. It relies heavily on the contentious claims about what figures are projectively equivalent, but anyone who could accept them met some very attractive mathematics. For example, Poncelet's proof of Pascal's theorem proceeds by taking six points on a conic and the candidate Pascal line (defined by two of its points) and projects this figure onto a circle with a line at infinity, when the proof is the very simple one given above.

To define pole and polar, Poncelet took a conic and four points on it A, B, C, D. He drew the lines AB and CD letting them meet at E, the lines AC and BD letting them meet at F, the lines AD and BC letting them meet at G, and announced that the line FG was the polar of the point E, that EG was the polar of F and EF was the polar of G, because this was the case when one of those lines (let us take EG) was sent to infinity and the conic to a circle. In this case the quadrilateral $ABCD$ is a rectangle and F is its centre. (See Figure 4.2.)

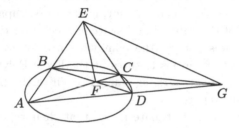

Figure 4.2 Poncelet on poles E, F, G and polars FG, EG, EF

4.2 Cauchy responds

How was Poncelet's work received? Were his strange arguments accepted or rejected? Was his work influential or dismissed, perhaps to achieve importance only several years later? What was the significance of the work in its day? Here we are in an unusually fortunate position. It was the custom at the time for the French to set up commissions to report on major works submitted for publication, and to comment on their quality and suitability for publication. In this case, a commission consisting of Arago, Poisson and Cauchy was set up in 1820, with Cauchy as chair, to report on a *Mémoire* by Poncelet which later appeared, possibly modified, as part of his book. A word about these people.

Arago entered the École Polytechnique with distinction in 1803, and left before the end of his second year to take up a career in the Bureau des Longitudes, which was a prestigious place to be. His speciality was applied analysis. In 1810 he took over Monge's courses at the École Polytechnique on "analysis applied to geometry" and descriptive geometry, and he also taught astronomy and geodesy. He never became a major creative mathematician or scientist, but he went on to wield considerable academic power behind the scenes. In 1815 political changes had helped install him on the Advisory Council of the École Polytechnique, replacing Monge. He then became admissions examiner for the military school at Metz. In 1830, another heavily political year when professors were required to take an oath of allegiance to the new monarchy, he became the perpetual secretary for mathematics in the Académie des Sciences in Paris, and he remained in that influential position until 1844.

Poisson was much more of a mathematician, a disciple of Laplace, and, with Cauchy and Fourier, one of the three leaders of his generation of mathematicians, those who followed Laplace, Lagrange and Legendre. And Cauchy, born in the revolutionary year of 1789, was emerging as the leading mathematician of his generation in France, a man of enormous energy, capable of working in almost all branches of the subject and switching between them with striking rapidity. In 1820 he was not yet the man who brought rigour to the calculus, that was to be set out in his lectures at the École Polytechnique in 1821 and 1823, but he was already a rising star. As chair of the commission, it was Cauchy who wrote the report.

Where we are unusually fortunate is not that their report was published, but in that Poncelet chose to reprint it, verbatim, in the *Traité*. As you might expect, the report gives a reasonable summary of the contents of the *Mémoire*, and therefore of the book, sometimes in more intelligible language. More interestingly, it comments on the quality. Here is the passage where they discuss Poncelet's method of continuity.

> The admission of the principle into geometry consists in supposing that, in the case where a figure composed of a system of lines or curves always has certain properties while the absolute or relative dimensions of its various parts vary in an arbitrary manner, between certain limits, the same properties necessarily persist when one goes beyond the dimensions which one had hitherto supposed were within these limits; and that, if some parts of the figure disappear on the second hypothesis those which remain continue to enjoy, the ones with respect to the others, the properties that they had in the primitive figure. This principle, it should be said, is only a bold induction, by means of which one can extend theorems, initially established with certain restrictions, to the case where these restrictions no longer hold. Applied to curves

of the second degree, it leads the author to exact results. Nonetheless, we think that it should not be admitted generally and applied indifferently to all sorts of questions in geometry, nor even in analysis. By placing too much confidence in it, one can be lead into manifest errors. One knows, for example, that in the determination of definite integrals, and consequently in the evaluation of lengths, areas, and volumes, one encounters a great number of formulae that are only true when the values of the quantities that they contain remain within certain limits.

For the rest, we distinguish happily between M. Poncelet's considerations of continuity and those which are directed towards the properties of lines to which he gave the name of ideal chords of conic sections. [202, pp. ix–xi]

Cauchy and his two colleagues then explained how complex numbers enter the story when one solves algebraically the equations for a conic and a line that does not meet the conic. The common solutions for x and y will be of the form $a + ib$ and $a - ib$. So the average of the xs and of the ys (which gives the coordinates of the midpoint) will be real, and their difference purely imaginary (of the form $2ib$). So one can construct a chord with a definite midpoint and a definite length. As they put it:

It therefore becomes useful to substitute for the imaginary chord which does not exist a fictive chord (of this length) drawn in the given direction and whose midpoint coincides with a point we shall speak about below. It is this fictive chord that one can give the name ideal chord by which M. Poncelet denotes sometimes the indefinite line that he considers and sometimes the imaginary chord intercepted by the curve, while he calls the centre of the ideal chord the real point which analysis indicates is the middle of the imaginary chord. The sense in which the author uses the word ideal is thus somewhat modified, in such a way that the ideal lengths remain real lengths constructible in geometry. [202, xi]

It is otherwise clear from the report that Cauchy and the others largely approved of the results that Poncelet had found, and they recommended that the *Mémoire* be published. In fact it was not, but was reworked and included in the *Traité*, where it appears as the first section. Poncelet said he had included the report in the book so that readers could see how much he had benefited from it. There is an unquantifiable element of irony in that remark.

As Cauchy admitted, the report did not confine itself to summarising the content of the *Mémoire*, but proposed a way of redefining some of its key ideas. For Poncelet, an ideal chord joins the points where a line meets a conic – but in a new sense of the word "meet". Cauchy sought to eliminate this novel language, while allowing that the ideal chord is a useful object. But it is now a much more algebraic object, whose principal properties are found by algebra and which are then displayed geometrically. Poncelet had set himself the task of finding a uniform geometrical language, and would surely not have welcomed this part of the report. Indeed, I presume that by republishing the report in his *Traité* but not altering his approach he was deliberately saying that he did not intend to benefit from this criticism; rather, it was Cauchy and the others who were wrong. To get a sense of how far Poncelet had moved from the mode of calculation that had kept him alive in Saratov, note that Chasles observed in his *Aperçu historique* that the *Traité* proceeds "without a word of calculation" [31, p. 215].

Late in his life, Poncelet also described a distressing meeting he had had with Cauchy in 1820.

> Fearing, and with good reasons, that twelve years of work and cease-less meditation would not clarify a deceptive problem and perhaps even make me the subject of ridicule in the eyes of my superiors, of my friends, and of everybody interested in geometry, though indifferent and a bit indulgent, I managed to approach my too-rigid judge at his residence at N. 7 rue Serpente. I caught him just as he was leaving for Saint-Sulpice. During this very short and very rapid walk, I quickly perceived that I had in no way earned his regards or his respect as a scientist, and that it might even be impossible to get him to understand me. Humble petitioner that I was, I thus restricted myself to respect-fully informing him that the objectionable points and difficulties that he believed he saw in the adaptation of the principle of continuity to geometry were essentially results of the insufficient attention that had heretofore been accorded to the law of signs, a law that had absorbed my attention since 1813, when I was in Russia, and especially since my return to France in 1814. I explained that the mathematical discussion of this law could have preceded my communication with the Académie, had the esteemed men on that body not dissuaded me from doing so. However, without allowing me to say anything else, he abruptly walked off, referring me to the forthcoming publication of his *Leçons à l'École Polytechnique*, where, according to him "the question would be very properly explored".[3]

---

[3] Taken from Belhoste [10, p. 55]; the original is in Poncelet, *Applications d'analyse*, 1864 [205, vol. 2, p. 564].

4.3 Other responses

Poncelet had some readers who appreciated him greatly. Chasles, in his *Aperçu historique*, recorded that some called him "the Monge of his century" [31, p. 88] for being the first to appreciate the power of projective geometry since Desargues, and recognised him as one of the founders of modern geometry. On the other hand, Chasles repeated Cauchy's opinion of the method of continuity without demur (p. 199). Indeed, Chasles was Poncelet's main rival in later years, and his account of Poncelet in the *Aperçu historique* is oddly cold and distant, possibly for that reason.

Poncelet's geometry was welcomed by other former students of Monge at the École Polytechnique, such as Malus, Dupin, Hachette and Lancret, and by other geometers, such as Gergonne, Brianchon and Olivier. They appreciated it for its generality and because, as Monge had taught, it rested on objects which you could imagine as physical objects: points, lines, etc. Perhaps not surprisingly, those who liked Poncelet's geometry were teachers of engineers at the advanced specialist écoles. Those who opposed it taught mathematics in the more general École Polytechnique but they were dominant in the prestigious Académie des Sciences.

Although Cauchy's rigorous analysis was regularly censured in the later 1820s by the external examiner (Gaspard de Prony) for its lack of geometry, Cauchy's influence was the more powerful. Monge's pupils, found themselves unable to keep geometry sufficiently alive, and the syllabus was determined by the teachers at the École Polytechnique. The successes of descriptive geometry turned out to be confined to techniques of use in engineering; the subject did not become a tool of great conceptual power. Even Monge's own work lacked good examples of what descriptive geometry could do in theoretical geometry. And pure geometry, such as Poncelet produced, never generated much in the way of applicable mathematics. It could have flourished as a pure subject, the revitalised heir of Euclidean geometry, but even this did not happen. Poncelet virtually abandoned it in the 1830s in favour of an experimentally driven study of machines, and others did too, Gergonne for example. Michel Chasles, its stoutest defender, lived to see it pass across the Rhine to Germany and noticed that it had become written in a language he could not speak.[4] And when in 1865 Poncelet, a member of the Académie des Sciences since 1834, brought out the second edition of his *Traité* he still did not back down, but in the notes added to the new edition he reaffirmed his belief in the merit of his ideas about ideal chords and imaginary points of intersections.

---

[4] It may be that Chasles did not read German, or wish to learn it, but it is possible that what he could not read was Steiner's papers, which are obscure in their very arguments. Cremona called him "this celebrated sphinx" [44, p. 3].

4.4 Poncelet's more conventional methods

It is not possible to go into many details here about Poncelet's methods, and it is historically more instructive to describe his most extreme, but a number of his central arguments were more mundane, and were to form the basis of Chasles' later rewriting of projective geometry. Here is a crucial one, leading up to the definition of a "supplementary" conic.[5]

Let ABC be a triangle whose sides, extended as necessary, meet a conic as follows: AB cuts the conic at R and R', BC cuts the conic at P and P', CA cuts the conic at Q and Q'. One can go round the triangle two ways: ABC or ACB. Either way, this equality results:

$$AR{\cdot}AR'{\cdot}BP{\cdot}BP'{\cdot}CQ{\cdot}CQ' = AQ{\cdot}AQ'{\cdot}CP{\cdot}CP'{\cdot}BR{\cdot}BR'.$$

The proof is very elegant, as Poncelet noted, and genuinely projective (the reader should check why). It is therefore enough to consider it for the circle, but for the circle it is true that $AR{\cdot}AR' = AQ{\cdot}AQ'$ by properties of secants proved in Euclid's *Elements* – use similar triangles. The equation therefore holds for the circle and therefore for any conic, because it is projectively invariant.

Now consider special cases, which put one firmly in Euclidean geometry. For example, when the point A is at infinity, and the lines RBR' and QCQ' are parallel, one finds $\frac{BR{\cdot}BR'}{BP{\cdot}BP'} = \frac{CQ{\cdot}CQ'}{CP{\cdot}CP'}$, which holds for all points on the line PP'. A theorem named after Carnot falls out if instead the sides of the triangle are tangents to the conic, so $P = P'$, etc.: $AR{\cdot}BP{\cdot}CQ = AQ{\cdot}CP{\cdot}BR$. With very little more work Poncelet showed that conics, other than the parabola, have centres, and thence that they have conjugate diameters and convenient properties that, written algebraically, would correspond to simple equations for them.

---

[5] Taken from [202, ch. I, section I].

5

Duality and the Duality Controversy

5.1 Pole and polar

We have seen that Poncelet used a way (in a sense, discovered by Brianchon) of pairing each point of the plane with a line, called the polar of the point, and each line in the plane with a point, called its pole. All that was needed for this was a conic. Any conic will do, and if a different conic is chosen the details of which point is paired with which line is changed, but nothing else. We also saw that the process of starting from a point (a pole) and producing a line (its polar) is the inverse of the process of starting from a line (a polar) and producing a point (its pole). We also saw that if three points lie on a line then their polars meet in a point and, conversely, if three lines meet in a point then their poles lie on a line. (See Figure 5.1.)

So Poncelet proclaimed a principle of duality, which said: given a conic, if you replace each line in a figure by its corresponding pole, and each point by its polar line, you obtain a new figure in which concurrent lines are replaced by collinear points and vice versa, and the original theorem becomes a theorem about the new figure on exchanging the words point and line, collinear and concurrent, throughout. (Such a pair of theorems are called dual theorems.) The simple act of dualising a theorem yields another theorem, this time about the dual figure.

Let us apply duality to Desargues' theorem (see Figure 5.2). What do we get?

J. Gray, *Worlds Out of Nothing*,
Springer Undergraduate Mathematics Series,
DOI 10.1007/978-0-85729-060-1_5, © Springer-Verlag London Limited 2011

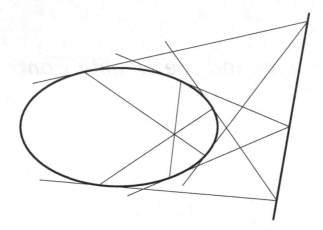

Figure 5.1 Poles and polars

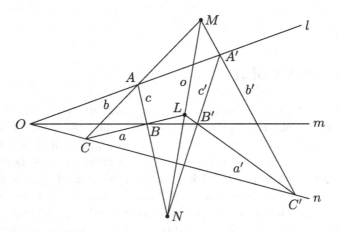

Figure 5.2 Desargues' dual

Answer: the converse! Notice that by cunningly choosing the names of the lines, the statement of the theorem and its converse involve nothing more than passing from upper- to lower-case letters.

Another amusing mathematical exercise is to dualise the construction of the fourth harmonic point (to construct what might be called a fourth harmonic line) and to check that it agrees with the construction of the fourth harmonic point.

Although Poncelet's reasoning is correct, he generated a typically Parisian controversy. The idea of duality had been around for some time. One can, for example, dualise Pascal's theorem, as a pupil of Monge, Charles Julien Brianchon, had already done in 1806 [27], to get this beautiful result about

any 6 tangents to a conic: 6 tangents to a conic meet in six points which lie
in pairs on three concurrent lines. More precisely: if a, b, c, d, e and f are
6 tangents to a conic, and a and b meet at A, b and c at B, and so on, until
f and a meet at F, then the three lines AD, BE and CF meet in a point.
Brianchon's path to his discovery was this. He knew from Monge's descriptive
geometry that given a fixed quadric surface Q and a point P (for convenience,
outside it) the lines through the point that touch the surface form a cone with
P as its vertex, and the cone touches the surface in a conic section, which, of
course, lies in a plane, Π_P. Moreover, as the vertex of the cone moves in a
plane Π, the corresponding plane Π_P always passes through the same point.
(We would say that the point P and the plane Π_P are pole and polar with
respect to the quadric Q.) Brianchon observed that when the point varies over
a second quadric surface Q', its corresponding plane Π_P envelops a new quadric
surface Q''. He then used this observation to produce the theorem that today
bears his name (see Figure 5.3).[1]

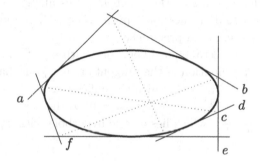

Figure 5.3 Brianchon's theorem

5.2 Gergonne versus Poncelet

But it is one thing to realise that dualising a figure is a good way to obtain new
theorems, which is what Poncelet did, and quite another thing to claim that
points and lines are interchangeable concepts which must logically be treated
on a par. This was the view that Gergonne put forward in 1825. Interpreted
in such generality, Gergonne's principle of duality is one of the most profound
and simple ideas to have enriched geometry since the time of the Greeks, but in
the resulting dispute, which concerned priority as well as mathematical depth,

---

[1] See Brianchon [27] and Chasles' *Aperçu historique* [31, pp. 370–371].

little justice was done to either side. The controversy, by the way, took the form in part of a fascinating polemic, listed as such in the contents pages of volume 18 of Gergonne's *Annales* (1827–1828) [89, pp. 125–156].

Poncelet disagreed forcefully with Gergonne. Part of the reason for this was that they had clashed over an application of the idea of duality to the study of curves other than conics. Gergonne had initiated this line of enquiry in a paper he printed in his own journal, the *Annales de Mathématiques Pures et Appliquées*, in 1827–1828, but he rather spoiled the effect by making a serious mistake. Poncelet pointed the mistake out, but was unable to see how to put it right [203].

Gergonne's crucial idea was this. Suppose you have a curve. Use duality to replace each point on the curve by a line. When you do this for every point on the curve you get a family of lines; Gergonne asked what this family looks like. It turns out that it envelops a curve, called the dual curve to the original one.

Equivalently, you can start with a curve and at a point P on the curve, take the tangent to the curve at that point, take the pole of that line, and find the locus of these points as the point P varies along the original curve. This method is easier to do (unless, and perhaps even if, you know how to find envelopes) so we shall use it in preference.

Let's apply it to the rectangular hyperbola, with equation $xy = 1$, parameterised by $(t, 1/t)$. The equation of the tangent at that point has slope $\frac{-1}{t^2}$, and is therefore $y - \frac{1}{t} = \left(\frac{-1}{t^2}\right)(x - t)$, which simplifies to $x + yt^2 - 2t = 0$, which I shall write in the form $\left(\frac{-1}{2t}\right)x + \left(\frac{-t}{2}\right)y + 1 = 0$, so the corresponding point in the dual space is $\left(\frac{-1}{2t}, \frac{-t}{2}\right)$, which lies on the hyperbola with equation $4xy = 1$.

5.2.1 Curves of higher degree

It's rather more interesting to attempt to apply this to a cubic curve. This introduces a new theme: the study of curves other than conic sections.

It's going to be worthwhile standing back and not getting lost in the details. If the equation of the curve is $F(x, y) = 0$, where F is of degree 3, then a line through a point (a, b) is almost always of the form $y = mx + c$. (Note that m and c are related by the equation $b = ma + c$, so once we know m we know c.) This line will be a tangent to the curve if $F(x, mx + c) = 0$ has repeated roots. These roots will be the x-coordinates of the points of tangency. For the equation to have repeated roots there will be a condition on m, determining the tangent lines themselves.

When does an equation of degree 3 have repeated roots? Consider a cubic equation $G(x) = 0$. It has a repeated root if and only if it is of the form $G(x) = (x - \alpha)^2 (x - \beta) = 0$. In this case, differentiation with respect to x gives $G'(x) = 2(x - \alpha)(x - \beta) + (x - \alpha)^2 = 0$, so $G'(x) = (x - \alpha)(2(x - \beta) + (x - \alpha)) = 0$, so G and G' have the repeated root in common. This argument reverses, so we deduce that a cubic equation has a repeated root if and only if it has a root in common with its first derivative. In that case the root will be the common solution of the equations $G(x) = 0$ and $G'(x) = 0$. Now, two equations of degrees k and m have km solutions, and here we have $k = 3$ and $m = 2$, so there are six common solutions.

Exercise 5.1

You might like to check this on the very familiar case of a quadratic. The equation $ax^2 + bx + c = 0$ has a repeated root if and only if $b^2 - 4ac = 0$, in which case the root is $\frac{-b}{2a}$. Let us verify this by the indicated method. We differentiate with respect to x and obtain $2ax + b = 0$, which only has the root $x = \frac{-b}{2a}$. That must be the common root with the quadratic equation, confirming that part of the claim, and the quadratic equation has that root if and only if $a(\frac{-b}{2a})^2 + b(\frac{-b}{2a}) + c = 0$ which simplifies to $\frac{-b^2}{4a} + c = 0$ or $b^2 - 4ac = 0$.

To return to the cubic curve. We have seen that through a point there will be 6 tangents to the cubic curve. Now let us dualise. Each tangent gives rise to a point on the dual curve. The configuration of 6 tangents through an arbitrary point dualises to 6 points on the dual curve that lie on an arbitrary line, or, better said, an arbitrary line meets the dual curve in 6 points. So the dual curve has degree 6.

If we had started with a curve of degree n, its derivative would be of degree $n - 1$ and so through an arbitrary point of the plane there will be $n(n - 1)$ tangents to the curve, so the dual curve would have degree $n(n-1)$. Note that when $n = 2$, $n(n - 1) = 2$, as we expect.

Exercise 5.2

Use duality to prove that the number of tangents common to two curves of degrees m and k is in general, and at most, $mk(m - 1)(k - 1)$ (no. 2 in Poncelet's list established in 1818).

5.3 Gergonne on the principle of duality

(From his *Annales de Mathématiques*, vol. 18, pp. 150–152 and 214–216, in Fauvel and Gray [74, 17.A3])

Duality

We have observed, not long ago, that at the point which mathematics has reached today, and encumbered as we are with theorems of which even the most intrepid memory cannot flatter itself it retains the statements, it would perhaps be less useful to science to seek new truths than to attempt to reduce the truths already discovered to a small number of guiding principles. In any case a science perhaps recommends itself less by the multitude of propositions which make it up than by the manner in which these propositions are related and connected to one another. Now, in each science there are certain elevated points of view where it is enough to stand there to embrace a great number of truths at a glance – which, from a less favourable position, one could believe were independent of one another – and which one realises accordingly are derived from a common principle, often even incomparably more easy to establish than the particular truths of which it is the abridged expression.

It is with a view to confirming these considerations with some quite remarkable examples, on common points and common tangents to plane curves (etc.), that we are proposing to establish here a small number of general theorems offering an infinity of corollaries, among which we restrict ourselves to pointing out the simplest or those most worthy of notice ...

Since we are not concerned at all here with metrical relations, all our theorems are double. To make this correspondence easier to grasp we place in two columns, the ones facing the others, the theorems which correspond to each other, as we have already done several times.

Properties of algebraic curves lying in a plane

Let there be a plane figure composed in any way one wishes of points, lines, and curves. Let us conceive that having drawn arbitrarily in the plane of this figure, an arbitrary curve of the second order, one then constructs in the same plane another figure, all of whose points and all of whose lines are the poles and polars of all of the lines and all the points of the first, with respect to this curve of the second order considered as directrix. The two figures thus drawn are said to be polar reciprocals the one of the other, since the first can be derived from the second just as it was supposed to be the origin of the other. Now, in consequence of the properties of poles and polars, well known today, here are the principal relations which are found to exist between these two figures.

1. If there is a system of a certain number of points lying on a line, then in the other figure there will be a system of exactly as many lines meeting in a point.
2. If there is a system of points lying on the same curve, then in the other figure there will be a system of exactly as many tangents to a curve of the same order.

1. If there is a system of a certain number of lines meeting in a point, then in the other figure there will be a system of exactly as many points lying on a line.
2. If there is a system of tangents to the same curve, then in the other figure there will be a system of exactly as many points lying on a curve of the same order.

A correction

Unfortunately for Gergonne, both of the statements numbered 2 above are false, as Poncelet speedily pointed out. Gergonne reprinted Poncelet's criticisms in the next edition of his own *Annales*, and replied with an article of his own, from which the following is taken.

We shall presently need two words, one to state that a curve is such that a line cuts it in m points (or that a curved surface is pierced by a line in m points) and the other to state that a curve is such that one can draw m tangents to it from a given point in its plane (or that a curved surface is such that one can find m tangent planes to it through a given line). But in order not to introduce new words here for which the repugnance of the public, however ill-founded it might be, would nonetheless be invincible, we shall adopt the word degree for the first case and the word class for the second, i.e. we introduce the following definitions:

Definition 1. A plane curve is said to be of the mth degree when it has m real or ideal intersections with a given line.

Definition 2. A plane curve is said to be of the mth class when m real or ideal tangents can be drawn to it from a given point in its plane.

[...]

That is what we should have said [in] our 17th volume, but the use of the word "order", which was entirely misplaced on that occasion, led us to commit an error, as we have already noted above, and we owe a sincere obligation to M. Poncelet, whose doubts, however vaguely expressed, made us re-examine our work and made us feel the necessity of putting it right.

However, the corrections to be introduced are neither very numerous nor very difficult. First of all in the entire memoir one may regard everything in the left hand column as exact since everything is deduced independently of the principle of duality by a very simple and rigorous analysis. It will be the same with the columns on the right (and they form the greater number) which related only to curves and curved surfaces of the second order, because these lines and surfaces are both of the second degree and the second class. But the entire correction of the memoir, according to the ideas which have been omitted, can perhaps be summed up in these few words: Replace the word order by the word degree in the column on the left and by the word class in the column on the right, understanding by these latter words what has been explained above.

Why did Gergonne introduce his study of duality, and what gains did he claim for it? His aim was to produce a guiding principle which would enable one to see how certain geometric ideas are related, and to deduce new ones, so unifying the study of geometry. It was as an example of what can be done that he wrote down two pairs of theorems about a curve and its dual. However, the second pair of theorems was wrong. The remedy was simply to change two words – seldom can so grave a mistake have been put right so deftly. But it will help us to see what the mistake was. It lies in the very last word. Gergonne claimed that the dual curve was of the same order as the original one. But this is only true if the original curve is a conic section, the case Gergonne understood best. When the curve is a cubic, say, the claim is generally false. The necessary mathematics had already been explained by Poncelet in his *Traité*, if it did not in fact go back to the handful of geometers of the 18th century.

There was worse to come. As the German mathematician Julius Plücker pointed out, the charm of duality was that if you did it twice you got back to where you started from. Dualising once replaced point with line and line with point. Dualising again should replace line with point and point with line in such a way that each point was returned to itself, each line to itself. But the fact of the matter was that if you dualised a curve (of degree higher than 2) the degree of the dual curve was higher than the degree of the original curve. That being the case, there was no way dualising twice could restore the original curve. How could the situation be put right, and Gergonne's delightful idea be made to yield correct results? Poncelet had some ideas in his long paper "Analyse des transversales" of 1832 [203], but they were not precise enough. He considered cubic curves and suspected, correctly, that cusps and double points on a curve had to be taken into consideration. The way forward was to be found by Plücker, who had followed the controversy between Gergonne and Poncelet, but brought a new, more algebraic, point of view to bear on the question of duality, but this is a matter for a future chapter.

6

Poncelet, Chasles, and the Early Years of Projective Geometry

6.1 What was done – differing opinions

To hear Poncelet tell it, what he invented was a uniform method of tackling problems in geometry, brought about by the use of ideal and imaginary elements, that reduced calculation to a minimum (and thereby made geometry easier to understand). To hear Cauchy tell it, Poncelet's methods were at best heuristic, and liable to mislead, and insofar as they led to correct conclusions it would be safer to use algebra (in particular, complex numbers) at least to validate the use of ideal methods if not to replace them entirely.

To hear Poncelet again, it would seem that this tremendous original achievement of his was somewhat neglected. To hear Chasles, as we shall in a few minutes, it would seem that the bulk of the credit should go to Monge anyway. Poncelet himself merely contested priority with the likes of Gergonne; Chasles' claim puts Poncelet entirely in the shadow of a man of the previous generation and makes one wonder about Chasles. What did he do? What were his motives in writing as he did?

And what would we say of the mathematics? Ideal chords seem very mysterious. The introduction of complex points undermines the intuitive, physical, visual element in the geometry. On the other hand, the idea that not only are all conics projectively equivalent, but also that pairs of conics are equivalent to pairs of circles is highly attractive. The idea of duality is surely delightful, but apparently it leads to paradoxes. What are we meant to make of all of this? The first point I wish to make is that that's history for you. It is a jumble of

J. Gray, *Worlds Out of Nothing*,
Springer Undergraduate Mathematics Series,
DOI 10.1007/978-0-85729-060-1_6, © Springer-Verlag London Limited 2011

this sort. It is not always, or even often, the case that a great mathematician does profound original work and the mathematical world immediately accepts and applauds it. There are occasions, and the early modern history of projective geometry is one of them, where the work itself was so confused that its profundity is still hidden.

Notice next that there are different ways to think about a complicated problem like this. We might choose to try to make sense of the thing by concentrating at least initially on the mathematics. Are we talking about good, if idiosyncratic, mathematics that people would eventually have to accept? Is it odd, or even flawed? Can we say precisely what the quality of Poncelet's work is, or was? That equivocation raises a genuine issue. Are we trying to say what Poncelet did that was so significant as seen from the 1820s, or what gives it its significance at some later date, such as today? We might agree that we would like a clear set of statements about what Poncelet did, in language faithful to his but not in his own words, just to be sure that we understand it. But what matters more is to tease out why it is important, and the answer to that question depends on when it is asked. Still, let us at least list the questions: what did Poncelet do? Why (or, to what extent) was it important in the 1820s and 1830s? Why is it worth talking about now?

We might choose to try to make sense of the matter in terms of personalities, academic rivalries and cooperations. This is to some extent the other side of the coin of evaluating the mathematics. If someone says something is important, does that tell us about the thing, or about the person making the judgement? In most walks of life, we learn to go either way on these issues – sometimes it's the thing, sometimes the person that the judgement illuminates. But that feels a little strange in the context of the history of mathematics, and I want to discuss the problem a little further. We are very used to saying that mathematics is correct (when it is), and only worth having if it is either right or can be put right. And when it is right then, somehow, it becomes valuable, because mathematics is, we all like to say, valuable.

Let me ask you, the reader, to reflect on such statements as these from your own experience for a minute. You made choices to get where you are, between mathematics and the sciences, or perhaps philosophy; in the balance of pure and applied mathematics that you have studied; in saying that this course or that one is not for you. Now you might allow that some courses that you didn't connect with are nonetheless valuable. But the world at large cannot afford to be so easy-going and tolerant. There are major issues to do with the value of mathematics that must be continually lived with and occasionally tackled: the balance of pure and applied, for example. The utilitarian argument is the easiest to follow. Sit down with the engineers, it says (or, these days, the computer scientists and bio-scientists), find out what they want and need by way of mathematics, and provide it. The number of these people is large, the work they do will benefit from your contribution, and so society will benefit. There are arguments along the lines of cohabitation: some topics have an intrinsic charm and a use (wavelets, differential geometry in string theory, and so on). The applied folk can apply, leaving the purer-minded to invent. There are arguments difficult to mount without sounding elitist which say that mathematics has its own, profound claim on our attention, and not to worry, it's amazing how it turns out to be useful (wavelets, differential geometry, ...).

I don't want to resolve these delicate issues for you, but to alert you to them and to take you back to the 1820s. If we see a disagreement about the importance of something, it can be possible to understand that disagreement by looking at the different perspectives of the protagonists in the dispute. Their motives might be base, they might reflect genuine, even major, differences in priorities. One thing that may be worth saying is that the present-day split between mathematics and science was nothing like so clear then. It is a creation of the (later) 19th century. There was a sense of the difference between useful and not-so-useful mathematics, and between precise and approximate uses of mathematics. But to give just one example, the major courses at the École Polytechnique used to run the two years of the student's life (first-year topics were followed in the second year by more advance presentations of the same topic), so naturally the instructor used to follow the students through, to ensure continuity. This in turn meant that the first-year calculus course, say, would be given by one person in one year and someone else in the next. Cauchy, whose name is so closely tied to the rigorisation of analysis, used to alternate in this fashion with Ampère, who is much better remembered for his work on current electricity (which is why his name survives in the term "amp"). Versatile though these people were, one cannot imagine such a thing today.

The point about the École Polytechnique was that it was the start of a student's higher education. After two years there, students left to go to one of the advanced and specialised écoles (such as mining engineering (École des Mines), the most prestigious of them all, or bridges and roads (École des Ponts et Chaussées)). There was a sort of selection and a sort of application process. So the point of the education at the École Polytechnique was to equip the young beginner with enough to go on to and benefit from the harder stuff that came later. This naturally strengthened the hand of those who advocated general methods, the basics, the fundamentals that everyone might be supposed to need, with the compromises appropriate to the skills of a beginner. But there was one remarkable feature of the École Polytechnique that more than counterbalanced this. There was only one École Polytechnique. It was the institution that every aspiring French mathematician, scientist, engineer wished to attend. It was the bottleneck through which everyone had to pass. Control that, and you control the education of the students; there was nowhere else for them to go. (It is not the case today, for example, that Oxford and Cambridge control higher education in Britain; in principle they could make a series of mistakes and become outclassed by places like Warwick.) Now the French Revolution had set itself the task of making a new kind of person: the citizen. What should go into the education of young people, including students, that makes them into citizens? Mathematics, of all subjects, was the most adaptable and seductive. To emphasise philosophy, or French literature, was to plunge into controversy. Who should be read, what views are to be acquired? But mathematics, when the ruling temper is if not atheist at least anti-clerical, has no need of the deity. And if times become more religious, why, mathematics has simply nothing to say about religion, and never did. Mathematics was at the centre of a Frenchman's education. It was therefore at the centre of the education provided at the apex of the system, the École Polytechnique itself. And mathematics is taught by mathematicians.

It was therefore mathematicians who had to decide what mathematics should be taught. Should they lean to the general, the practical, the readily intelligible if imprecise? What compromise should be struck with their military masters?

6.2 Institutions and careers

In the case of the École Polytechnique, the balance was slowly shifted away from the useful stuff and towards the rigorous material. When Cauchy introduced his much more rigorous analysis, the students were so ill-disciplined in their response that they had to be reprimanded and Cauchy was ordered to tone it down, but the point was established nonetheless. Leave mathematicians alone, as perforce military men must, and slowly they make the subject more rigorous and abstract and general. It is not a wrong or foolish interpretation of their mission, but it is not the only one. One of the subjects that was squeezed to make way for analysis was Monge's descriptive geometry. It didn't disappear; it lasted at the École Polytechnique until 1919, but it was cut back.[1]

As for Poncelet's projective geometry, Poncelet was among those who were based elsewhere, in his case at the School of Applied Artillery and Engineering in Metz, a position Arago had persuaded him to take in 1824. He became very interested in this, in particular in the experimental side, and this took him away

---

[1] I am indebted to one of the referees for pointing out that it was still part of the written entrance exam for both the École Polytechnique and the École Normale Supérieure until 1955.

from the topic of projective geometry. He was also hindered by the cumbersome publication system in France. His memoir on duality (reciprocal polars) was submitted in 1824, but Cauchy failed to report on it until 1828, and it was only published in 1829, by which time some of the key ideas had been published by Gergonne (not altogether accurately, as we saw). Poncelet wrote vigorously to Gergonne proclaiming his priority, and some of the correspondence was published, but the very public priority dispute wounded Poncelet greatly, and he suffered from a lack of interest in his work among the leading mathematicians in Paris. He took the sensible way out, transferred his papers to a different journal, the new *Journal für die reine und angewandte Mathematik*, (*Journal for Pure and Applied Mathematics*) edited in Berlin by Leopold Crelle, and gradually immersed himself in his work on machines, where he went on to have a very distinguished career. That left the subject of geometry in the hands of the likes of Gergonne and Chasles.

6.3 Chasles

Figure 6.1 Chasles

Gergonne was the editor of his journal, set up to avoid the publication problems just alluded to, and he published prolifically, but he folded his journal in 1831 on reaching the age of 60. Chasles then emerged as the principal geometer of the next generation. He had money from his father, who wished him to be a stockbroker, but this did not suit him and he retired in his thirties to study mathematics and its history. In 1829 the Belgian Academy of Sciences announced a prize for a philosophical examination of the different methods of modern geometry, in particular the theory of reciprocal polars. Chasles wrote a long essay arguing that the theory of projective geometry rests essentially on the notion of cross-ratio, which is preserved by any projective transformation. The work won the academy's prize and, before publishing it, Chasles asked if he could expand both the historical introduction and the notes on recent work. This was done, and his famous *Aperçu historique* [31] from which we have already quoted was published in 1837. Now the notion of cross-ratio was known to Brianchon, and is mentioned as such (without the name cross-ratio) in Poncelet's *Traité* [202]. But Poncelet preferred to rely on the concept of a harmonic division, which is the special case when the cross-ratio $= -1$, and the pairs of points A, B, and C, D are related this way: $\frac{AC}{BC} = \frac{AD}{DB}$. This is the cross-ratio that occurs naturally, so to speak, for example in poles and polars and complete quadrilaterals.

Chasles' *Aperçu historique* became and remains the first thing to consult when thinking about the history of projective geometry, although it is not without its faults. But it brought Chasles nothing but success. He was elected to the French Académie des Sciences in 1839 as a corresponding (i.e. junior) member. In 1841 he was appointed to the École Polytechnique, where he taught geodesy, astronomy and applied mechanics until 1851, and from 1848 he had a personal chair at the Sorbonne where he taught higher geometry. He was not just a historian. In fact, his subsequent historical works have not worn well, but he did interesting work on a very subtle branch of geometry called enumerative geometry, which counts the number of configurations of particular kinds. This proved to be an interesting subject, although proofs in it were often less than rigorous. Its most famous expositor was Hermann Schubert, and I have taken its most eloquent result from him (not Chasles): the number of cubic space curves touching 12 arbitrarily given quadric surfaces is 5,819,539,873,680. There's something you probably didn't know! That the number is actually correct was proved only in the last 20 years.

The most interesting thing about Chasles' *Aperçu historique* in the present context, and one of the biggest problems with his work, was the fact that he could not read German. Had he been able to do so, he would have known that the very concept of cross-ratio was already in the literature (as we shall see shortly). He could also have given a richer account, especially of the more recent history of the surface. But in the French context, his use of cross-ratio, developed at length in his *Traité de géométrie supérieure* (1852) [32] with nothing more exotic than imaginary points, did a lot to establish that projective geometry was a legitimate, rigorous discipline in no way dependent on the principle of continuity, the theory of ideal points, and other mind-stretching devices used by Poncelet.

6.4 What was done?

Who, then, brought projective geometry to life in France? There is no doubt that the modern consensus says Poncelet. But as I hope I have made plain, the technical content of Poncelet's projective geometry was often found to be obscure, if not downright unacceptable, and later mathematicians simply wrote it out of the subject. Talk of ideal chords never caught on. The projective geometry that succeeded is at least as much Chasles' creation. Is the subject then really his vision of a non-metrical geometry based on transformations, and so on, a way of reasoning about figures that is broader than Euclid's? As we have seen, Chasles was among those who would reply that the essential step was due to Monge. It was Monge who said: study a figure by studying its orthogonal projections. All Poncelet did was introduce central projections, which consummate Monge's programme by being the most general transformations that send lines to lines.

Such a reply would be unfair. Monge's achievements were in a restricted context, the highly utilitarian but not mathematically very profound theory of descriptive geometry. He did much else besides, but Poncelet's work is not, so to speak, a mere footnote to that of Monge. There is more truth in the idea that it is Chasles who has lost out, that without his rewriting of Poncelet's projective geometry the subject would have been done in a much more algebraic style. But this is a discussion we can have more profitably in a few chapters' time.

It would be reasonable to say that it was Poncelet and only Poncelet who had the vision of a new theory of geometry, as general in its methods as algebra, and which did not need algebra to proceed. This revivified geometry had its own important problems, and a productive method of dealing with them, based on the ingenious use of transformations to reduce the general figure to a simple one. It was also related, in a way I have not had time to discuss, to the familiar, metrical, Euclidean geometry. This vision exceeded that of Monge, but it did not succeed until it was domesticated, a process due in France to Chasles and to others in Germany.

6.5 Chasles, Steiner and cross-ratio

The broad thrust of Poncelet's approach to projective geometry was generally taken to establish that there was a new topic in geometry, or at least a new approach to it, the one he had referred to as non-metrical geometry. But the details of Poncelet's approach were not generally adopted. Instead, his theory was rewritten in terms of the concept of cross-ratio. This was done in one of two ways.

In the first way, one accepts the definitions of conic section already in place, and specifies a certain class of non-metrical transformations of the plane, the perspectivities. These transformations are somehow extended to be defined for points at infinity. A sequence of perspectivities is a projectivity, and one shows that a perspectivity, and hence a projectivity, maps straight lines to straight lines, conics to conics and preserves cross-ratio.

Chasles' later approach is in some ways more radical. He began his *Traité de géométrie supérieure* of 1852 [32] with the concept of cross-ratio (which he called the anharmonic ratio of four points), and defined a transformation of the plane (extended with a line at infinity) to be a homography if it mapped straight lines to straight lines and preserved cross-ratio (in fact, it can be proved that a continuous map sending lines to lines must preserve cross-ratio). He gave a construction for a homography taking any four points (no three on a line) to any four points (no three on a line) and then showed that this homography is unique, because its effect on any other point is completely determined. But he did not show how a homography can be realised as a sequence of perspectivities.

In more detail, Chasles argued as follows. He defined the concept of cross-ratio of the four points A, B, C, D on a line to be $\frac{AC}{AD} : \frac{BC}{BD}$ (which is the same as $\frac{AC \cdot BD}{AD \cdot BC}$); and he defined the concept of the cross-ratio of four lines OA, OB, OC, OD through a point O as $\frac{\sin(AOB) \cdot \sin(COD)}{\sin(AOD) \cdot \sin(COB)}$ (we shall write this expression as $CR(O; A, B, C, D)$). He showed that if the points A, B, C, D lie on a line then $CR(O; A, B, C, D) = CR(A, B, C, D)$ and therefore the cross-ratio of four lines through a point is a projective invariant. (See Figure 6.2.)

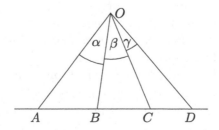

Figure 6.2 Cross-ratio in terms of angles

His terminology was a little different: what we have called the cross-ratio he called the anharmonic ratio of four points, because, he explained, when it was -1 the four points were in a relationship the Greek geometers had called "harmonic".

He showed that the uniqueness of the fourth harmonic point means that if D is the fourth harmonic point to points A, B and C then $CR(A, B, C, D) = -1$, and extended this uniqueness property to show that, given points A, B and C then if two points D and E are such that the cross-ratios are equal, $CR(A, B, C, D) = CR(A, B, C, E)$, then $D = E$. It follows that a similar property holds for lines through a common point. He gave a construction for the fourth point, D, given three collinear points A, B and C and a value λ for the anharmonic ratio of A, B, C and D. As Chasles showed, it follows from this that a homography of one line to another is uniquely determined by its effect on three distinct points. It is clear that there is a homography taking three points A, B, C on a line ℓ to three points A', B', C' on a line ℓ', because this can be done by a sequence of two perspectivities. Uniqueness of the homography doing this follows from the result just mentioned.

In his *Traité des sections coniques* [33], which was intended to follow the *Traité de géométrie supérieure*, Chasles defined conic sections via the concept of cross-ratio, and checked that conics on his definition were precisely the conics on the usual definition. It now followed that conics were mapped to conics by homographies.

Chasles based his study of conics on the following result about four points A, B, C, D on a conic and the cross-ratio $CR(P; A, B, C, D)$, where P is an arbitrary fifth point on the conic.

Theorem 6.1 (The projective definition of a conic)

(i) If Q lies on the conic C then

$$CR(P; A, B, C, D) = CR(Q; A, B, C, D),$$

and

(ii) if

$$CR(P; A, B, C, D) = CR(Q; A, B, C, D)$$

then Q lies on the conic C.

He showed that this cross-ratio is independent of the position of the point P by transforming the conic to a circle (which he observed can always be done) and then observing that it was trivially true that the cross-ratio $CR(P; A, B, C, D)$ is a constant when the points A, B, C, D and P lie on a circle. This is because the angles at P do not depend on the position of P but only on the fixed arcs AB, etc. With this result behind him, Chasles could define a conic as the curve through four points A, B, C, D with the property that for any point P on the curve the cross-ratio $CR(P; A, B, C, D)$ has the same constant value, and if tangents are drawn to the curve at the four points they cut any fifth tangent in four points that have the same cross-ratio. It is necessary to prove that the conic thus defined consists of exactly the points on the usual conic through A, B, C, D and P, and this Chasles also did (it forms part (ii) of Theorem 6.1).

To prove part (ii) of the theorem, suppose that Q is a point not on the circle but such that

$$CR(Q; A, B, C, D) = CR(P; A, B, C, D),$$

where P is an arbitrary point on the circle (other than A, B, C, D). Let QA meet the circle at P, and draw the lines PB, PC and PD. Draw the line BC and let it meet the lines PA and PD at the points A' and D' respectively, and the line QD at the point D''. Then we have, by the uniqueness of the fourth point making a given cross-ratio with three given points:

$$
\begin{aligned}
CR(A', B, C, D') &= CR(P; A, B, C, D) \\
&= CR(Q; A, B, C, D) \\
&= CR(A', B, C, D''),
\end{aligned}
$$

which implies that $D' = D''$, which is impossible. We deduce that Q lies on the circle.

Chasles also gave (independently, it seems) Steiner's nice construction of a conic. One is given three points, A, B and C, and a further two points O and O' on none of the lines AB, BC, CA. Consider the three lines OA, OB and OC and a fourth line ℓ through O, and also the three lines $O'A$, $O'B$ and $O'C$ and a fourth line ℓ' through O', where ℓ is arbitrary and ℓ' is so chosen that the two cross-ratios are equal: $CR(OA, OB, OC, \ell) = CR(O'A', O'B', O'C', \ell')$. Chasles showed (in [33, ch. XXV]) that as the line ℓ varies, the common point of ℓ and ℓ' traces out a conic, indeed the conic through the five points A, B, C, O and O'. Early in the book he had established the special case that if A, B and C lie on a line then the locus is a straight line.

To see this, note first of all that, given the points O, A, B and C, for each value λ of the cross-ratio there is a unique position of the line ℓ such that the cross-ratio of the four lines OA, OB, OC and ℓ is λ. The same is of course true of the lines through the point O', so it follows that for each value λ there is a unique position of the lines ℓ and ℓ' such that the two sets of cross-ratios are equal. It follows that for each value of λ there is a point P on the lines ℓ and ℓ' such that $CR(O; A, B, C, P) = CR(O'; A, B, C, P)$.

To show that this point P lies on the conic through O, O', A, B and C, observe that, by the equalities of the cross-ratios and the definition of a conic, the conic through A, B, C, P and O passes through O', and the conic through A, B, C, P and O' passes through O. But these conics must be the same (because the points A, B and C do not lie on a line), so the point P lies on the conic through A, B, C, O and O'. Therefore the locus of the point P is a subset of the conic through the points A, B, C, O and O'. To establish the converse, consider the conic through the points A, B, C, O and O', and let P be a point on it. From the definition of a conic it follows that the cross-ratios are equal: $CR(O; A, B, C, P) = CR(O'; A, B, C, P)$. So P lies on the appropriate intersection of the lines through O and O', as required. This result gives a method of constructing arbitrarily many points on a conic by ruler alone.

Such an account should really be supplemented by a derivation of the fundamental theorems of linear projective geometry – the theorems of Pappus and Desargues – and an account of duality, to give an account of what the fundamental transformations can do. Chasles seems to have felt that he had done justice to the theorems of Pappus and Desargues in his earlier *Aperçu*. As for duality, he discussed poles and polars and what he called correlative transformations (that send points and lines in one figure to lines and points in another, exchange collinearity and concurrence, and preserve cross-ratio) but he did not attach as much importance to it as later writers were inclined to do.

6.6 Extracts from Chasles' *Aperçu historique*

6.6.1 Chasles on descriptive geometry

In recent times, after a rest of almost a century, pure Geometry has been enriched by a new doctrine, descriptive geometry, which is the necessary complement to the analytic geometry of Descartes and which, like it, must have immense results and mark a new era in the history of geometry.

This science is due to the creative genius of Monge. It embraces two objects:

The first is to represent all bodies of a definite form on a plane area, and thus to transform into plane constructions graphical operations which it would be impossible to execute in space.

The second is to deduce from this representation of the bodies their mathematical relationships resulting from their forms and their relative positions.

This beautiful creation, which was initially intended for practical geometry and the arts which depend on it, really constitutes a general
theory, because it reduces to a small number of abstract and invariant
principles and to easy and always correct constructions, all the geometric operations which can be involved in stone cutting, carpentry,
perspective, fortifications, gnomonies, etc, and which apparently can
only be executed by mutually incoherent processes, which are uncertain and often scarcely rigorous.

But besides the importance due to its first intention, which gives
a character of rationality and precision to all the constructive arts,
descriptive geometry has another great importance due to the real
services which it renders rational geometry, in several ways, and to
the mathematical sciences in general. [For this reason] geometry thus
reaches a state where it can most easily lend its generality and its intuitive evidence to mechanics and the physico-mathematical sciences. [31,
pp. 189–190]

6.6.2 Chasles on Monge and his school

Monge gave us, in his *Traité de géométrie descriptive*, the first examples of the utility of the intimate and systematic alliance between
figures in three dimensions and plane figures. It is by such considerations that he proved, with rare elegance and perfect evidence, the
beautiful theorems which constitute the theory of poles of curves of
the second degree; the properties of centres of similitude of three circles taken two by two whose centres lie three by three on a line, and
various other figures of plane geometry.

Since then, the pupils of Monge have cultivated this truly new kind
of geometry with success, so that one has often, and with reason, given
them the name of the school of Monge, and it consists, as we shall say,
in introducing into plane geometry considerations of the geometry of
three dimensions. [31, p. 191]

Although declining to give details for reasons of space, Chasles in a footnote
specifically mentioned the following as members of the school.

Brianchon, who in a memoir of 1810 presented new and extensive
reflexions on the subject to which, Poncelet tells us, he owes his own
initial idea about the numerous beautiful geometrical researches contained in his *Traité des propriétés projectives* and Gergonne who per-
formed the useful service of writing his own works, always imprinted

with profound philosophical insight, and of founding his *Annales de Mathématiques* for the productions of former pupils of the École Polytechnique.

6.6.3 Chasles on Monge's work

Ancient geometry is a slew of figures. The reason is simple. Because one then lacked general and abstract principles, each question could only be treated in a concrete state, with the figure which was the object in question, and about which the only way forward could be to discover the elements necessary for the proof or the sought-for solution. [...]

This lack in ancient geometry was one of the relative advantages of analytic geometry where it was avoided in a most happy manner. So one could ask accordingly if there was not also a way of reasoning without the continual assistance of figures in pure and speculative geometry, for even when their construction is easy they are a real inconvenience and quite tire the spirit and exhaust the mind.

The writings of Monge and teaching of this illustrious master, whose style has been preserved by one of his most famous disciples (Arago), have resolved this question. They have taught us that it is enough, now that the elements of science have been created and are very extensive, to introduce into our language and our ideas of geometry, some general principles and transformations analogous to those of analysis, which, in making a truth known to us in its primitive purity and in all its facets, provide easy and fertile deductions by means of which one arrives naturally at one's goal. Such is the spirit of Monge's doctrines; and although his descriptive geometry, which provides us with examples of it, by its very nature makes essential use of figures, it is only in its effective and mechanical applications, where it plays the role of a tool, that it operates like this; but no one more than Monge could conceive of and do geometry without figures. It is a tradition in the École Polytechnique that Monge knew to an extraordinary degree how to conceive of the most complicated forms in space, to penetrate to their general relation and their most hidden properties with no other help than his hands, whose movements accorded most admirably with his words, sometimes difficult but always with a true eloquence appropriate to the subject, in the neatness and precision, the richness and profundity of ideas. [31, pp. 208–209]

6.7 A quick introduction to modern projective geometry

6.7.1 The real projective plane

The beginner may be forgiven for finding all this talk of points and even a line "at infinity" dangerously sloppy. One way to make rigorous sense of it is to return to the idea of a point-like eye looking in all directions. This "eye" sees all the points of the projective plane; it "sees" points in space lying on the same line of sight as the same point, it even "sees" real points lying in diametrically opposite directions as the same. To remove the inverted commas, we formalise the real projective plane as the set of all Euclidean lines through a fixed point in three-dimensional Euclidean space. More precisely (and with respect to a fixed but arbitrary origin in $\mathbb{R}^3$) the points of the real projective plane are the one-dimensional vector subspaces of the vector space $\mathbb{R}^3$ from which the origin has been deleted. We denote the real projective plane by $\mathbb{RP}^2$.

So a point of the real projective plane, $\mathbb{RP}^2$, is an equivalence class of non-zero vectors $\mathbf{v}$, where $\mathbf{v}$ is equivalent to $\mathbf{w}$ if and only if there is a non-zero k such that $\mathbf{v} = k\mathbf{w}$. In terms of coordinates, a vector $\mathbf{v}$ in $\mathbb{R}^3$ with components (x, y, z) is equivalent to any vector of the form (kx, ky, kz) with $k \neq 0$. If, as is usual we denote an equivalence class by square brackets, a projective point is something of the form $[x, y, z]$ where not all of x, y, z are zero (the origin has been deleted) and $[x, y, z]$ and $[kx, ky, kz]$ denote the same projective point. A projective line in the real projective plane is a set of projective points whose associated vectors form a plane in $\mathbb{R}^3$.

The transformations of projective geometry are those that are induced from the linear transformations of the real vector space $\mathbb{R}^3$. These linear transformations map one-dimensional vector spaces to one-dimensional vector spaces, and two-dimensional ones to two-dimensional ones, so they map projective points to projective points and projective lines to projective lines. For completeness, it would have to be shown that every map of the projective plane mapping projective points to projective points and projective lines to projective lines is induced from a suitable linear transformation; I omit this.

Before generalising this concept, it's worth looking at the projective plane in a little detail. Indeed, how do we look at it? A good way is to observe that every projective point, thought of as a one-dimensional vector subspace of $\mathbb{R}^3$, meets the unit sphere in two diametrically opposite points. So we may think of the projective plane as the unit sphere with diametrically opposite points identified. To see this, we consider just the northern hemisphere with diametrically opposite points on the equator identified. We can almost see all of this, but we must remember that when a moving point sails south of the equator it reappears in the northern hemisphere but "round the other side". An amusing exercise is to follow the strip around the 0 and 180 lines of longitude up from one part of the equator, over the north pole, and down the other side and satisfy yourself that in the projective plane the strip becomes a Möbius band.

Another way of looking at the projective plane is to surround the origin in $\mathbb{R}^3$ with a cubical half-box made up of the planes with equations $x = 1$, $y = 1$, $z = 1$. Lines through the origin meet the plane with equation $z = 1$ unless they are parallel to that plane. These are the lines composed of the vectors $(x, y, 0)$. The other lines are the lines composed of the vectors (kx_0, ky_0, kz_0), $z_0 \neq 0, k \neq 0$, and they meet the plane $z = 1$ in points with coordinates $\left(\frac{x_0}{z_0}, \frac{y_0}{z_0}, 1 \right)$. So the plane $z = 1$ is a picture of the projective plane in which projective points either appear as points or they don't appear at all. However, it is easy to see that every projective point makes an appearance on at least one of the three specified planes. Thus we re-establish the transfer of information between Cartesian coordinates (x, y) and projective coordinates $[x, y, 1]$.

As for the transformations of projective geometry, it is an interesting exercise to check that there is always exactly one projective transformation taking any ordered set of four given projective points to any other such set, provided that no three of these points in either case lie on a projective line.

The key things to note here are that linear algebra immediately gives us a transformation taking three points (not on a line) to three given non-collinear points. We may, for example, choose the first set of points to be a basis $\{e_1, e_2, e_3\}$ of $\mathbb{R}^3$, and map them to any given linearly independent set $\{a_1, a_2, a_3\}$. It is elementary to write down an invertible matrix doing this: its ith column is the column vector corresponding to a_1. But at the level of projective geometry, we may replace each a_1 by $k_i a_1$, and this allows us to adjust the matrix and send a fourth projective point, such as $[1, 1, 1]$, to an arbitrary fourth point. So to map four projective points to four projective points (no three on a line) we map the first set to $\{e_1, e_2, e_3, [1, 1, 1]\}$ by the inverse of the transformation just found, and compose with a transformation mapping that set to the second given set of points.

You can also check that an ordered set of four points on a projective line may be sent to any other such set if and only if the cross-ratio of the two ordered sets is the same. For this, a definition of cross-ratio is needed. The cross-ratio of four collinear projective points a_i, $i = 1, 2, 3, 4$, is obtained as follows. By collinearity, $a_3 = \lambda_1 a_1 + \lambda_2 a_2$ and $a_4 = \mu_1 a_1 + \mu_2 a_2$. The cross-ratio is then defined to be $\frac{\lambda_2 \cdot \mu_1}{\lambda_1 \cdot \mu_2}$. It can be checked that this quantity is well defined (it does not depend on the choice of vectors representing the projective points) and it is invariant under projective transformations.

6.7.2 Projective spaces

Now that we have a definition of the real projective plane, it is easy to see what
the definition of n-dimensional real projective space must be – just follow the
above argument starting with $\mathbb{R}^{n+1}$. Similarly, we may start with the complex
numbers and a complex vector space instead of the real numbers, and obtain
the complex projective n-space $\mathbb{CP}^n$. We may even start with an arbitrary
field.[2]

While that is enough for many fascinating topics in mathematics, geometers
want to go further. As this book indicates, one could start with an entirely
axiomatic account of projective space in terms of some undefined entities called
points and lines and some incidence properties. The question then arises of how
closely related the concepts are of a projective space defined over some field and
an axiomatically defined projective space. Is every projective space defined by a
set of axioms in fact a projective space defined over a field? Spaces defined over
a field have coordinates in the way described above, which makes them easy
to understand. However, the answer to the question is very definitely no: there
are axiomatically defined projective spaces that cannot be coordinatised at all.
A hint of this emerges with the issue of the relationship between Desargues'
theorem and Pappus's theorem.

In between there are projective spaces whose coordinates lie in a skew field
(a non-commutative system) but not in a field. It turned out that Pappus's
theorem is true in a projective geometry if and only if the coordinates are
drawn from a commutative number system, and Desargues' theorem is true
if and only if the number system is associative, so Pappus's theorem implies
Desargues', and there can be projective geometries in which both theorems are
true, in which Desargues' theorem is true but Pappus's is false, and geometries
in which they are both false. However, as is briefly discussed below, Desargues'
theorem is automatically true in any projective geometry of dimension three or
more, so one has a limited and for that reason a special set of possibilities. For
example, Pappus's theorem is not true in the quaternionic plane, and there is
a Cayley plane obtained by starting with Cayley's octonions, but there is no
Cayley space of dimension three or more.[3] I cannot resist borrowing a comment
from one of the most helpful of the referees: the Cayley plane is one of the most
beautiful geometric objects "the *panda* of geometry".

---

[2] For a good account, see P. Samuel [216].
[3] Quaternions and octonions are described in the book by H.-D. Ebbinghaus et al.,
Numbers, 1991 [58, ch. 7].

Euclidean Geometry, the Parallel Postulate, and the Work of Lambert and Legendre

Euclid's *Elements* is a set of books of great antiquity, written around 300 BC. Our knowledge of them derives from a few copies of copies from shortly before 1000 AD – nearer to us in the present day than to Euclid himself. We rely, and have relied for four hundred years, on a succession of editions and commentaries by many people. This transmission with commentaries was reasonably well established by 1600. The best modern edition was published just before 1900. It is the work of a tireless Danish scholar, J. L. Heiberg, and is the basis for the standard English edition, that of Sir T. L. Heath.[1]

The structure of Euclid's *Elements* is complicated. There are 13 books on a variety of topics. Typically, they open with definitions. There are some general rules of deduction (called Common Notions) and five postulates of a more geometric nature, of which the fifth (the parallel postulate) says

> That, if a straight line falling on two straight lines make the interior angles on the same side less than two right angles, the two straight lines, if produced indefinitely, meet on that side on which are the angles less than two right angles.

The first figure in Figure 7.1 depicts the parallel postulate according to Euclid: when $\alpha + \beta < \pi$, the lines k and m meet where they appear to.

---

[1] Currently a Dover paperback, recently reissued by Green Lion Press in a lightly revised edition [70].

J. Gray, *Worlds Out of Nothing*,
Springer Undergraduate Mathematics Series,
DOI 10.1007/978-0-85729-060-1_7, © Springer-Verlag London Limited 2011

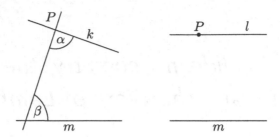

Figure 7.1 The parallel postulate

The second figure in Figure 7.1 depicts the parallel postulate as given by Playfair's axiom: in a plane, given a line m and a point P not on the line, there is a unique line through P that never meets m. This line is called the parallel to m through P.

(These postulates do not say the same thing, but in the presence of Euclid's other axioms each implies the other, so they are equivalent.)

The parallel postulate (in either form) permits one to prove such basic theorems as these: the angle sum of a triangle is π, and Pythagoras's theorem.

In fact, without it, there's apparently not much one can say. Euclid's *Elements* manage some 28 theorems, including results about congruence and the isosceles triangle theorem, but, significantly, neither of the results just quoted above. But, it is a strange thing to assume. As the sum of the two angles gets closer to π, it is clear that the point where the lines meet gets further and further away, in principle, beyond the galaxy. It is not a statement that seems either logically forced upon one, or true on impeccable empirical grounds.

In fact, it is a profound remark about straight lines. Aristotle has been quoted as saying that "If this (apparently he was waving a rod around) is what we mean by a straight line, then the angles in a triangle make two right angles" [237]. The significance of this remark will gradually become apparent.

Because the parallel postulate is not inevitably or transparently true (but in the belief that it is true) attempts were made to deduce it. The practice was: start with Euclid's *Elements*, remove the parallel postulate, and try using only what remains to prove the parallel postulate. Success in this enterprise would mean that you knew the parallel postulate was true not because you had had to assume it, but because it followed from assumptions that no one doubted.

There were many such attempts:[2] Ptolemy made one, as did Proclus, one of our best surviving sources for Greek mathematics, also numerous writers in Arabic too, all of whom, perforce, we leap over to arrive at the Oxford

---

[2] Among other accounts, see Gray [100].

mathematician John Wallis, who, in the second of his two lectures given on the evening of 11 July 1663, argued as follows [246]: take any two points A and B on a straight line c, and lines a through B and b through A making angles α and β, such that $\alpha + \beta < \pi$. To construct a point C on a and b thereby showing that they meet, draw through B a line b' enclosing an angle α with the line c. Slide b' along c until it overlaps a, keeping it all the time at an angle α to c. Clearly at some point it meets a; we exhibit one such line b'', and call the appropriate points on c and a, A_1 and C_1 respectively. Now draw a triangle ABC on AB similar to A_1BC_1, and then C must lie on a and b as required.

On the face of it this is a direct proof of Euclid's fifth postulate, formulated as it was in the original, and so it is, if we are willing to incorporate Wallis's assumption that one can draw arbitrary similar triangles as an axiom. However, the concept of form independent of size is no more self-evident than the concept of parallels, and mathematicians, Wallis among them, recognised that they must accept a weaker conclusion: the existence of similar triangles is equivalent to the existence of parallel lines. In other words, if there is to be a geometry in which the parallel postulate fails, then it can contain no figures of the same shape but arbitrarily different sizes. Figures cannot shrink or expand without distortion.

All Wallis's argument actually shows is that: Euclid's *Elements* − parallel postulate + arbitrary scaling $\Leftrightarrow$ Euclid's *Elements* with the parallel postulate.

7.1 Saccheri

In 1733, the year of his death, a Jesuit priest, Girolamo Saccheri, published a book *Euclides ab omni naevo vindicatus* (*Euclid freed of every flaw*) [214] in which he took Euclid's *Elements* − parallel postulate, and showed that:

HAA:[3] if there is a triangle with angle sum $< \pi$, then every triangle has angle sum $< \pi$;

HRA: if there is a triangle with angle sum $= \pi$, then every triangle has angle sum $= \pi$;

HOA: if there is a triangle with angle sum $> \pi$, then every triangle has angle sum $> \pi$.

Of these, HRA is the case in Euclid's *Elements*.

---

[3] The abbreviations derive from Saccheri's name for the hypotheses of the acute angle, the right angle and the obtuse angle respectively.

HOA, he showed, led to a contradiction.[4] That left him with HAA. If he could show that it too led to a contradiction then he would have shown that Euclid's was the only consistent geometry. Before looking at what he did in that direction, ask: what is wrong with spherical geometry? It exemplifies the HOA. The answer is that spherical geometry is not Euclid's *Elements* − parallel postulate + HOA, because it does not permit the indefinite extension of lines. That said, it should be clear that the existence of spherical geometry, if only in the form of the spherical trigonometry used by astronomers, must have some bearing on the enterprise of sorting out Euclid's *Elements*.

Saccheri's attempt on the HAA is interesting. It contains many true, proved and unexpected results. For example, it introduces the concept of the asymptotic parallel. To explain this, observe that the HAA geometry is one in which given a line l and a point P not on the line, there are infinitely many lines through P that never meet l. All these lines may be called parallels to l through P. Saccheri showed that each one of them has a common perpendicular with the base line l, except for one in each direction, which separates the parallels from the lines that meet l. These lines he called the asymptotic parallels through the point P to the line l.

However, Saccheri's book climaxes in Prop. XXXII with an attempt to argue that asymptotic parallels have a common perpendicular at their point of intersection at infinity, which is an indefensible position given Saccheri's other assumptions about geometry.

Here, I should say, arises one of those interesting disagreements between historians. I originally dismissed Saccheri's claim as 'meaningless nonsense', without realising that it had been given a much more sympathetic reading by Greenberg in his *Euclidean and non-Euclidean geometries*, who claims on p. 218 that "There is no very serious error in Saccheri's treatise." In support of this position, he quotes Saccheri's words (from the Scholion on p. 233) on the contrast between the refutations of the HOA and the HAA, which are (in full):

> It is well to consider here a notable difference between the foregoing
> refutations of the two hypotheses. For in regard to the hypothesis of ob-
> tuse angle the thing is clearer than midday light; since from it assumed
> as true is demonstrated the absolute universal truth of the controverted
> Euclidean postulate, from which afterward is demonstrated the abso-
> lute falsity of this hypothesis; as is established from P. XIII. and P.
> XIV. But on the contrary I do not attain to proving the falsity of the
> other hypothesis, that of acute angle, without previously proving; that
> the line, all of whose points are equidistant from an assumed straight

---

[4] This is not exactly obvious. For proofs, see Gray [100] or, for example, Lobachevskii [153].

line lying in the same plane with it, is equal to this straight, which itself finally I do not appear to demonstrate from the viscera of the very hypothesis, as must be done for a perfect refutation.

This requires a little explanation. Saccheri is here admitting to a defect in his second proof Prop. XXXII, namely that it is less than visceral in its force. In fact, after the argument that two lines have a common perpendicular at their common point at infinity, Saccheri continued to address the HAA for many more pages. He was not happy with his first proof of Prop. XXXII, and gave another, in Part II of the book, which he introduced by saying that

And here I might safely stop. But I do not wish to leave any stone unturned, that I may show the hostile hypothesis of acute angle, torn out by the very roots, contradictory to itself.
However this will be the single aim of the subsequent theorems of this Book.

Unfortunately for Saccheri, his argument in this part of the book is simply wrong: the crucial error lies in his attempt to show that in a quadrilateral $CABD$ with equal sides CA and DB and right angles at A and B the length of the curve joining C and D that is everywhere equidistant from AB is equal in length to AB. Now, if this attempt to tear out the HAA by its roots was not sufficiently visceral for Saccheri, nonetheless he concluded that it was conclusive, and the "hypothesis indeed of acute angle from this was demonstrated false" (p. 241).

If we set aside this aspect of his conviction as resting on a flawed argument, what can we make of the first refutation (concerning common perpendiculars at infinity)? After proving, to his satisfaction, in Prop. XXXII that "The hypothesis of acute angle is absolutely false; because repugnant to the nature of the straight line" because it allowed two lines to have a common perpendicular at their common point at infinity, Saccheri embarked on a series of lemmas to produce "the most exact demonstration". Lemma 1 is an argument (missing from Euclid's *Elements*) that two straight lines cannot enclose an area. Then he remarked that a straight line can always be drawn joining any two points, and then (in Lemma 2) he proved that two straight lines cannot have one and the same segment in common. The argument is long; it was difficult for him because it is one of those topics that are implicit in the *Elements* but which cannot be deduced from it without some discussion of what a straight line is taken to be. In Lemma 3 he showed that two straight lines that have a point in common do not touch each other but cross there. Lemma 4 states that "Every diameter bisects its circle, and the circumference of it." Lemma 5 shows that "Among rectilinear angles, all right angles are exactly equal to one another,

without any deviation even infinitely small." Now Saccheri drew all this together and deduced the existence of a unique straight line perpendicular to
a given straight line at a given point, and observed that it cannot split into
two. This enabled him to spell out exactly why the HAA was repugnant to the
nature of the straight line.

My reading of this is that Saccheri was completely convinced by his first
proof, although he could see that some additional remarks, about matters one
presumes Euclid had taken for granted, were in order. But recognising that
the additional remarks, even if completely valid, lack the simplicity and evident logical force of the refutation of the HOA, he then gave a second proof,
which also lacks the simplicity he would have liked. He found both proofs convincing, and a modern reader must find both proofs are flawed. Since there is
no possibility of there being two lines with a common perpendicular at their
common point at infinity in a geometry based on the HAA, I would have to
conclude that there is, after all, a serious error in Saccheri's work. This does
not rule out novel geometries that depart further from Euclid's and in which
Saccheri's words can be redeemed, but the only interpretation I can place on
his words would be to admit the possibility that two distinct lines through a
point can, at least initially, point in the same direction. This is not a prospect
an 18th-century geometer would have wanted to contemplate.

Enter Johann Heinrich Lambert.

7.2 Lambert

The next mathematician of interest in the study of the problem of parallels
is the Swiss mathematician Johann Lambert.[5] He was a wide-ranging thinker,
one of the first to consider our galaxy finite and only one island universe out
of many – a distinction he shares with Kant (see Lambert's *Cosmologische
Briefe* (*Cosmological letters*), 1761 [146]). He was an early member of the Berlin
Academy of Sciences with Euler, and contributed to many areas of knowledge:
to photometry and optics (1760), where he established Lambert's cosine law for
the intensity of a refracted beam of light as well as the inverse square law for
the intensity of illumination at a distance; to statistics; to pure mathematics,
where among other things he was the first to prove that π was irrational (1766);
to logic, where he extended Leibniz's symbolic studies; to astronomy, where he

---

[5] Lambert is well worth a full-length study. For now, see the article on him in the
Dictionary of scientific biography [90] and the article by Gray and Tilling [99].
Kirsti Andersen tells me that Lambert did not want a picture of him to appear in
his lifetime and the picture of him comes from a caricature that his friend Johann
Bernoulli III published in 1786, saying it was a good likeness.

Figure 7.2 Lambert

established Lambert's law for the parabolic orbit of a comet (1761). In his *Freye Perspektive* (1759, 1774) [145] he not only discussed perspective drawing but also the intriguing problem of what geometric constructions can be made with a straight edge alone and only a fixed compass. Like Kant he sought to reintroduce the a priori into science, and he wrote purely philosophical works which now meet with differing assessments of their merits:

His *Theory of parallels* [147] was not published in his lifetime, unlike Saccheri's work, which had attracted considerable attention for a short while and of which Lambert was probably aware. However, roles were thereupon reversed, and Saccheri's book was largely forgotten almost throughout the 19th century, while Lambert's remained known to the experts at least. Lambert was possibly unsatisfied with his own study, which often happens with good work that cannot be carried through to its conclusion. The book was published after his death by Johann Bernoulli III in 1786.

Lambert became interested in the topic when he learned of a thesis on 28 attempts on the parallel postulate by Klügel, a student of Kaestner's, a mathematics professor at Göttingen. This inspired him to take up the subject himself. He endorsed Saccheri's trichotomy, and re-derived a contradiction on the HOA. But he could find no fault with the HAA. He could see that if this or that alternative proposition was true the HAA would self-destruct, but no fatal flaw emerged.

He did notice that if the HAA was true there was what he called an absolute measure of length. To explain that, observe that we can define angles over the radio (as fractions of one complete circle). But we cannot define lengths without previously agreeing on a unit. So we say angles have an absolute measure of length, but lengths do not. However, to each length, Lambert noticed, one can associate a unique angle, the angle in the equilateral triangle with that side. That triangle is unique because on the HAA similar triangles are congruent (Wallis's theorem). For a time he thought an absolute measure of length was a logical absurdity, but finally he decided that it wasn't.

Lambert also noticed that the area of a non-Euclidean geometry triangle with angles α, β and γ is proportional to $\pi - (\alpha + \beta + \gamma)$, and as a result proclaimed that the geometry would be true on an imaginary sphere. To explain this, notice that if the radius of a sphere is R, then the area of a triangle with angles α, β and γ is $R^2 (\alpha + \beta + \gamma - \pi)$ and that, in the case of the HAA, formulae for the area like $R^2 (\pi - (\alpha + \beta + \gamma)) = -R^2 (\alpha + \beta + \gamma - \pi) = (iR)^2 (\alpha + \beta + \gamma - \pi)$ occur. Accordingly, Lambert remarked [147, §82]: "From this I should almost conclude that the third hypothesis would occur in the case of the imaginary sphere."

7.3 Legendre

Legendre took the opportunities presented by the restructuring of French education during and after the revolution to reintroduce something like Euclidean geometry into the syllabus. This forced him to confront the parallel postulate, and he tried in various ways to prove it. His book [149] first appeared in 1794, numerous subsequent editions attest both to its success and its author's attempts to vindicate the Euclidean theory of parallels. All failed, in various ways, including this one (which may be called the doubling argument, and appeared in the 3rd to the 8th editions).[6]

Let ABC be a triangle with angles α, β, γ such that $\alpha + \beta + \gamma < \pi$; for definiteness let the defect, which is defined to be $\pi - (\alpha + \beta + \gamma) = \delta$. Locate A' symmetrically situated to A with respect to BC (this can be done by rotating ABC through 180° around the midpoint of BC) and extend AB and AC. Draw through A' a line meeting AB at B' and AC at C' (not necessarily a line "parallel" to BC, which would beg the question). By symmetry the defect of triangle $A'BC$ is also δ (see Figure 7.3).

Since the defect of the angle sum of the large triangle $AB'C'$ is the sum of the defects in the angle sum of the four triangles separately, we have in $AB'C'$

---

[6] For much more information on Legendre's geometry, see [206].

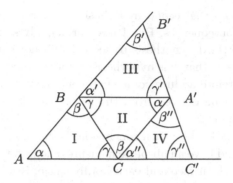

Figure 7.3 Legendre's figure

a triangle with defect $> 2\delta$. Indeed, the first defect is

$$2R - (\alpha + \beta' + \gamma'') = 2R - (\alpha + \beta' + \gamma'')$$
$$+ (2R - (\alpha + \beta'' + \gamma'))$$
$$+ (2R - (\alpha' + \beta + \gamma)) + (2R - (\alpha'' + \beta + \gamma)),$$

on adding the angles at A', B and C separately and subtracting $2R$ each time. But the sum on the right can be rearranged, on putting the angles in each smaller triangle together, as

$$2R - (\alpha + \beta' + \gamma'') = 2R - (\alpha + \beta + \gamma) + (2R - (\alpha + \beta + \gamma))$$
$$+ (2R - (\alpha' + \beta' + \gamma')) + (2R - (\alpha'' + \beta'' + \gamma''))$$
$$= 2(2R - (\alpha + \beta + \gamma)) + (2R - (\alpha' + \beta' + \gamma'))$$
$$+ (2R - (\alpha'' + \beta'' + \gamma'')).$$

Since the last two terms in brackets are positive, by assumption, it follows that $2R - (\alpha + \beta' + \gamma'') > 2(2R - (\alpha + \beta + \gamma))$, as claimed.

Continuing in this manner we obtain triangles with defects δ, 2δ, 4δ, 8δ, and so on, triangle $AB^{(n)}C^{(n)}$ having defect $> 2^n\delta$. Therefore, for n sufficiently large we have a triangle whose defect is greater than π. This is evidently impossible, and so Legendre has apparently proved that the angle sum of a triangle is 0, from which the ordinary theory of parallels follows.

Where is the error? To be revealed in a later chapter!

7.4 Lambert on the consequences of a non-Euclidean parallel postulate

One easily sees that one can go much further with the third hypothesis [i.e. the HAA] in this way, and that similar theorems can be found to those derived on the second hypothesis [i.e. the HOA], albeit with quite opposite consequences. But I have principally sought such consequences of the third hypothesis to see if it did not contradict itself. From them all I saw that this hypothesis would not destroy itself at all easily. I will therefore adduce some such consequences without seeing how far they can be derived from the second hypothesis by making corresponding changes.

The most remarkable of such conclusions is that if the third hypothesis holds we would have an absolute measure of length for each line, for the content of each surface and each bodily space. Now this overturns a theorem that one can unhesitatingly count amongst the fundamentals of Geometry, and which up to now no one has doubted, namely that there is no such absolute measure. Indeed, Wolf makes it into a theorem which he derives from the definition of quantity, and which can be stated as follows: *quantitas dari sed non per se intelligi potest* [there can not be a quantity known in itself]. However, this theorem, like the definition, must be amended, because unquestionably there are quantities that are intelligible in themselves and have a definite unit.

For lines, surfaces, and bodily spaces the same is also true; and I don't believe that one should introduce a definition into Geometry in order to put it right.

Now, in order to prove the first-mentioned corollary, let A, B, C, D, E be right angles, and, assuming the third hypothesis, let G, F, H, J be acute, and indeed $H < G$, and $J < H$, and likewise $F < G$ and $J < F$. Now I say that the angle G is the measure of the quadrilateral $ADGB$, if indeed $AB = AD$, and likewise the angle J shall be the measure of the quadrilateral $ACJE$, if $AC = AE$.

For, from considering the equality of the sides $AB = AD$ and the right angles A, B, D, the acute angle G can fit no other quadrilateral except those whose sides AB, AD have the absolute lengths of AB, AD. If, e.g., one takes the greater sides $AE = AC$ and puts right angles at E, C then on the third hypothesis the angle $J < G$. So G does not fit on J. If $AE = AC$ were to be taken smaller than $AD = AB$ then would J be $> G$ and so G would not fit on J in any case.

Therefore the angle G is the absolute measure of the quadrilateral $ADGB$. Since the angle has a measure intelligible in itself, if one took e.g. $AB = AD$ as a Paris foot and then the angle G was 80° this is only to say that if one should make the quadrilateral $ADGB$ so big that the angle G was 80°: then one would have the absolute measure of a Paris foot on $AB = AD$.

This consequence is somewhat surprising, which inclines one to want the third hypothesis to be true! However, this advantage not withstanding, I still do not want it, because innumerable other inconveniences would thereby come about. Trigonometric tables would have to be infinitely extended; the similarity and proportionality of figures would entirely lapse; no figure could be presented except in its absolute size; Astronomy would be an evil task; etc.

But these are *argumenta ab amore et invidia ducta* [arguments drawn from love and hate] which Geometry, like all the sciences, must leave entirely on one side. I therefore return to the third hypothesis. According to it, it is not only the case that in every triangle the angle sum is less than 180°, as we have already seen, but also that the difference from 180° increases directly with the area of the triangle. So I want to say: if of two triangles one has a greater area than the other then the angle sum of the first triangle is smaller than that of the other.

I shall not prove this theorem completely here, as I state it, rather I shall give only so much of the proof as will enable the rest of it to be understood overall.[7]

Now it seems to me to be remarkable that the second hypothesis holds when one takes spherical instead of plane triangles, because for them not only is the sum of the angles greater than 180° but the excess is proportional to the area of the triangle.

It seems even more remarkable that what I am saying here about spherical triangles can be deduced without regard for the problem of parallel lines, and based on no other idea than that each plane surface passing through the centre of a sphere divides it into two equal parts.

I should almost therefore put forward the proposal that the third hypothesis holds on the surface of an imaginary sphere. At least there must always be something which does not allow it to be overturned by plane surfaces as easily as the second hypothesis can be. [147, §78–82]

---

[7] I omit this passage, JJG.

8

Gauss (Schweikart and Taurinus) and Gauss's Differential Geometry

Figure 8.1 Gauss

8.1 Gauss

Carl Friedrich Gauss (1777–1855) is one of the few truly brilliant mathematicians to deserve the label "genius".[1] In later life his mother and he used to like to say that he taught himself to read with little instruction beyond learning the individual letters and that he had more or less taught himself arithmetic. These stories have some plausibility, because as an adult Gauss learned several languages, including English and Russian, and was a phenomenal calculator. His prodigious gifts as a child brought him to the attention of Martin Bartels, himself a good mathematician, and through him to the Duke of Brunswick, who

---

[1] The best single source on Gauss's life remains the biography by Dunnington, recently reissued [55], but Mittler [164] is a fine new resource in German.

was happy to sponsor the child's education at the impressive local Collegium Carolinum. Gauss amply repaid them for their support, and all his life was ever the loyal subject.

He published his first discovery in 1797, which was that the regular 17-sided figure is constructible by ruler and compass alone. This may not strike you as interesting; it did not impress the mathematician Gauss took it to (the by then 80-year-old Kaestner) but the point is not simply that the Greeks and everyone since them had missed this but that Gauss has seen that the problem reduces to finding the 16 complex roots of the equation $\frac{x^{17}-1}{x-1} = 0$ (which is clear once you know about complex numbers), and that the combination of the facts that 17 is prime and $17 - 1 = 16$ is a power of 2 means that the equation can be solved by solving a succession of quadratic equations, and solving a quadratic equation is a ruler-and-compass construction. Today this result is regarded as an impressive piece of Galois theory – but Gauss was 35 years ahead of Galois.

In 1799 he published his doctoral thesis, on the fundamental theorem of algebra. This is the claim that every polynomial equation of degree n with real or complex coefficients has n solutions (possibly repeated). But what brought him fame were two remarkable achievements in 1801. One was his book, the *Disquisitiones arithmeticae* [83], the first book that established the hitherto obscure subject of number theory as a major field for mathematicians and where, among other things, the theory of the 17-gon was explained. The other, much more newsworthy discovery was the rediscovery of the asteroid Ceres. This was the first asteroid to be discovered, but soon after Piazzi in Italy had found it it disappeared behind the sun. Piazzi published his observations and challenged the community of astronomers to find where it would be when it next emerged into view. Gauss won. The key to his method was his discovery of the statistical method of least squares. But Gauss did not publish his method until 1809, and by then Legendre had found it and published it independently, in 1805. There was a nasty priority dispute in 1809 as a result.

Gauss had been a student at the University of Göttingen, and while there one of his friends was Wolfgang Bolyai, with whom he discussed the foundations of geometry. Bolyai later recalled how he got to know Gauss:

> Even today I am a friend of his, although I am far from being able to compare myself with him. He was very modest and didn't make much showing; not three days as in the case of Plato, but for *years* one could be with him without recognising his greatness. What a shame, that I didn't understand how to open up this silent "book without a title" and read it! I didn't know how much he knew, and after he saw my temperament, he regarded me highly without knowing how insignificant I am. The passion for mathematics (not externally manifested) and our moral agreement bound us together so that while often out

walking we were silent for hours at a time, each occupied with his own thoughts. [55, p. 27]

One wonders if Bolyai was not a little intimidated by the brilliance of his friend. And friendship it was, although after 1799 they parted, never to see each other again. A mutual friend described Bolyai that year as someone who takes part in things "only as a philosopher, who on such occasions finds material with which to institute observations on human follies. This is so much his rule of habit, as I have discovered from several cases, that it is difficult for him to miss any of these worldly affairs, not that he wants to enjoy them along with the others, but in order to strengthen his tranquillity of mind." [55, p. 31] Yet Bolyai himself described his parting with Gauss in these gorgeously overwritten words:

I accompanied him [Gauss] in the morning, Saturday, 25 May 1799, to the peak of a small mountain toward Brunswick. That feeling of seeing each other for the last time is indescribable. Even a word about tears is ineffective. The Book of the Future is closed. Then we parted with a dying, farewell handshake, almost without words, with the distinction that he, led by angels of the temple of fame and glory, returns to Brunswick, and I, much less worthy, yet calm with a good conscience, go to Göttingen, although pursued afterward as a martyr of truth by an army of the many less worthy ones. [55, p. 31]

Gauss was quick to spot errors in purported proofs of the parallel postulate, and willing to discuss them in correspondence, but reluctant to publish. This makes it hard to know what he knew. In 1817, Gauss wrote to a fellow astronomer called Olbers that:

I am becoming more and more convinced that the necessity of our (Euclidean) geometry cannot be proved, at least not by human reason nor for human reason. Perhaps in another life we will be able to obtain insight into the nature of space, which is now unattainable. Until then we must place geometry not in the same class with arithmetic, which is purely a priori, but with mechanics. [55, p. 180]

8.2 Schweikart and Taurinus

Two amateur investigators, F. K. Schweikart and his nephew F. A. Taurinus, also thought the problem was worthwhile. In 1818 Schweikart, who was a professor of law at Marburg, communicated a remarkable theorem to Gauss via their mutual acquaintance, the astronomer Gerling. It was based on the HAA (known to him from Lambert's work), and is as good a place to start as any. He wrote:

> There are two kinds of geometry: a geometry in the strict sense – the Euclidean; and an astral geometry. Triangles in the latter have the property that the sum of their three angles is not equal to two right angles. This being assumed, we can prove rigorously:
> (a) That the sum of the three angles of a triangle is less than two right angles;
> (b) that the sum becomes ever less, the greater the area of the triangle;
> (c) that the altitude of an isosceles right-angled triangle continually grows, as the sides increase, but it can never become greater than a certain length, which I call the Constant.
>
> Squares have, therefore, the following form [see Figure 8.2].

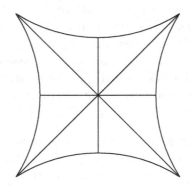

Figure 8.2 Schweikart's square

> If this Constant were for us the Radius of the Earth, (so that every line drawn in the universe from one fixed star to another, distant 90° from the first, would be a tangent to the surface of the earth), it would be infinitely great in comparison with the spaces which occur in daily life.
>
> The Euclidean geometry holds only on the assumption that the Constant is infinite. Only in this case is it true that the angles of every triangle are equal to two right angles: and this can easily be proved, as soon as we admit that the Constant is infinite. [88, pp. 180–181]

The confidence with which Schweikart was willing to build on a geometrical assumption that was not one of Euclid's is striking. He was not looking for a contradiction in his new geometry, but accepted the new assumption and explored its consequences. He even speculated that it might provide a suitable geometry for studying physical space. Gauss replied to Gerling that he agreed with Schweikart's theorem, adding that he could solve all the problems in this "Astral geometry" once the Constant spoken of by Schweikart was given [88, p. 182].

In his two books on geometry published in 1825 and 1826 Taurinus went further than his uncle, and produced a thorough study of geometry based on trigonometrical formulae.[2] Although he always believed in the truth of Euclid's geometry and the parallel postulate, and although his endeavours were directed at understanding geometry better in order to prove the parallel postulate, his researches yielded an important set of formulae which were to prove important in the years ahead. This shift of approach is highly significant. Until the 1810s and 1820s geometers interested in the parallel postulate used classical techniques and a style modelled on that of Euclid himself; Taurinus broke with this style and considered trigonometrical relations between angles. His entirely formal methods were to prove productive in other hands, and so it is necessary to understand something of his overall approach.

It had been known for many years that the following formulae connect angles on the surface of a sphere with angles subtended at the sphere's centre. (Note that we are using A, B, C for angles on the surface of the sphere; and a/k, b/k, c/k for angles at the centre of the sphere, whose radius is k, subtended by the lengths a, b, c.)

(1) $\cos(a/k) = \cos(b/k)\cos(c/k) + \sin(b/k)\sin(c/k)\cos A$

(2) $\cos A = -\cos B \cos C + \sin B \sin C \cos a/k$

Taurinus asked what would happen to these fundamental formulae of spherical trigonometry if one considered a sphere of imaginary radius (such as Lambert had hinted at, fifty years before) or, to put it another way, if one made the purely formal substitution of replacing k by $\sqrt{-1}\,k$.

The effect is rather startling. What one gets is this:

(1′) $\cosh(a/k) = \cosh(b/k)\cosh(c/k) - \sinh(b/k)\sinh(c/k)\cos A$;

(2′) $\cos A = -\cos B \cos C + \sin B \sin C \cosh a/k$,

where cosh and sinh are the hyperbolic equivalents of the usual spherical cosine and sine. Recall that $\cos\theta = \frac{e^{i\theta}+e^{-i\theta}}{2}$, $\sin\theta = \frac{e^{i\theta}-e^{-i\theta}}{2i}$, and $\cosh\theta = \frac{e^{\theta}+e^{-\theta}}{2}$, $\sinh\theta = \frac{e^{\theta}-e^{-\theta}}{2}$.

Thus, just as the spherical trigonometrical formulae (1) and (2) are used to study the properties of triangles on spherical surfaces, so the new formulae

---

[2] Partially reproduced in Engel and Stäckel [64].

produced by Taurinus, $(1')$ and $(2')$, could be used to study the properties of triangles on surfaces with a new geometry, akin to that hinted at by Lambert.

Taurinus now had a "geometry" in which triangles with sides a, b, c, and corresponding angles A, B, C, are related by the formulae $(1')$ and $(2')$. He called it a "log-spherical" geometry because of the role played by the hyperbolic functions cosh and sinh. He showed that it agreed with the "Astral geometry" of his uncle, and even found Schweikart's Constant as a function of the radius k.

But Taurinus really wanted Euclid's geometry to be true, so in his first book he gave ten reasons why log-spherical geometry could not be a geometry of space. These reasons are far from convincing; for example, "Were the log-spherical geometry true, Euclidean geometry could not be, the possibility of which cannot be doubted." [64, p. 259] He simply did not want to accept it. In his second book, however, he relented sufficiently to allow that such a geometry could exist. It was the geometry of some surface or other, but he could go no further with it.

Taurinus communicated some of his discoveries to Gauss in 1824. Gauss must have found them distressingly incoherent, for he replied that he found them incomplete, leaving something to be desired geometrically. He was also not inclined to believe that Taurinus had found a contradiction in the idea of a geometry in which triangles have angle sums less than π when he, Gauss, had looked for one unsuccessfully for 30 years. Indeed, while he stopped short of wholly endorsing the possibility of a non-Euclidean geometry, he spoke rather warmly of the properties of such a geometry. However, he was emphatic that his reply should not be published [64, pp. 249–250]. While Gauss moved to protect his reputation, Taurinus took the risky step of publishing at his own expense, but the book was a flop, and finally he had all the unsold copies burned.

8.3 What Gauss knew

What, indeed, did Gauss actually know about the subject? In his own work, he had developed an elementary theory of directed asymptotic parallels. This was an inconclusive exercise in conventional geometry, albeit in a new context. The 1817 letter to Olbers, for example, was about the work of his former student Wachter, who had sent Gauss a defence of the parallel postulate, based on the idea that any four points in space, that do not all lie in a common plane, lie on a sphere. Evidently they met to discuss the problem, for Wachter then wrote to Gauss setting down what he recalled of their conver-

sation [88, pp. 175–176]. In this letter he mentions the profound idea that if
the parallel postulate is false then it will be necessary to study geometry on
a sphere of infinite radius, but he did not spell out what such a mysterious
object could be. It is not clear if Wachter or Gauss came up with the idea
originally.

In his letter to Taurinus, Gauss spoke of a geometry quite different from
Euclid's, that was self-consistent (*in sich selbst durchaus consequent*) that he
found entirely satisfactory. He could solve all the problems in it once a value
was assigned to a parameter that could not be determined a priori, and the
larger this parameter became the more the new geometry approximated Euclidean geometry. The theorems in the new geometry might seem paradoxical;
for example, no matter how large the sides of a triangle become the area of the
triangle remains less than a fixed constant. The only thing that reason seemed
unable to accept was that in the new geometry there was an absolute measure of
length, but even this idea, he said, he sometimes wished was true. "But despite
the Nothing Saying word-wisdom of the metaphysicians we know too little, even
nothing at all about the true nature of space, and we may confuse something
which seems unnatural to us with the Absolutely Impossible." [64, p. 250] The
letter is curiously ambiguous, coming close to declaring outright that another
geometry is possible, only to withdraw at the last minute. Was Gauss saying
that he too could still be confused, or was he trying to get Taurinus to see that
he was?

In 1829 Gauss wrote to Bessel that "my conviction that we cannot base
geometry completely a priori has, if anything, become even stronger", but that
he probably would not publish his ideas in his lifetime because he feared the
howls of the Boeotians (a tribe the ancient Greeks had regarded as particularly
stupid) [88, p. 200]. Bessel replied that he regretted this modesty on Gauss's
part, and indeed that he could believe that space was slightly non-Euclidean.
In April 1830 Gauss wrote back to say that "It is my inner conviction that the
study of space occupies a quite different place in our a priori knowledge than
the study of quantity ...We must humbly admit that if Number is the pure
product of our mind, space has a reality outside of our minds and we cannot
completely prescribe its laws a priori." [88, p. 201]

So it seems that Gauss was willing to contemplate a new geometry but not
to give a flat-out, head-on description of it, or even to work out the details to
his own satisfaction. That was quite contrary to his normal practice; he died
leaving behind a wealth of discoveries he had not bothered to disclose. This
would suggest that Gauss did not want to investigate the new geometry, and
maybe even had reservations about it.

8.3.1 Gaussian curvature

The most interesting case is Gauss's work on differential geometry, described
in his *Disquisitiones generales circa superficies curvas* (*General investigations
of curved surfaces*) of 1827.[3] This is a subject he transformed entirely. He
was involved in the survey of the estates of Hanover (a good activity for a
loyal subject) and he threw himself into every detail, going on the field trips,
assisting with the measurements – he invented the helioscope, which greatly
increased their accuracy – and doing a great deal of trigonometry and error
estimations. As a result, he took up the study of the geometry of surfaces. A
surface in space (e.g. a sphere) acquires its geometry from the ambient space.
Thus: lengths along curves, shortest curves (geodesics), angles, etc., are inferred
from the primitive concepts in the surrounding three-dimensional space.

To deal with the obvious fact that a surface in space, other than the plane,
is curved, and the effect this has on its geometry, Gauss took an arbitrary
surface and studied the map from it to the unit sphere, obtained by looking
along normals to the celestial sphere. The local effect on area proved crucial.
Here one takes a piece of surface, U say, enclosing a point P, say, and finds
the ratio of the area of the image of this piece on the sphere to the area of the
piece of surface. If the surface is denoted Σ, the sphere is denoted S and the
map along the normals is denoted g (for Gauss), so $g : \Sigma \to S$, then Gauss
considered

$$\frac{\text{area}\,(g\,(U))}{\text{area}\,(U)}.$$

Gauss did this for smaller and smaller pieces of surface, and worked out a way
of talking about the limit of the ratio as the piece U tends to the point P. In
symbols, $\lim_{U \to P} \frac{\text{area}(g(U))}{\text{area}(U)}$. He called this quantity the curvature of the surface
at the point P. It has become known as the Gaussian curvature of the surface
at the point P.

If the surface is itself a sphere, the effect is just a scaling that depends on
the radius of the sphere. In fact, if the sphere has radius R then the Gaussian
curvature is $\frac{1}{R^2}$. But if the original surface is a plane, all the normals point
to the same point on the celestial sphere, and the map maps the whole plane
to a point: area is shrunk to zero. If the original surface is a cylinder, all the
normals point along a circular arc on the celestial sphere, and the map maps
the whole cylinder to a line: area is again shrunk to zero. Finally, if the surface
is a saddle, a region on the saddle is mapped to a region on the sphere, but as

---

[3] See Gauss, *Werke*, IV [87, pp. 217–258] for the original, and the edition by P.
Dombrowski, in *Astérisque* 62, 1978 [84] for the Latin original, a reprint of the
English translation by A. Hiltebeitel and J. Morehead of 1902 [84], and a valuable
commentary.

you go one way on the saddle the normals go the other way on the sphere, so the map sends positive area to negative area.

Gauss did a lengthy calculation and came up with a result that surprised even him. It surprised him so much he called it the theorema egregium or exceptional theorem: Gaussian curvature is intrinsic. This means that it depends on quantities which are measurable in the surface alone, and do not involve the third dimension. If two-dimensional creatures measured Gaussian curvature on a sphere, they could, in principle, decide that all talk of the sphere being embedded in a three-dimensional space was mathematically superfluous!

Surfaces with different Gaussian curvatures cannot be isometric (mapped exactly one onto another). But surfaces with the same Gaussian curvature are locally indistinguishable geometrically. The famous example is the plane and the cylinder – the map which realises the isometry between them is printing off a cylindrical drum.

To get an impression of the Gauss map, consider the ellipse in the (y, z)-plane with equation $\frac{y^2}{3} + z^2 = 1$. Spin this curve around the z-axis and the ellipsoid with equation $\frac{x^2}{3} + \frac{y^2}{3} + z^2 = 1$ is obtained, but the point to be made presently is clearer if we stick to the curve. The ellipse is described parametrically by the points $(y, z) = (\sqrt{3}\cos\theta, \sin\theta)$. A simple calculation finds that the slope of the normal at that point is $\frac{\sqrt{3}\sin\theta}{\cos\theta}$. So the normal vector to the ellipse at that point (and based at that point) is $(\cos\theta, \sqrt{3}\sin\theta)$ and the unit normal vector there is $\frac{1}{(\cos^2\theta + 3\sin^2\theta)^{1/2}}(\cos\theta, \sqrt{3}\sin\theta)$.

To understand this pictorially, consider the region of the ellipse where θ runs from 0 to $\pi/3$ (shown in grey in Figure 8.3), and the region where θ runs from $\pi/3$ to $\pi/2$ (shown in black in Figure 8.3), where the ellipse is the outer curve. The grey region is quite small. The Gauss map maps each point on the ellipse to the point on the unit circle defined by the corresponding unit normal vector, and the relevant part of the unit circle is the inner curve in the figure. On this circular arc, the image of the grey region is quite large, and that of the black region much smaller, corresponding to the fact that the curvature of the ellipse is greater where θ is smaller.

To see the behaviour of the Gauss map on the corresponding ellipsoid, rotate the ellipse around the z-axis.

For what it's worth, the Gaussian curvature of the point with coordinates (u, v) on the ellipse parameterised by $(a\cos u\sin v, b\sin u\sin v, c\cos v)$ is given by

$$g(u, v) = \frac{a^2 b^2 c^2}{\left(c^2\sin^2 v(b^2\cos^2 u + a^2\sin^2 u) + a^2 b^2\cos^2 v\right)^2}$$

and you can check that when $a = b = c = R$ and the ellipsoid is simply the sphere of radius R, the curvature everywhere is indeed $\frac{1}{R^2}$, as stated above.

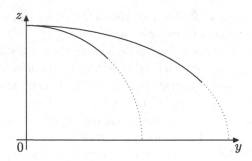

Figure 8.3   An impression of the Gauss map

So Gauss had a deep insight into the geometry of surfaces, and evidently could have embarked upon a crude classification into surfaces with positive curvature, zero curvature and negative curvature. Let's say constant positive curvature, zero curvature and constant negative curvature. That's the sphere (and maybe others, but actually there are no others), the plane (essentially) and – what? Pursuing this analogy – this trichotomy – the corresponding trigonometry is: spherical, plane and – what?

Gauss did not investigate, except possibly after the work of Bolyai and Lobachevskii was known to him. For this reason I believe that credit goes to the two men who come next in these chapters: Bolyai in Hungary/Romania and Lobachevskii in Russia.

9.1 János and Wolfgang Bolyai

János Bolyai, who was born in Klausenburg, Transylvania, Hungary[1] on 15 December 1802, was educated at home by his father Wolfgang (Farkas). Wolfgang had become the professor of mathematics at the Evangelical Reformed College in Maros-Vásérhely (now Târgu-Mures, Romania) in April 1804. The college dated back to 1557, the town was a pleasant one in the wine district, and Wolfgang showed himself to be a widely educated man. He had a picture of Gauss on the wall, alongside one of Shakespeare, and one of Schiller. He also wrote plays himself and translated several English and German works into Hungarian.[2]

In keeping with the Romantic sweep of his tendencies he adopted Rousseau's philosophy of education as set forth in *Émile* (1762) [212]. This promoted the importance of play and naturalness in contrast to the prevailing authoritarianism of the age. The father taught his son the first six books of Euclid's *Elements* [70], after which he moved on to Euler's *Algebra* [71]. When János was 12 he was allowed to attend lectures at the college. In due course he graduated top of the class in 1817 with prizes for Latin and a reputation as a violinist. He had even taught the other students in mathematics and physics. Unlike his father, however, he had no taste for poetry.

His father worried that he studied too much, at some cost to his health.

---

[1] Now Cluj in Romania.
[2] On the life and work of Wolfgang Bolyai, see Gray [103] and the references there, especially Engel and Stäckel [65].

From 1818 to 1823 he was packed off to the Royal Engineering Academy in Vienna, where he trained as a cadet and then he served in the Austrian army for 10 years as an engineer. This seems to have given him time for mathematics, and in 1833 he retired on a pension as a semi-invalid to Maros-Vásérhely – so there may have been something to his father's worries after all.

He would have learned of the parallel postulate from his father, but his interest in it was sharpened by an acquaintance he made in Vienna, Otto Szász, who had a new way of defining parallels. He suggested to János that the line through a point P parallel to a line m should be considered as the limiting position, a'', of a line a through P as it rotates (see Figure 9.1), with the property that every line below it meets the line m. Bolyai later called it an asymptotic parallel.

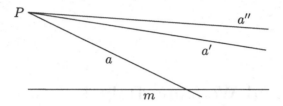

Figure 9.1   A parallel line as a limiting case

In 1820 János began to think that his failures to prove the parallel postulate might be due to the fact that the parallel postulate was not actually true. He switched direction and began to consider a geometry independent of the parallel postulate. He wrote to his father, in words that recall his Rousseauist upbringing, that:

> ... one must do no violence to nature, nor model it in conformity to any blindly formed chimera; that on the other hand, one must regard nature reasonably and naturally, as one would the truth, and be contented only with a representation of it which errs to the smallest possible extent.

This alarmed his father, who wrote back:[3]

> You must not attempt this approach to parallels. I know this way to the very end. I have traversed this bottomless night, which extinguished all light and joy of my life. I entreat you, leave the science of parallels alone ... I thought I would sacrifice myself for the sake of the truth.

---

[3] These and other quotations from the Bolyais are taken from Meschkowski [161, pp. 31–34].

I was ready to become a martyr who would remove the flaw from geometry and return it purified to mankind. I accomplished monstrous, enormous labours; my creations are far better than those of others and yet I have not achieved complete satisfaction ... I turned back when I saw that no man can reach the bottom of this night. I turned back unconsoled, pitying myself and all mankind. Learn from my example: I wanted to know about parallels, I remain ignorant, this has taken all the flowers of my life and all my time from me.

And yet again:

I admit that I expect nothing from the deviation of your lines. It seems to me that I have been in these regions; that I have travelled past all reefs of this infernal Dead Sea and have always come back with broken mast and torn sail. The ruin of my disposition and my fall date to this time. I thoughtlessly risked my life and happiness – aut Caesar aut nihil [either Caesar or nothing].

The son did not listen to his father, perhaps because his approach to the problem was aimed at creating a new geometry, not at finding a flaw in it, and on 3 November 1823 he wrote to him to say that he was succeeding:

I am determined to publish a work on parallels as soon as I can put it in order, complete it, and the opportunity arises. I have not yet made the discovery but the path that I am following is almost certain to lead to my goal, provided this goal is possible. I do not yet have it but I have found things so magnificent that I was astounded. It would be an eternal pity if these things were lost as you, my dear father, are bound to admit when you see them. All I can say now is that I have created a new and different world out of nothing. All that I have sent you thus far is like a house of cards compared with a tower. I am as convinced now that it will bring me no less honour, as if I had already discovered it.

His father was now sympathetic, and urged his son to publish his results as soon as possible. He suggested that a good place would be as an appendix to a work on geometry that he had been writing for some time. János later commented that:

He advised me that, if I was really successful, then there were two reasons why I should speedily make a public announcement. Firstly because the ideas might easily pass to someone else who would then publish them. Secondly there is some truth in this, that certain things

ripen at the same time and then appear in different places in the manner of violets coming to light in early spring. And since all scientific
striving is only a great war and one does not know when it will be replaced by peace one must win, if possible; for here pre-eminence comes
to him who is first.

However, when János visited his father on leave in February 1825, he was
unable to convince him. The problem was an arbitrary constant that entered
the formulae his son had found. This could lead to a contradiction if it were
to turn out not to be arbitrary but to have two distinct values simultaneously.
Their disagreement continued until 1829, and the father continued to advise his
son not to waste his life as one hundred geometers before him had. Finally, they
agreed to publish it anyway. The two-volume work, entitled *Tentamen juventutem studiosam in elementa matheosis purae...* was published by the college
in Maros-Vásérhely in 1832. János's ideas formed a 28-page *Appendix* [19] – to
what effect, we shall see in a later chapter.

9.2 János Bolyai's new geometry

János Bolyai's new world from nothing began with the definition of parallels
that Otto Szász had suggested. If lines AM and BN lie in the same plane and
AM is not cut by BN but every line in the angle ABN cuts AM then Bolyai
said the line BN is parallel to the line AM. There is a unique parallel to AM
through the point B to the right, and there is a unique parallel to the left
(shown as BN' in Figure 9.2, where the angles at A are right angles). As a
result the angles NBA and $N'BA$ are equal to an amount δ, say, which Bolyai
assumed is less that a right angle. Later Bolyai investigated how the angle δ
depends on the length of the perpendicular BA.

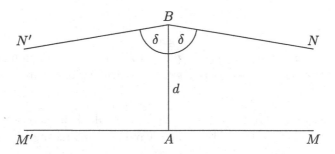

Figure 9.2 Parallels in Bolyai's geometry

Bolyai proceeded to draw out some basic properties of parallel lines from his new definition. In particular he showed that if a and b are parallel and A is a fixed point on a, then there is a unique point B on b such that the angles $MAB = \alpha$ and $NBA = \beta$ are equal (see Figure 9.3).

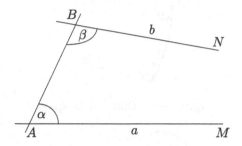

Figure 9.3 Parallels and equal angles in Bolyai's geometry

Bolyai now made a vital shift into three dimensions. No previous investigations of the parallel postulate had taken this route, and it shows that Bolyai was concerned that his new geometry be taken as a possible geometry of physical space. It also allowed him to develop crucial results that gave his results any chance of convincing a sceptical public. It also makes the exposition quite unlike Euclid's *Elements*.

To follow the march of his argument, recall that when we wish to study the geometry on a sphere, we naturally imagine the surface of a sphere sitting in three-dimensional Euclidean space. Distances between pairs of points on this sphere are measured along the surface of the sphere (no tunnelling is allowed). The resulting geometry on the sphere is called the "induced" geometry on the sphere, because it has inherited its concept of distance from the surrounding Euclidean space.

Bolyai proceeded in much the same way, but starting with non-Euclidean three-dimensional space. He introduced a special surface, which he denoted by the letter F, obtained (if I may simplify a little) as follows. He took a point, A, a straight line a through A, and in each plane containing the line a he considered all the lines parallel to a (in the same direction, which is to the right in Figure 9.4). On each parallel line, b, he picked out the point B with the property that the angles MAB and NBA are equal (their size depends on the position of the point B). This gave him a curve, which he denoted L, in the plane containing the lines a and b. The set of all these curves, as the plane through a is varied, defined the surface F (see Figure 9.5). It is most likely that the names L and F were derived from the German Linie for line and Fläche for surface, even though the *Appendix* was written in Latin.

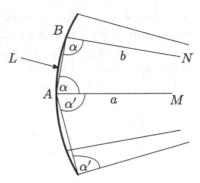

Figure 9.4 Bolyai's *L*-curve

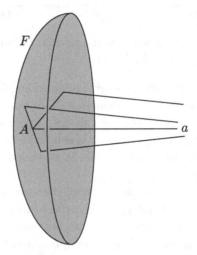

Figure 9.5 Bolyai's *F*-surface

The surface *F* is bowl shaped (see Figure 9.5). The original line *a* appears as an axis meeting the curve *L* at right angles, and the surface *F* can be thought of as swept out by the curve *L* as it is rotated about the axis *a*.

As Bolyai observed, if the parallel postulate is true then the angle δ in Figure 9.2 is a right angle, the curve *L* is just the straight line through *A* perpendicular to *a* and the surface *F* is just the plane through *A* perpendicular to *a*. However, if the parallel postulate is false, then the angle δ in Figure 9.2 is less than a right angle, the curve *L* is not a straight line but a curve perpendicular to *a* and the surface *F* is not a plane but a surface perpendicular to *a*.

Bolyai then set out to discover as many theorems as possible which are true whether the parallel postulate is true or false, noting along the way those theorems that are only true when the postulate is false. Results which are true on either assumption Bolyai called "absolute" theorems. Among the theorems he discovered are these:

1. all the lines b parallel to the axis a meet the surface F at right angles (if the parallel postulate is false this identifies the surface F as a candidate for the sphere of infinite radius discussed by Gauss and Wachter);

2. any plane containing the axis a meets the surface F in a curve L.

These theorems are also true when the parallel postulate is true, but they are trivial because the surface F is a plane in that case. Bolyai also showed that

3. any plane not containing the axis a or a line b meets the surface F in a circle. (This theorem is only true when the parallel postulate is false.)

The most important result established by Bolyai is true whether the parallel postulate is true or false, but it is much more interesting, even surprising, when the parallel postulate is false:

4. on any surface F, if two curves L cross a third and the sum of the interior angles is less than two right angles, then the two L curves intersect.

This result implies that the parallel postulate holds for curves L on a surface F. It follows that if the L curves are taken as straight lines, then Euclidean geometry holds on the surface F whether or not the parallel postulate is assumed. It meant that Bolyai could now compare figures in two-dimensional non-Euclidean geometry with other figures in two-dimensional Euclidean geometry, much as we can pass between figures on a sphere and corresponding figures on a plane.

The discovery that in a space in which the parallel postulate is false there is a surface on which the induced geometry is Euclidean was remarkable. After all, mathematicians had failed to find a geometry in Euclidean three-dimensional space upon which the induced geometry was non-Euclidean. It means that the geometer studying three-dimensional non-Euclidean space can bring in two-dimensional Euclidean geometry just as we study geometry on the sphere by thinking of it in a three-dimensional Euclidean space.

The question of finding a surface in three-dimensional Euclidean space on which the induced geometry would be non-Euclidean geometry remained open. Had such a surface been easy to find, the discovery of non-Euclidean geometry would doubtless have been made early on. However, it was to turn out that no such surface exists.[4] That does not mean that non-Euclidean geometry does not exist, only that there is no intuitively accessible model of it as a surface in Euclidean space, but this is a subtle distinction to make. Bolyai's discovery

---

[4] Strictly speaking, no surface of this kind smooth enough to admit the formalism of differential geometry exists.

means that a three-dimensional non-Euclidean geometer would find Euclidean geometry almost as easy to study as we find spherical geometry (the infinitude of the Euclidean plane would surely cause some residual difficulty).

To make use of the F-surface and its Euclidean geometry Bolyai compared triangles in non-Euclidean geometry with those on the F-surface. He passed between them by considering a (non-Euclidean) triangle, ABC, right-angled at B, and the lines AM, BN and CP perpendicular to ABC and parallel to AM (see Figure 9.6).

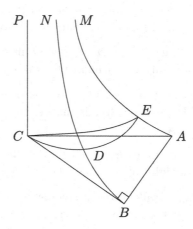

Figure 9.6 Euclidean and non-Euclidean triangles in space

Then, in a move that took him even further from Euclid's way of thinking about geometry, he introduced trigonometric formulae. He considered the surface consisting of all the lines parallel to BN and CP through points of BC, and showed that it is perpendicular to the similar surface consisting of all the lines parallel to BN and AM through points of AB. He then took the F-surface through C perpendicular to CP, and supposed that BN meets it at D and AM meets it at E. He then deduced that the angle between the L-curves CD and DE at D is a right angle. It is also clear that the angle between the L-curves DC and EC at C is equal to the angle between the non-Euclidean lines BC and AC at C. Moreover, rotating the figure about the axis CP shows that the ratio of the circumferences of circles whose radii are CD and CE in the F-surface is equal to the ratio of the circumferences of the circles whose radii are CB and CA in the non-Euclidean plane. This means that the elementary trigonometric formulae connecting the sides and angles of the F-triangle CDE can be transformed into results about the sides and angles of the corresponding non-Euclidean triangle.

Bolyai next considered the formulae that relate the sides and angles of a triangle on a (Euclidean) sphere, which in his day were taught to every land surveyor and military engineer. He showed, strikingly, that they too are absolute theorems. It meant that he could also use spherical trigonometry in the new setting.

To find the appropriate trigonometric formulae in a geometry where the parallel postulate is false Bolyai began with the fundamental figure in such a geometry. This consists of a straight line AB of length y, meeting a line AM at right angles and a line BN at an acute angle u, where the lines AM and BN are parallel. The length y determines the angle u and vice versa. The angle u is called the angle of parallelism corresponding to the length y, and is sometimes written $u = \Pi(y)$. Bolyai now sought (in [19, §29]) the expression for the length of the line segment y as a function of the angle u, shown in Figure 9.7.

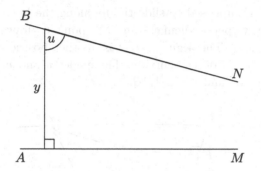

Figure 9.7 The angle of parallelism

His argument was a long and difficult one. He first considered an arbitrary point C on the line a and the L-curve through it. He called this L-curve L'. (See Figure 9.8.)

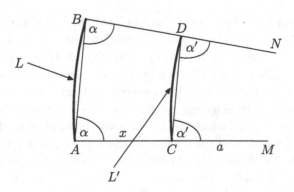

Figure 9.8 The curve L'

He showed that the ratio $AB\!:\!CD$ is independent of AB and depends only on the length $AC = x$, by observing that if AB is doubled, so is CD, and more generally if AB is increased by a factor of k then so is CD. He denoted this ratio X and set himself the task of evaluating it. His answer is a formula relating u and Y, where Y is the same function of y that X is of x, and u is the angle of parallelism corresponding to the length y. First, he showed by a simple scaling argument that $Y^{1/y} = X^{1/x}$, from which it follows, although Bolyai did not say so immediately, that $X = e^{kx}$ for some arbitrary constant k.

To find X he considered the curve which is everywhere equidistant from a straight line. In Euclidean geometry this is another straight line, but if the parallel postulate is false it will not be. Bolyai imagined it was swept out by the tip of a line segment that moves perpendicular to the given line. In Figure 9.9 the triangle ABC is supposed to slide rigidly along the line a, with the edge AC remaining always perpendicular to a. The point C draws the equidistant curve, c, to the line a. The segment BD is another position of the segment AC, so the point D lies on the curve c. The angles u and v are as shown in Figure 9.9: $u = \angle CAD$ and $v = \angle ADB$.

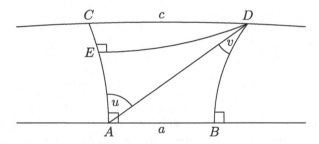

Figure 9.9 The angles u and v

Bolyai showed that the ratio of the length of the line segment AB to the length of the segment CD of the curve c is equal to $\sin v/\sin u$ by dropping the perpendicular DE from D to AC and first applying his trigonometric formulae to the triangles AED and ABD. Then he used a limiting argument to find the ratio of the lengths AB and CD. He observed that since the ratio AB/CD is clearly a constant (it depends only on the height AC not the width AB) this ratio can be evaluated in the limit as AC moves off infinitely far, when the angle u tends to a right angle and the angle v to the angle δ (see Figure 9.9 above).

However, he also showed that in Figure 9.10, where $x = AC$, the ratio X he was interested in before was equal to $\sin u/\sin v$. From this he could deduce the result he wanted: the angle of parallelism is given by the formula $Y = \cot(u/2)$. If we use the result that $Y = e^{ky}$ this can be rewritten as the formula $\sinh ky = \cot u$ or as $\sinh ky = \cot \Pi(y)$.

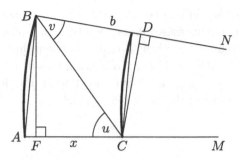

Figure 9.10 The ratio $X = AB/CD$

This formula allowed Bolyai rapidly to deduce a number of useful results. He found the formula for the perimeter of a circle of radius r, and then the formulae connecting the sides and angles of triangles when the parallel postulate is false. It is these formulae that contain an arbitrary constant and that had accordingly worried his father. Bolyai observed that when the parallel postulate is false any three pieces of information about a triangle suffice to determine the rest, and he gave all the formulae in the easy case when the triangle has a right angle (formulae for the general triangle follow with a little routine work). This is not quite the case in Euclidean geometry: knowing the angles of a triangle determines its shape, but not its size (one might of course reply that only two pieces of information are required to determine the angles of a Euclidean triangle, because the angle sum must be two right angles).

Bolyai also showed that the new formulae reduce to their Euclidean equivalents when the arbitrary constant increases to infinity. In particular, the formula corresponding to Pythagoras's theorem in non-Euclidean geometry reduces to the usual expression for Pythagoras's theorem in Euclidean geometry in this way. In all these formulae, the non-Euclidean lengths (x, say) always enter divided by the arbitrary constant k, so an approximation that holds when x/k is very small may be said to hold when k is very large and equally when x is very small. So to a very high degree of approximation, the Euclidean version of Pythagoras's theorem holds for small triangles in non-Euclidean geometry, showing that small regions of non-Euclidean space are approximately Euclidean.

Bolyai went on to show that in the new geometry the area of a triangle in the plane (not on the special surface F) is equal to its angular defect, and then to show that if the parallel postulate is false one can construct a square equal in area to a given circle, provided the radius of the circle is chosen with care.[5] Thus Bolyai squared the circle. Squaring the circle is synonymous with

---

[5] I thank Marvin Greenberg for making this clear to me. Bolyai's construction is described in detail in [103, pp. 69–75].

achieving the impossible and had been regarded that way since at least 450 BC when the Greek playwright Aristophanes wrote *The Birds*. So it must have been with some excitement that Bolyai realised that in his new geometry one can indeed square the circle, and he addressed this result (in [19, §33]) to "the friends of truth" who, he supposed, would not find it unwelcome. His method occupies the final sections of the book (§34–§43).[6]

9.3 János Bolyai's section 32

In section 32 of the *Appendix* [19], Bolyai made his strongest claim that he could do anything in the new geometry that could be done in Euclidean geometry, thus defying, as it were, anyone to find fault with it. He produced a method of resolving problems in the new geometry, "which being accomplished (through the more obvious examples), finally will be candidly said what this theory shows". Most commentators seem to have failed to appreciate the significance of this remark.

Bolyai, and his readers, were familiar with maps in an atlas. When the earth's surface is represented on a flat, Euclidean, plane, there are formulae for the area of a region of the sphere in terms of latitude and longitude, and for the length of a curve on the sphere. A glance at Mercator's projection shows that the formulae cannot be as simple as they are in ordinary plane geometry: equal increments of y on the map clearly correspond (as one goes north) to smaller and smaller steps on the surface of the earth. So the formulae are different, but the idea is the same.

Bolyai seems to have supposed that his readers would recognise these arguments and appreciate them in the altered setting of his new geometry.

He drew a picture of a curve ABG in the familiar Cartesian plane with x- and y-axes (see Figure 9.11) and indicated how it was to be interpreted as a picture of non-Euclidean geometry drawn in a Euclidean plane. The x-axis is to represent a non-Euclidean straight line; the perpendiculars to the x-axis through C and F on the x-axis meet a curve at points B and G respectively. Through the point B he then drew the ray BH parallel to the x-axis, making equal angles FCB and HBC.

What Bolyai did not say, is that the perpendiculars CB and FG must be L-lines, in which case he should have explained how they can be represented as Euclidean straight lines. Bolyai then considered what happens to the triangle BGH when the distance CF becomes very small. The arc length BG along the curve becomes very small, and better and better approximated by the non-

---

[6] See the *Appendix* [19] and also the account in Gray [103].

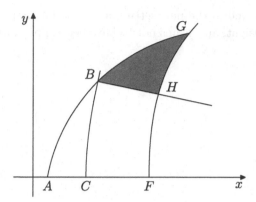

Figure 9.11 The fundamental figure in Bolyai's geometry

Euclidean length of the non-Euclidean line segment BG. Knowledge of very small lengths will then yield knowledge of arbitrary finite lengths along the curve by means of integration. Bolyai was confident that the trigonometric formulae connecting the sides of the very small triangle BGH are obtained from the formulae for an arbitrary finite non-Euclidean triangle, which he had established in the previous paragraph. This is true, although not elementary, however Bolyai escaped the pedagogic problem, not for the first or only time in the *Appendix* by saying (correctly): "It can be demonstrated" that the requisite approximation is $BG^2 = dy^2 + BH^2$. He then used this formula to deduce formulae for the areas and volumes of some non-Euclidean figures.

Here, Bolyai missed a great opportunity to start again. He could have used the above argument to give a much more convincing account of non-Euclidean geometry. With a bit of extra work, he could have shown that the entire picture of non-Euclidean two-dimensional geometry could appear in the right half-plane (the region defined by $x > 0$); that straight lines appeared as curves of a certain appearance; that two lines were parallel if (in a typical case) they met on the y-axis; and that the trigonometric formulae for triangles were those he had derived earlier. In these pictures (as in conventional cartography) in the usual (x, y)-plane very small distances were to be measured according to his new rule, and arbitrary finite distances determined from these by integration. As a result, equal non-Euclidean distances would appear to shrink as one neared the y-axis (the exact opposite of Mercator's projection as one heads away from the equator).

Most likely, the world would not have been convinced. This suggested interpretation makes Bolyai into Riemann (as we shall see below). But the fact that Bolyai got as close as he did to formulating the elements of his new geometry in terms of the calculus, and so being able to tackle all the problems of differential geometry as well in the new setting as in the familiar Euclidean one, is testimony to his insight, and seems not to have been appreciated sufficiently in his day or since.

10
Lobachevskii

10.1 Lobachevskii and Kasan

The new mathematical world of a geometry other than Euclid's, was the independent discovery of Nikolai Ivanovich Lobachevskii, born in Kasan in Russia in 1792 and his Hungarian contemporary, János Bolyai (1802–1860). Although little understood in their lifetimes, their work eventually helped to overturn almost every belief about the mathematics of space, and to open the way to the modern geometries of Hilbert and Einstein. After his death, Lobachevskii became known as the Copernicus of geometry. On the occasion of his centennial

J. Gray, *Worlds Out of Nothing*,   115
Springer Undergraduate Mathematics Series,
DOI 10.1007/978-0-85729-060-1_10, © Springer-Verlag London Limited 2011

he was taken up by Malevich and the Russian futurists, and if his bicentennial was a quieter affair it was nonetheless worth noting.[1]

Lobachevskii came from a poor background, but his mother was able to have him enrolled at the local Gymnasium (or high school) on a scholarship in 1800. In 1805 the Gymnasium was made the kernel of the new University of Kasan, and in 1807 Lobachevskii began to study there, in the footsteps of his brother Alexei, who was to become a successful chemist. Nikolai was fortunate, for the university, unable to find a Russian of sufficient merit, had appointed Martin Bartels as professor of mathematics. Bartels was the mathematician who had taught Gauss when Gauss was in his teens, so he knew talent when he saw it. By 1807, when Bartels came to Kasan, Gauss has become both the leading pure mathematician and the leading astronomer of his day. If, as the joke of the day had it, Bartels was the best mathematician in Germany (because Gauss was the best mathematician in the world) still more was he the best mathematician in Russia. Bartels not only trained Lobachevskii well, but protected him from the excesses of his youth, for the younger man was often in trouble with the authorities. On various occasions he lost holiday privileges, and was even suspected of atheism. When he should have graduated, the university agreed that his work merited the degree even if his achievements in mathematics were ignored, but found that his conduct had been too bad to overlook, and refused to pass him. Bartels lobbied the other professors, and three days later Lobachevskii was awarded not the ordinary degree but the Master's qualification. His career as a professional mathematician could begin.

But the university was improperly run, and soon the governing body was split into cliques around the director. Each side took to leaking information about the misdeeds of the others to the Ministry of Education in St Petersburg, until the scandal could not be ignored. The man dispatched to tidy up the mess was Magnitsky. He conducted a whirlwind tour of the university and the local schools, supported by those professors who saw him either as the much-needed new broom or as their own chance to get on the winning side. Within a month he had seen all he needed to see, and he reported back to the Minister that the professors were generally very able, but the university was underfunded (he singled out the medical faculty especially) and that education in the whole region was sadly amiss. The province was perhaps half a century behind Europe, which was not, he added, what the Emperor wanted. Thus alerted, the Ministry took the steps Magnitsky desired. He became the director of the university, with responsibility for education in the whole region of Kasan.

Once in office, he ruled with a single-minded sense of purpose his earlier reports had tended to conceal. The main aim of education, sadly neglected these

---

[1] The biographical information on Lobachevskii that follows is taken from Engel's account in Lobatschefskij [155], Kagan [126] and Rosenfeld [211].

past few years, was to inculcate a knowledge of the true faith and a sense of humility and obedience. Prizes in science were replaced with prizes only in theology. Students who broke the rules were labelled sinners and made to spend time in solitary confinement contemplating pictures of the Last Judgement. Every university appointment, every school teacher, was scrutinised not just for their teaching ability but their moral character. Non-Russians, and especially Germans, were thought to be dangerous free thinkers and either left or were dismissed. Past records were examined and reports sent back to St Petersburg on everybody. Lobachevskii, mindful no doubt of what the records could say about him, was complicit in this, signing the reports of the Education Committee on which he served, and of the University Council. He took up the cause of the library, in urgent need of reform, but was frustrated at every step by the librarian and resigned. He carried a large teaching load, not just in mathematics but in theoretical physics and astronomy, and, as he was required to do, filled in monthly reports on the students and their intellectual and moral progress.

He did manage to leave to a colleague the burden of showing how mathematics too played its part in revealing the truths of religion, and when he wrote up his own lectures on geometry as a book, Magnitsky sent it to the academician Fuss in St Petersburg for review. Fuss decided it should not be published. He was concerned that the book used the metre as a unit of measurement and divided the quarter circle into one hundred degrees, a division "proposed at the time of the French Revolution, when the rage to destroy extended even to the calendars and the division of the circle." [240, p. 17] Lobachevskii asked that his manuscript be returned; it was found in 1898 and published in 1909. In it he notes that Euclidean geometry has not yet been established rigorously, but does not suggest that there might be an alternative.

Magnitsky's reign lasted for seven years, until 1826. Financial corruption was again souring the university's reputation, but his downfall was an imprudent letter addressed to the Emperor Alexander accusing the Grand Dukes, Nicolas among them, of liberalism. Alexander's early and unexpected death made Nicolas the new emperor, and although he was in fact no friend of liberty in education, he found the letter unforgivable. Within days, Magnitsky was banished and his puppet the Rector dismissed. Three days later, Lobachevskii gave a public lecture "On the principles of geometry, with a rigorous demonstration of the theory of parallels". The manuscript of this talk is now lost, but later references suggest that it marks the start of Lobachevskii's awareness of a non-Euclidean geometry.

The new Director of Education for the Kasan region restored elections for the rectorship of the university, and Lobachevskii was elected. Under his leadership the library, allowed to drift under Magnitsky, was finally put on a proper footing. New professors were appointed, new buildings built. Under his rectorship, an observatory and an anatomical theatre were built. In 1830 cholera swept through the country. When it reached Kasan, Lobachevskii sealed off the university, allowing only the doctors in and out. Strict measures, albeit of doubtful efficacy, were taken to prevent the disease entering the university in the supplies of food and drink. All mail was fumigated on arrival, washed in a chlorine solution and set aside until the epidemic was over. In line with the prevailing theory of the miasmatic transmission of illnesses, the university walls were wreathed in boiling vinegar. The isolation worked, and what had been a terrifying epidemic outside resulted in only forty cases and sixteen deaths inside. Even the Emperor wrote to express his appreciation of Lobachevskii's efforts.

The 1830s were the high point of his life, for in addition to all his heavy teaching duties and his demanding work as the Rector, Lobachevskii found the energy and time to publish his original work for the first time. What he had discovered, perhaps by 1826 and certainly by 1829 when he published an article in the *Kasan Messenger*, was an alternative geometry to Euclid's. In this geometry the theorems are different and often contradict those of Euclidean geometry, but the definitions are equally plausible. It becomes, as a result, necessary to use some empirical means to determine which of the two geometries is the true one. It does not need much imagination to see how such researches would have fared under Magnitsky, but we have no documents to establish how far Lobachevskii had gone before the liberalisation of 1826.

10.2 Lobachevskii's new geometry

How can there be another geometry, and how did Lobachevskii come to discover it? It is at the least ironic that, in his time as Rector, Lobachevskii was deeply involved in the energetic building programme of the university, overseeing it to the smallest detail. For Lobachevskii deeply believed that the Euclidean foundations were flawed. In his view terms like "line", "surface" and "position" were highly abstract and obscure. What one knows about, he was to argue on several occasions, are bodies and the motion of bodies. It is by thinking about these much clearer ideas that the concepts of geometry are to be built up. In this he was echoing the ideas of d'Alembert as expressed in the *Encyclopédie*, whose philosophy of geometry was well adapted to the application of geometry and the calculus to Newtonian mechanics. It was Lobachevskii's view that it is from a consideration of bodies that one derives ideas about straight lines, and that these, carefully considered, need not be as Euclid had said they were.

Lobachevskii offered his readers the example of the geometry on the surface of a sphere. Here the curve of shortest length joining two points is the arc of a great circle between them (the arc cut out on the sphere by a plane through the centre of the sphere). This is a geometry familiar to navigators; Lobachevskii proposed to put it on a par with Euclid's, and having done so to set up yet another, which was his own discovery. It afforded a charming trinity.

In Euclidean geometry it is a fundamental theorem that the angles of any triangle add up to two right angles. As noted earlier, this follows from the assumption, known as the parallel postulate, which asserts (in a simpler but equivalent form) that given any line and any point P, not on the line, there is a unique line through the point P that does not meet the given line.

In the geometry on the sphere the parallel postulate is false, because any two great circles meet, and moreover the angle sum of any triangle is greater than two right angles.

In Lobachevskii's new geometry the parallel postulate is again false, but in a different way. In this geometry it is true that given any line and any point P, not on the line, there is a line through the point P that does not meet the given line, but it is no longer unique. It becomes unique once a direction is specified for each line. Lobachevskii could have decided to call any pair of non-intersecting lines parallel, but had he done so it would not have been the case that if line a is (on this definition) parallel to line b and line b is parallel to line c then line a is parallel to line c. For this reason, he preferred to have three classes of line with respect to a given line: those that cut it, those that never meet it, and the two lines, one in each direction which he called the parallels.

Sadly for Lobachevskii, his skills as a speaker deserted him as a writer. His papers were unconvincing. In 1837 he attempted to reach a mainstream European audience by writing in French in the new German journal, Crelle's *Journal (Journal für die reine und angewandte Mathematik)*, but the article was top-heavy with formulae and very little of their geometrical meaning came through because he wrote as if the audience had access to his previous articles in Russian periodicals. When he published his ideas in a self-contained booklet in Berlin in 1840, it was only too obviously the case that his argument ran in this fashion: let lines be like this, then the theorems you get are these. To which it could still be replied that, perhaps, the theorems you will get will turn out to be self-contradictory and your new geometry an illusion. Indeed, this had been for over a century the traditional way to defend Euclidean geometry. Several eminent geometers had even claimed that from assumptions like Lobachevskii's they had actually derived the requisite contradiction. Lobachevskii himself had begun his most thorough defence of his discoveries by refuting some of the most famous these flawed defences of Euclid.

What gave Lobachevskii his conviction that his geometry was not self-contradictory, and that no proposition in it could be simultaneously true and false, was a mixture of two things. One was his starting point in the analysis of the motion of bodies, concealed though it was from his European readers. The other was his finishing point, for Lobachevskii expressed his theorems in the language of trigonometry and the calculus. He deliberately sought out formulae because he deeply believed that geometry was about measurement, and that measurements, numbers, are related to one another by formulae. In turn, the validity of these formulae was a matter of algebra, whatever might be their geometrical significance. But he confined these thoughts to his Russian-language publications. So Lobachevskii inferred (and this, logically, is a flaw in his argument) that his new geometry made sense because he had trigonometric formulae, and moreover these formulae made sense if the parameter they contained became indefinitely large, for then the formulae described a geometry indistinguishable from Euclidean geometry. One can add that if the parameter was made purely imaginary the formulae of spherical trigonometry resulted.

In fact, his papers and books closely resemble János Bolyai's. Both worked on a geometry in three dimensions. Both expressed their fundamental results in terms of new trigonometric formulae. Lobachevskii was clearer in some respects, and more concerned to find theorems in the new geometry than to find theorems common to both Euclidean and the new geometry. He had the sensible habit of giving names to important features of the new geometry (which he called the imaginary geometry) whereas Bolyai merely used letters (and was criticised on this account by Gauss). He called the curve perpendicular to a family of parallel lines a horocycle, and the corresponding surface a horosphere. Like Bolyai, he showed that the geometry on the horosphere is Euclidean.

Lobachevskii's best single attempt to explain his discoveries that had any chance of being read outside Russia was his little booklet of 1840 [153], and the reader is recommended to read the extracts below and draw the accompanying figures. In this work Lobachevskii gave a clear account of the implications of the new assumption about parallels, one remarkably similar to, but clearer than, Bolyai's. The similarities mean that there is no need to go over the details of the exposition again, but it is worth saying that once one grants the new definition, the rest follows logically. What was missing was the motivation for considering such a definition and, perhaps more sensibly, the astronomical considerations Lobachevskii had already taken up and found to be inconclusive.

10.2.1 Lobachevskii's first foundations of geometry

I should look briefly at Lobachevskii's first presentation of his new geometry, called in the German translation "Ueber die Anfangsgründe der Geometrie" (On the foundations of geometry), which forms a 66-page article spread over five issues of the Kasan *Messenger* for 1829 and 1830. In it he described a three-dimensional geometry based on the primitive ideas of distance and congruence. This allowed him to talk of spheres, and of congruent spheres (those with equal radii). He then fixed two points and looked at all possible pairs of congruent spheres with those points as centres; when the spheres meet they meet in a circle. This enabled him to define a plane as all the points lying on the circles obtained in this fashion. A similar argument one dimension lower allowed him to define a straight line.

He then considered the planar configuration consisting of a point A, a line ℓ that does not pass through A, and the perpendicular AB from A to ℓ. He proposed that there could be three kinds of line through the point A that lie in this plane: those that meet the line ℓ, those that do not, and among the lines that do not meet ℓ, two that separate the lines that meet ℓ from those that do not: these he called the parallels to ℓ through A. The parallels, he said, make an angle at A with the perpendicular AB that depends on the length of AB, and if this length is a then the angle is, in Lobachevskii's notation in this article, $F(a)$. In later work Lobachevskii called this function the "angle of parallelism".

The book is full of derivations of trigonometric formulae for triangles. They are obtained first for right-angled triangles, and they make considerable use of the angle of parallelism function. A typical example (§13, Eq. 14) of the trigonometric formulae Lobachevskii obtained is this: in a right-angled triangle with sides a and b opposite the angles A and B respectively

$$\tan A = \cos F(a) \tan F(b).$$

It is characteristic of all these formulae that they involve not the side lengths of the triangles but their F-values. In the course of these derivations Lobachevskii obtained (§12, Eq. 12) this expression for the function F (where an arbitrary parameter that enters the argument has been set equal to 1, corresponding to a choice of unit of length in this geometry):

$$\tan \frac{1}{2} F(x) = e^{-x}.$$

It follows from this, although Lobachevskii did not say so explicitly, that in terms of the hyperbolic functions,

$$\cos F(x) = \tanh(x), \sin F(x) = \frac{1}{\cosh x}, \text{ and } \tan F(x) = \frac{1}{\sinh x}.$$

Curiously, although Lobachevskii went on to introduce expressions of the form $e^t + e^{-t}$ he never used the hyperbolic functions themselves. Perhaps he did not know them; more likely, because he never used them in any of his writings on geometry, he knew them but did not think they were helpful.

The significant novelty in this work is the derivation of formulae for arc lengths and areas of curved figures; Lobachevskii gave quite a thorough account of the intrinsic differential geometry of his new space. This differential geometry disappears from his work in the 1830s and is not in his booklet of 1840 upon which most accounts – including the one in this book – are based. This omission may be justified in view of the complete lack of influence the "Anfangsgründe" had, but it is still interesting to look at what Lobachevskii wrote.

Lobachevskii used what he called right-angled coordinates, by which he meant that a straight line is taken as the x-axis and the straight lines at right angles to this line are labelled y. In this scheme a point P has coordinates (x, y) if the perpendicular from P to the x-axis has length y and the distance along the x-axis from the foot of the perpendicular to the origin is x. He applied his trigonometric formulae to a triangle with coordinates $(-x, y), (+x, y)$, and $(+x, y')$, with $y' > y$, and by then allowing the sides to be very small obtained an expression for the element of arc length along a curve (I have squared the expression in §21, Eq. 34 to bring it into line with normal usage):

$$ds^2 = dy^2 + \frac{dx^2}{\sin^2 F(y)}.$$

It is very tempting to replace $\sin F(y)$ in this expression by the value given above and write this formula as

$$ds^2 = dy^2 + \cosh^2 y\, dx^2,$$

when the analogy with the formula for arc length on the sphere in latitude (y) and longitude (x) coordinates leaps out at us:

$$ds^2 = dy^2 + \cos^2 y\, dx^2,$$

but Lobachevskii seems to have missed this, even though he was keen to note the analogies between his trigonometric formulae and those in spherical trigonometry.

Lobachevskii then developed length and area formulae for figures in the new geometry. I shall not follow him here, and refer the reader to Papadopoulos's commentary on the *Pangéométrie* of 1855, where Lobachevskii returned to this theme. But it worth noting another topic that Lobachevskii also then set aside for a while: in §15 he raised the question of whether investigations into stellar parallax would reveal the nature of space, and found, using observations on the stars Rigel, Sirius, and epsilon Eridani that the question was very unlikely ever to be resolved.

It is intriguing that Lobachevskii's final attempt to interest the mathematical world in his discoveries reverts in many ways to his first. The derivation of the trigonometry is rather different, the differential geometrical formulae are put to greater use, but the sense that the new geometry is coherent at every level from the infinitesimal to the astronomical is just as clear. It is hard to imagine that he could have done more, or better than he did; the fate of his work in his lifetime says much more about his readers than his writings.

10.2.2 Astronomical evidence

Lobachevskii showed that on this new definition of parallels the angle sum of a triangle was now always less than two right angles. Moreover, the angle sum gets less as the triangle gets bigger, and this suggested to Lobachevskii that one could attempt to see if space was in fact non-Euclidean. His idea was to consider the parallax of stars.[2] But the parallax of stars is almost negligible, and all Lobachevskii could conclude in 1829 was that ordinary Euclidean geometry was accurate enough for all practical purposes [155, §15]. In his last work, the *Pangéométrie* [154], Lobachevskii gave a different argument involving two measurements on a star arbitrarily far away, from which the arbitrary constant in the formula for the angle of parallelism (defined below) can be determined [154, §9, pp. 76–78]. This argument rests implicitly on the fact that in Euclidean geometry the parallax gets smaller as the star is further away, and it may be arbitrarily small if stars are to be found far enough away. But in non-Euclidean geometry the parallax cannot fall below a certain level, determined by the diameter of the earth's orbit. If measurements of parallax seemed always to exceed a certain non-zero amount, this would be evidence in favour of the validity of non-Euclidean geometry.

10.3 From Lobachevskii's *Geometrische Untersuchungen*, 1840 [153]

10.3.1 Opening remarks

In geometry I find certain imperfections which I hold to be the reason why this science, apart from transition into analytics, can as

---

[2] A task that was in fact beyond the astronomers of Lobachevskii's day until Bessel a few years later, so the observations Lobachevskii used were in fact hopelessly inaccurate.

yet make no advance from that state in which it has come to us from Euclid.

As belonging to these imperfections, I consider the obscurity in the fundamental concepts of the geometrical magnitudes and in the manner and method of representing the measuring of these magnitudes, and finally the momentous gap in the theory of parallels, to fill, which all efforts of mathematicians have been so far in vain.

For this theory Legendre's endeavours have done nothing, since he was forced to leave the only rigid way to turn into a side path and take refuge in auxiliary theorems which he illogically strove to exhibit as necessary axioms. My first essay on the foundations of geometry I published in the *Kasan Messenger* for the year 1829. In the hope of having satisfied all requirements, I undertook hereupon a treatment of the whole of this science, and published my work in separate parts in the *Gelehrten Schriften der Universität Kasan* for the years 1836, 1837, 1838, under the title "New Elements of Geometry, with a Complete Theory of Parallels". The extent of this work perhaps hindered my countrymen from following such a subject, which since Legendre had lost its interest. Yet I am of the opinion that the Theory of Parallels should not lose its claim to the attention of geometers, and therefore I aim to give here the substance of my investigations, remarking beforehand that contrary to the opinion of Legendre, all other imperfections – for example, the definition of a straight line – show themselves foreign here and without any real influence on the theory of parallels.

[...]

All straight lines which in a plane go out from a point can, with reference to a given straight line in the same plane, be divided into two classes – into cutting and not-cutting.

The boundary lines of the one and the other class of those lines will be called parallel to the given line.

From the point A let fall upon the line BC the perpendicular AD, to which again draw the perpendicular AE.

In the right angle EAD either will all straight lines which go out from the point A meet the line DC, as for example AF, or some of them, like the perpendicular AE, will not meet the line DC. In the uncertainty whether the perpendicular AE is the only line which does not meet DC, we will assume it may be possible that there are still other lines, for example AG, which do not cut DC, how far soever they may be prolonged. In passing over from the cutting lines, as AF, to the not-cutting lines, as AG, we must come upon a line AH, parallel to DC, a boundary line, upon one side of which all lines AG are such as do not meet the line DC, while upon the other side every straight line AF cuts the line DC.

The angle HAD between the parallel HA and the perpendicular AD is called the parallel angle (angle of parallelism), which we will here designate by $\Pi(p)$ for $AD = p$.

If $\Pi(p)$ is a right angle, so will the prolongation AE' of the perpendicular AE likewise be parallel to the prolongation DB of the line DC, in addition to which we remark that in regard to the four right angles, which are made at the point A by the perpendiculars AE and AD, and their prolongations AE' and AD', every straight line which goes out from the point A, either itself or at least its prolongation, lies in one of the two right angles which are turned toward BC, so that except the parallel EE' all others, if they are sufficiently produced both ways, must intersect the line BC.

If $\Pi(p) < \pi/2$, then upon the other side of AD, making the same angle $DAK = \Pi(p)$ will lie also a line AK, parallel to the prolongation DB of the line DC, so that under this assumption we must also make a distinction of sides in parallelism.

All remaining lines or their prolongations within the two right angles turned toward BC pertain to those that intersect, if they lie within the angle $HAK = 2\Pi(p)$ between the parallels; they pertain on the other hand to the non-intersecting AG, if they lie upon the other sides of the parallels AH and AK, in the opening of the two angles $EAH = \pi/2 - \Pi(p)$, $E'AK = \pi/2 - \Pi(p)$, between the parallels and EE' the perpendicular to AD. Upon the other side of the perpendicular EE' will in like manner the prolongations AH' and AK' of the parallels AH and AK likewise be parallel to BC; the remaining lines pertain, if in the angle $K'AH'$, to the intersecting, but if in the angles $K'AE$, $H'AE'$ to the non-intersecting.

In accordance with this, for the assumption $\Pi(p) = \pi/2$, the lines can be only intersecting or parallel; but if we assume that $\Pi(p) < \pi/2$, then we must allow two parallels, one on the one and one on the other side; in addition we must distinguish the remaining lines into non-intersecting and intersecting.

For both assumptions it serves as the mark of parallelism that the line becomes intersecting for the smallest deviation toward the side where lies the parallel, so that if AH is parallel to DC, every line AF cuts DC, how small soever the angle $\angle HAF$ may be.

10.3.2 Concluding remarks

All four equations for the interdependence of the sides a, b, c, and the opposite angles A, B, C, in the rectilineal triangle will therefore be (equations (1)):

$$\sin A \tan \varPi (a) = \sin B \tan \varPi (b)$$
$$\cos A \cos \varPi (b) \cos \varPi (c) + \frac{\sin \varPi(b) \sin \varPi(c)}{\sin \varPi(a)} = 1$$
$$\cot A \sin C \sin \varPi (b) + \cos C = \frac{\cos \varPi(b)}{\cos \varPi(a)}$$
$$\cos A + \cos B \cos C = \frac{\sin B \sin C}{\sin \varPi(a)}$$

If the sides a, b, c, of the triangle are very small, we may content ourselves with the approximate determinations

$$\cot \varPi (a) = a,$$
$$\sin \varPi (a) = 1 - \tfrac{1}{2}a^2,$$
$$\cos \varPi (a) = a,$$

and in like manner also for the other sides b and c.

The equations (1) pass over for such triangles into the following:

$$b \sin A = a \sin B$$
$$a^2 = b^2 + c^2 - 2bc \cos A$$
$$a \sin (A + C) = b \sin A$$
$$\cos A + \cos (B + C) = 0$$

Of these equations the first two are assumed in the ordinary geometry; the last two lead, with the help of the first, to the conclusion

$$A + B + C = \pi.$$

Therefore the imaginary geometry passes over into the ordinary, when we suppose that the sides of a rectilineal triangle are very small.

I have, in the scientific bulletins of the University of Kasan, published certain researches in regard to the measurement of curved lines, of plane figures, of the surfaces and the volumes of solids, as well as in relation to the application of imaginary geometry to analysis.

The equations (1) attain for themselves already a sufficient foundation for considering the assumption of imaginary geometry as possible. Hence there is no means, other than astronomical observations, to use for judging of the exactitude which pertains to the calculations of the ordinary geometry.

This exactitude is very far-reaching, as I have shown in one of my investigations, so that, for example, in triangles whose sides are attainable for our measurement, the sum of the three angles is not indeed different from the two right angles by the hundredth part of a second.

In addition, it is worthy of notice that the four equations (1) of plane geometry pass over into the equations for spherical triangles, if we put $a\sqrt{-1}$, $b\sqrt{-1}$, $c\sqrt{-1}$, instead of the sides a, b, c; with this change, however, we must also put

$$\sin \Pi(a) = \frac{1}{\cos(a)}$$
$$\cos \Pi(a) = \sqrt{-1}\,\tan(a)$$
$$\tan \Pi(a) = \frac{1}{\sin a \sqrt{-1}}$$

and similarly also for the sides b and c.

In this manner we pass over from equations (1) to the following:

$$\sin A \sin b = \sin B \sin a$$
$$\cos A = \cos b \cos c + \sin b \sin c \cos A$$
$$\cot A \sin C + \cos C \cos b = \sin b \cot a$$
$$\cos A = \cos a \sin B \sin C - \cos B \cos C.$$

11

Publication and Non-Reception up to 1855

11.1 Minding's surface

An interesting surface was discussed by H. F. Minding in the 1830s in papers he published in Crelle's *Journal* [162] although these papers were then largely forgotten for 30 years; this was a surface of constant negative curvature with the property that geodesics between points are not necessarily unique. Minding's surface is formed by rotating a tractrix about its vertical axis. A tractrix – the name is due to Huyghens – is the curve of the obstinate dog. A point Q on a line ℓ is attached to a point P by a line segment of fixed length. P, the dog, is dragged behind Q, the owner, who walks along ℓ, and the path of P is called the tractrix. It is conventional for the walk to start with PQ perpendicular to ℓ. Minding's surface, which is sometimes called the pseudosphere, is formed by rotating the tractrix about ℓ; the initial position of the point P then generates a circle of singular points (see Figure 11.1).

At any point, P, the surface looks locally like a saddle. Its curvature is everywhere negative, and, moreover, everywhere constant (that's a deduction from the nature of the tractrix) since for that curve, as one rises upwards the principal vertical curvature decreases and the principal horizontal curvature increases so as to preserve the value of the product, and moving sideways obviously changes nothing.

The geometry of Minding's surface is described by figures made up of geodesic segments – think of strings laid upon it and pulled taut. They must, of course, be constrained to lie always on the surface and not to jump across it

J. Gray, *Worlds Out of Nothing*,
Springer Undergraduate Mathematics Series,
DOI 10.1007/978-0-85729-060-1_11, © Springer-Verlag London Limited 2011

Figure 11.1   A pseudosphere

like tunnels in the earth. Minding's interest was in surfaces which, over small regions at least, were metrically the same – such as the cylinder and the plane. He noticed a property of the surfaces of constant curvature which will be important in what follows. Any figure drawn upon such a surface (plane, sphere, or pseudosphere) may be slid about without changing its shape. To see this, draw a figure on a surface and try and move it around. If the surface is curved, its curvature is like a force that prevents you drawing a perfect square. But variable curvature will snap open any figure you are moving around. Therefore the only surfaces which admit the sliding of congruent figures are those of constant curvature.

The tractrix and its axis resemble the picture of asymptotic parallels due to Saccheri and Gauss; this resemblance is not accidental. The surface S was also known to Gauss; he referred to it in an unpublished note written between 1823 and 1827 as the surface of revolution which is the opposite of the sphere, but he does not appear to have connected it with anything in non-Euclidean geometry [88, p. 264]. Minding also showed that the pseudosphere is not the only surface of revolution to have constant negative curvature, he gave other examples, but the tractrix is the simplest [162].

11.2 The Bolyais read Lobachevskii

The Bolyais eventually heard of Lobachevskii's work in 1844 from a Hungarian mathematician and physicist, Franz Mentovich, who met Gauss in Göttingen. Unable to obtain any copies of it, Wolfgang Bolyai wrote to Gauss in 1848, who replied that they might do better trying to locate the German booklet Lobachevskii had published in 1840 [153]. They got a copy (on 17 October 1848) and found numerous small points to criticise. Some of their arguments were preferable, notably the proof of the exponential relationship between lengths along two L-curves with the same axes. But they were rightly impressed by the

Russian's derivation of the trigonometric formulae. At first the two Hungarians thought of publishing a reply, but their efforts petered out, and it must be a sad reflection on their state that by now they were willing to give up. They had discovered a new geometry, one of the most momentous discoveries ever made, and the world simply ignored them.

11.3 Final years of János Bolyai

A copy of the Bolyais' book was sent to Gauss, but it was lost in the chaos of a local cholera epidemic, and another was sent. Gauss was impressed, writing to Gerling on 14 February 1832 that "I consider this young geometer von Bolyai to be a genius of the first order." [88, p. 220] Unhappily, the letter to the Bolyais gave a distinctly different impression. On 6 March 1832, Gauss wrote to the father, Wolfgang:

> If I commenced by saying that I am unable to praise this work, you would certainly be surprised for a moment. But I cannot say otherwise. To praise it, would be to praise myself. Indeed the whole contents of the work, the path taken by your son, the results to which he is led, coincide almost entirely with my meditations, which have occupied my mind partly for the last 30 or 35 years. So I remained quite stupefied. So far as my own work is concerned, of which up till now I have put little on paper, my intention was not to let it be published during my lifetime. Indeed the majority of people have not clear ideas upon the questions of which we are speaking, and I have found very few people who could regard with any special interest what I communicated to them on this subject. To be able to take such an interest it is first of all necessary to have devoted careful thought to the real nature of what is wanted and upon this matter almost all are most uncertain. On the other hand it was my idea to write down all this later so that at least it should not perish with me. It is therefore a pleasant surprise for me that I am spared this trouble, and I am very glad that it is just the son of my old friend, who takes the precedence of me in such a remarkable manner. [88, pp. 220–224]

It is an extraordinary letter, all the more so when one considers how little evidence there is that Gauss did know everything János Bolyai had found to say. Perhaps it can best be read as a statement of approval, signalling that the young man's work is consistent with everything Gauss knows about the subject, and even hinting that Gauss will not seek to publish on the subject. Wolfgang at least was pleased that the great geometer had endorsed his son's discoveries, but the son was appalled. It was to be almost a decade before he could be convinced that his father had not confided the ideas in the *Appendix* to Gauss, who then dishonestly claimed priority, just as his father had warned him that someone might. A long period ensued when father and son did not speak to each other. Wolfgang disapproved of the fact that János lived unmarried with a woman (Rosalie von Orban) by whom he had three children. Father and son then resumed an uneasy relationship which lasted until Wolfgang died in 1856, and János's relationship with Rosalie ended about the same time. János died in 1860.

11.4 Final years of Lobachevskii

Russian authorities like Ostrogradskii denigrated Lobachevskii's work, which they simply did not understand. His little German book got only one review, which was a travesty. One of the very few people to appreciate Lobachevskii's work when it came out was Petr Ivanovich Kotelnikov (1809–1879), who was a professor of mathematics at Lobachevskii's own university, Kasan. He delivered a public lecture at the university there in 1842 on the subject of the popular dislike of mathematics in the course of which he said:

> In this connection I cannot pass over in silence that the futile millennial attempts to prove with all mathematical rigor one of the fundamental theorems of geometry, to the effect that the angle sum in a rectilinear triangle is equal to two right angles, inspired Mr. Lobachevskii, a revered and meritorious professor of our university, to undertake the prodigious task of building a whole science, a geometry based on the new assumption that the angle sum in a triangle is less than two right angles a task that is bound to gain recognition sooner or later.[1]

But Lobachevskii had also sent a copy of his *Geometrische Untersuchungen* [153] to Gauss in Göttingen. He did not know of Gauss's interest in the subject, because Bartels seems not to have told him of it and Gauss was at pains not to publish on the subject. But Gauss's interest was profound and lifelong, and by 1840 he seems only to have lacked the one central organising perception on the topic that he sought and found in all of his best work. So he immediately acclaimed Lobachevskii's work, and since he had been learning Russian since the spring of 1839 (after finding he had no taste for Sanskrit[2])

---

[1] Quoted in Rosenfeld [211, p. 219].
[2] See Peters [189, vol. 2, p. 242, letter no. 641].

he sought out Lobachevskii's earlier papers, and in 1842 had him made a corresponding member of the Göttingen Academy of Sciences. This was to be the only acclaim Lobachevskii was to receive in his lifetime.

To quote from Rosenfeld's account (the direct quotations are from his book [211]), in February of 1841 Gauss wrote to J. F. Encke:

> I am making reasonable progress in learning to read Russian and this gives me a great deal of pleasure. Mr. Knorre sent me a small memoir of Lobachevskii (in Kasan), written in Russian, and this memoir, as well as his small German book on parallel lines (an absurd note about it has appeared in *Gersdorff's Repertorium*) have awakened in me the desire to find out more about this clever mathematician. As Knorre told me, many of his papers are in the Russian Proceedings of Kasan University.

And on 28 September 1846, Gauss wrote to Schumacher:

> Lately I had reason to reread the small work of Lobachevskii [*Geometrische Untersuchungen zur Theorie der Parallellinien...*]. This work contains the foundations of the geometry that would obtain, and form a coherent whole, if Euclidean geometry were not true. A certain Schweikart called this geometry *Astral geometrie*. Lobachevskii calls it imaginary geometry. You know that for 54 years (since 1792) I have shared the same views with some additional development of them that I do not wish to go into here; thus I have found nothing actually new for myself in Lobachevskii's work. But in developing the subject the author followed a road different from the one I took; Lobachevskii carried out the task in a masterly fashion and in a truly geometric spirit. I see it as my duty to call your attention to this work that is bound to give you truly exceptional pleasure.

At the risk of repetition, recall that there is very little (and arguably no) evidence in Gauss's work that he came to the most significant of the conclusions of Lobachevskii or Bolyai before reading their work.

By now the formerly energetic Russian's life was entering on a period of decline. A terrible fire in 1841, which had left half of Kasan in ashes, destroyed the observatory, although students, under Lobachevskii's direction, saved the equipment and were able to protect the library. After 30 years his rectorship came to an end and, keen to keep such democracy as they had, the university professors preferred to elect a new man. This cost Lobachevskii not only income, but also his university residence. An unsuccessful marriage lurched into financial crisis as attempts to reorganise his wife's considerable dowry simply put the money in the hands of an unscrupulous rogue. Such was the chaos of the household that Lobachevskii's biographers have been unable to establish the number of children born into it, but it may well have been 15, even 18. Seven survived to their teens. His even temperament subsided into fatigue, his mental powers diminished, and a kind of senility seems to have set in. He died in 1856.

11.5 Gauss's death, Gauss's *Nachlass*

Gauss had died the previous year, 1855, and when mathematicians came to work through the mass of unpublished papers he had left, they found to their astonishment that he had gone quite a way to discovering non-Euclidean geometry. His confidence in it had spread to a select few of his colleagues, such as the astronomer Bessel, who had written to Gauss that he was prepared to believe that space might be better described in terms of non-Euclidean geometry although small regions would be adequately described in Euclidean terms. It emerged that Gauss had corresponded with an old friend from his student days on the subject, and that in due course this man's son had written a paper on non-Euclidean geometry. Indeed, János Bolyai's work was remarkably like Lobachevskii's: the same definitions, the same emphasis from the start on describing a geometry of three dimensions, the same observation that empirical measurements would henceforth have to determine which of the two geometries was the physically valid one. The same formulae, which necessarily depended on an arbitrary constant in the same way that the radius of a sphere enters the formulae of spherical trigonometry. Faced with this profusion of work, ignored, disparaged, neglected, sometimes even hidden, going back over 30 years, the new leaders of the mathematics profession started to shift their ground. Even so, it was to be another generation before a way was found to put right the logical flaw in Lobachevskii's work, and to give a new foundation for geometry that mathematicians, at least, could accept.

Why does it matter? It matters as a piece of physics, for physical theories up to the time of Einstein took space as a theatre, a stage, in which bodies move, what Newton had famously called God's sensorium. To change the very nature of the theatre was therefore quite remarkable, akin to the even more profound changes in our concepts of space and time that were ushered in by Einstein. It also matters because it severs the cherished link between mathematics and physics. If mathematics is thought of as a body of logical truths, of which geometry is a central part, then the existence of a non-Euclidean geometry means that not all purely logical mathematics can be true. It exists in at least two conflicting forms, only one of which can be true, and empirical work is needed to decide which. Soon it fractured into infinitely many geometries, and the empirical task of finding the correct one became hopeless, but even Lobachevskii showed that geometry could no longer be true in any simple sense. This realisation seems to have alternately excited and depressed every one who came to it. To Dostoyevsky, in *The brothers Karamazov* [54], it was a source of incomprehension akin to the way in which the idea of God exceeds our understanding. To 19th-century British mathematicians and philosophers it was a source of polemics, as humanists sought to stress and theologians to diminish its impact on the mind's capacity to find significant a priori truths.[3] It matters to this day to mathematicians, who have continued to find striking applications of non-Euclidean geometry within their own subject. Perhaps Lobachevskii was a Copernicus, dethroning Euclid much as the Pole had dethroned Ptolemy. Perhaps he was a Columbus. János Bolyai said of his own work that he had "discovered a whole new world out of nothing". But it is unlikely that mathematicians have yet exhausted the richnesses of the world that Lobachevskii discovered in the relative isolation of Kasan.

---

[3] See Richards [208].

On Writing the History of Geometry – 1

12.1 Assessment and advice

This is an unusual chapter, one of three, that offers advice on how to tackle a piece of assessed work.[1] An example of suitable assessment questions will be found at the end of this chapter, along with yet more advice.

12.1.1 Reading and writing the history of mathematics

The opinion that good historians simply tell us what happened in the past is hopelessly naive. Arguments, opinions and judgements play an inescapably large role in the writing of history; one cannot look at the different views of historians and pretend that somehow one will transcend these and reach the true and correct opinion.

The necessity of argument arises in the first place because the historical record is imperfect. Much of it is lost, much of what exists is often difficult to access (written in different languages, stored in remote archives) or hidden among a morass of too much information. What survives and is read may be wrong, even deliberately wrong, unduly partial, and misleading in many ways. Arguments about what happened in Greek mathematics show this only too well, as tenuous fragments of evidence are pressed into whatever cause the

---

[1] The accompanying lecture was a review of the story so far.

J. Gray, *Worlds Out of Nothing*,
Springer Undergraduate Mathematics Series,
DOI 10.1007/978-0-85729-060-1_12, © Springer-Verlag London Limited 2011

author espouses.[2] The problem of evidence is a particular problem for the non-specialist, the beginner and the student, who simply do not have the time to browse among whole piles of material and depend, naturally and rightly, on the works of established historians. But in relying on the expert, another problem opens up: what is, and can only be, an opinion, becomes an established fact. What one historian once asserted (let us say carefully and after much thought) becomes repeated by others, and reaches the student as the truth of the matter.

For example, consider the claim, much disputed in the literature, that Gauss surveyed three mountains in northern Germany with a view to deciding if space was Euclidean or non-Euclidean. This simple statement is two things at once. It is a claim that Gauss surveyed a specific region of Germany, and it is a claim that he did this with a specific purpose in mind (we might allow that he could have had other reasons as well). The first claim is a matter of fact. And it is, moreover, true: there is a huge amount of documentation filling pages and pages of Gauss's *Werke* [87, 88] and doubtless archives as well. But the evidence about his motives is quite different: a reminiscence of Gauss published the year after he died. There is in fact no evidence at all that he organised the long and costly surveying work, which took up several years in the 1820s, with this aim in mind. What seems to have been the case is that Gauss was interested in the question of the Euclidean or non-Euclidean nature of space (we know that from many letters he wrote) and he may well have tried to investigate it, to the limits of accuracy he was working at – which were an order of magnitude better than his contemporaries. He may well have discovered that even the largest terrestrial triangle he could find was not big enough to resolve the matter either way. The best case for Gauss having done this work is Scholz's [219]. He argues that this was the start of a Göttingen tradition that saw the final determination of the nature of space as being out of reach, traces of which are visible in, for example, Riemann's Habilitation[3] lecture (see Chapter 18 in this book and [209, §18.5, p. 192]) where Riemann said that astronomical measurements establish that the curvature of space is certainly very close to zero.

Several points can be deduced from this little story. It shows, or rather, a look at the historical literature would show, that for some people the idea that Gauss did this work with this aim in mind has become established as a fact, and so they feel there is no need to reopen the documents upon which it was based. But this "fact", the supposed motivation, had to be argued for, and in fact the arguments are weak. Importantly, belief in the motivation cannot be allowed to rest on the undisputed fact that the survey took place, which was done for quite other reasons. The beginner cannot be expected to query

---

[2] Helpfully surveyed in Grattan-Guinness [98], and discussed more provocatively in Fowler [77].

[3] A German post-doctoral qualification.

every claim made by a historian, but he or she can notice when claims have no evidence, or when the supposed evidence is no evidence at all. And it may also, regrettably, need to be said that there are facts in history, and we argue with them at our peril. In this case the fact is that the survey did take place.

The problem is twofold. Facts are sometimes hard to recognise, statements purporting to be facts may rest on more guesswork than one expects (for example, an estimate of the population of a certain place at a certain date), some statements of "fact" are congealed judgements, there is fraud and dishonesty in the world. Even the best-attested historical fact has no greater claim on us than the true statements by which we live our lives. One does not need the services of a philosopher here, the work of a good mystery writer is enough to convince us that almost all facts have a provisional character about them. Secondly, the most interesting aspect of history is not the facts, but how they are put together. Every time historians attempt to explain something they are making a judgement based on an argument. We like to accept the conclusions of good arguments as facts, but they are judgements, well-argued opinions. When we respond, however critically, to the work of historians, we too are making up arguments, and reaching conclusions. We may improve on their work, but we cannot transcend the limitations of the historian.

That being the case, what should we do about it? We may hope to discover a persistent tendency of a historian we are reading. This need not be simple prejudice (although at times it is). All of us have a tendency to talk up "our side" be it the imperialist or the conquered, the rulers or the ruled. Like mathematicians, historians of mathematics have preferences for some branches of mathematics over others, some languages over others, some mathematicians over others. Like historians in other fields, some historians of mathematics have causes: they want you to like this group of people who, they might suggest, have been hitherto misunderstood. Historians of mathematics have a tendency to talk up the great mathematicians, amplifying the historical record by reading between the lines. The give-away word is "would", as in sentences like "Gauss would surely have noticed ...". It is not that such inferences are wrong; they are inevitable, but they need to be seen as inferences, and subject to analysis. One can ask if there is any detectable consequence of his noticing this, or any problem in assuming that he did not (or, that he did). It is one thing if Gauss is "bound" to have noticed an implication of an argument, another if he is "bound" to have read something. Even when no tendency is apparent, but certainly when it is, we should notice what sort of evidence is required for the various assertions historians make. Assertions about what happened are one thing, assertions about why it happened are another.

Another thing we can do is think about the quality of the arguments we are reading. This applies both at the level of historians' arguments, and those of the historical protagonists themselves. In a book such as this one we have both, but similar rules apply. It can be useful to run through the sorts of adjective that apply to arguments: strong, weak, convincing, persuasive, . . . A mathematical argument may strive to be totally logically compelling, a philosophical argument may hope to be irresistible. There are many occasions in life where persuasive is all one can hope for, where conflicting views have to be acknowledged. At the other extreme, arguments can be woeful, fatally flawed, have only a kernel of truth. There can be poor arguments for conclusions one wants to accept on other grounds, arguments that aren't arguments at all, strong arguments with, nonetheless, a gap, arguments that are attractive only at first sight. It is important in studying the history of mathematics to realise that mathematicians' arguments can be treated in this way too: one thing the history of mathematics shows very clearly is that the best and most original arguments mathematicians come up with are rather far from the carefully written ("gapless") chains of reasoning of modern textbooks. Invention and perfection can lie rather far apart.

The problem becomes most urgent when it becomes personal, when we too have to become historians and write some history. As historians, we have to produce arguments. We need facts in support of some opinions, and we may want to shift our position as we weigh the evidence. Then, when we have reached our conclusion – which is likely to be an opinion as to why something happened the way it did – we need to marshal our evidence, organise the presentation of our case, admit to complexity and the possibility of other interpretations, and then argue our way from beginning to end. Particular care is to be spent on the single word "because". Without it and its synonyms there can be no explanation, so its absence suggests that we have allowed facts (and of what sort?) to substitute for arguments. But when we use it we must check that the evidence supports it, the logic of the argument supports it, that other explanations are not as good (and may even be open to grave objection).

Writing history and reading it critically can be itemised, doubtless too simplistically, under five headings: argument, evidence, interpretation, opinion and understanding (AEIOU). The two that I have not already discussed are interpretation and understanding. An argument that, as it might be, Poncelet's work on projective geometry met with a difficult reception because of the strange way it was presented is supported by historical evidence that people did indeed find it difficult and uncongenial to read. But it is even stronger if you can show that you have interpreted what Poncelet wrote, and what critics, commentators and later authors wrote, correctly, and that your understanding of all this illuminates why his work got the reception it did. If you can explain why it got the various different receptions it did, so much the better.

12.1.2 Practice questions

We can try out the above advice on some examples. Quite deliberately, the questions call for personal judgements. Answers can be tentative, there is room for disagreement here. Reasoned discussion, amicable agreements to differ, changing one's mind in the light of new information or better arguments, are part of what life is about – mathematical life too.

Example 12.1

The following statements all concern Monge's achievement in making descriptive geometry a major topic in the École Polytechnique around 1800–1810.

Rank them in order of importance from 1 (most important) to 5 (not important at all). Statements which you consider to be (more or less) false should be separated out and marked with an F. You can rank some statements joint equal, but don't do it more than you feel you must. You won't have quite all the information you need, so as you rank them make a note of any extra assumptions you make.

(a) Descriptive geometry is essential in the work of a military engineer.

(b) Descriptive geometry is important in the education of a mathematician.

(c) Monge was an influential man behind the scenes, with a personal connection to Napoleon.

(d) Monge had a strong background in practical mathematics, engineering and the sciences.

(e) Geometry was widely taught in schools throughout France in this period.

(f) Napoleon's ministers told him to do it.

Let's see what we can do. First, a crude ranking. Statements (a), (c) and (d) strike me as relevant. I'm not convinced by (b) as it's a view, popular in its day, but not universally shared even then, as we can see in Chapter 2. Statement (f) is false. You can at least recognise that you have no evidence for it from this book. That leaves (e). Geometry was widely taught because the École Polytechnique entrance exam required it, but the central position of geometry at the École Polytechnique is the reason geometry is prominent in schools, not the other way round (A-levels are the same today in English schools). School education at the top levels is subordinate to university-level education, then and now.

Now to refine the ranking. Statement (d) is true, but doesn't carry much weight. What if he had been an ineffectual person, although technically very able? What if he had been a very skilled advocate of descriptive geometry but the subject was universally regarded as a waste of time? That leaves (a) and (c). It's not clear that Napoleon was involved in the syllabus of the École Polytechnique, but his advisers and ministers were, so the fact that Monge was "on side" must have helped. I'd like to know more, but my hunch is that the combination of (a) and (c) mattered just a bit more than (d), so the top ranks are (a), (c) and (d).

Down at the other end, I put (b) well ahead of (e), and (f) gets an F. So here's my answer:

(a) Descriptive geometry is essential in the work of a military engineer.

(c) Monge was an influential man behind the scenes, with a personal connection to Napoleon.

(d) Monge had a strong background in practical mathematics, engineering and the sciences.

(b) Descriptive geometry is important in the education of a mathematician.

(e) Geometry was widely taught in schools throughout France in this period.

(f) Napoleon's ministers told him to do it.

Note I'm assuming that (c) was strong enough for the government to trust that (a) was true. It then follows that the government will believe that teaching descriptive geometry in a big way will deliver their priorities.

Example 12.2

The following statements all concern Poncelet's version of projective geometry. Rank them in order of plausibility from 1 (highly plausible) to 7 (highly implausible, if not indeed false). You can rank some statements joint equal, but don't do it more than you feel you must. You won't have quite all the information you need, so as you rank them make a note of any extra assumptions you make.

(a) The geometry is rigorous, but hard to understand at first.

(b) The geometry is not rigorous, and hard to understand at first.

(c) The geometry is a powerful way to prove theorems.

(d) The geometry is a powerful way to discover theorems.

(e) The geometry is easy to use (with practice) but not rigorous.

(f) The geometry makes claims that are false.

(g) The geometry is not rigorous, but can be made rigorous with more work.

Here's my answer. A lot hangs on the evaluation of (f). The opinion of contemporaries was mixed. Some thought that claims were being made which were false, others that the dubious claims could be rescued by changing or extending the meanings of the terms involved. Here I've sided with those who think that Poncelet's version does contain false statements that cannot be rescued. That means that (g) is false, and (c) is ambiguous. The geometry is a powerful way to prove some theorems, and on other occasions it's misleading (which is alright as long as you know which occasion is which).

Statement (d) has a lot going for it: the geometry is a very good discovery method because it reduces complicated configurations to simple ones. Statement (b), sadly, is plain true. Statement (e) is true, although the geometry may be easier for uncritical minds than critical ones.

So I would rank the statements this way:

(d) The geometry is a powerful way to discover theorems.

(b) The geometry is not rigorous, and hard to understand at first.

(e) The geometry is easy to use (with practice) but not rigorous.

(f) The geometry makes claims that are false.

(g) The geometry is not rigorous, but can be made rigorous with more work.

(c) The geometry is a powerful way to prove theorems.

(a) The geometry is rigorous, but hard to understand at first.

Example 12.3

The following statements all concern Bolyai's and Lobachevskii's accounts of non-Euclidean geometry and its reception in the period 1800–1845. Rank them in order of plausibility from 1 (highly plausible) to 8 (highly implausible, if not indeed false). Again, you can rank some statements joint equal, but don't do it more than you feel you must. You won't have quite all the information you need, so as you rank them make a note of any extra assumptions you make.

(a) Bolyai's work was so poorly published as to be inaccessible.

(b) Lobachevskii's work was so poorly published as to be inaccessible.

(c) Their work was so badly written as to be incomprehensible.

(d) Their work contained fundamental flaws necessitating its rejection.

(e) Their work was too novel in its conclusions to be acceptable.

(f) Their work was dramatic and occasioned further, deep study.

(g) Their work was good enough for Gauss.

(h) If it's good enough for Gauss it's good enough for anyone!

Here are my opinions about these statements. Statement (a) is undoubtedly true, but (b) is not, Lobachevskii did after all publish in Crelle's *Journal* and he also wrote a booklet in German. Even so, crucial details were left in Russian. Statement (c) is an overstatement with a large kernel of truth in it, and truer of Bolyai than of Lobachevskii. Statement (d) is more tricky: their work is logically flawed, but automatic rejection does not follow. Why not attempt to rescue it instead? Statement (e) is a romantic value judgement. Probably many people would want to reject it on first hearing, but less conservative souls might have wanted to accept it. Does the historical classification "ahead of its time" make any sense? Or would one say that a superficial reading and an instant rejection is about what Bolyai and Lobachevskii should have expected. Statement (f) sadly was to prove false in their lifetimes, and certainly in the period 1800–1845. But (g) is true, and throws an odd light on (c), (d) and (e) because Gauss was indeed a conservative soul not seeking novelty for its own sake, and because (h) is pretty much true! There was, in the period, no sterner critic of mathematical arguments than Gauss at his best.

That gives me three blocks:

very plausible: (a), (g), (h);

somewhat plausible: (b), (c);

not convincing: (d), (f).

Where you put (e) is up to you (but in an essay argue your reasons).

This exercise of compiling and ranking opinions in the light of evidence is exactly what doing history is all about. Try it on the following questions.

Write notes for an essay on each of the questions below. Use not more than 150 words (imagine that the essay will be no more than 500 words).

Question 12.4

(i) What was the controversy about duality? Your answer should explain what is meant by duality in projective geometry, and explain how the controversy was resolved.

And/or:

(ii) Describe some of the advances made by German mathematicians in the study of projective geometry. In what ways did their work advance beyond the work their French predecessors had done?

Question 12.5

How did non-Euclidean geometry come to be accepted? Your answer should discuss improvements in the mathematical accounts of non-Euclidean geometry, and consider extra mathematical factors as well.

12.2 References and footnotes

In the references section you list all the items you've referred to in your essay, and nothing you haven't. Entries should be listed under a heading "References" at the end of your work. The entries in your list of references should appear as follows.

A book:
Non, A. 2001 *The book of famous names*, Celebrity Press, Washington.

An article in a book:
Nonentity, A. 2003 Songs to die for, pp. 49–64, in *Opera Singing for Beginners*, D.A. Capo (ed.), Covent Garden Press, London.

An article in a journal:
Stroyd, D. 2004 Born and razed, *Journal of Ancient Civilisations* 9, 64–100.

A Web item:
These should be referenced like a book, but give the full http address (the url).

List the entries alphabetically by author's last name (Non, Nonentity, Stroyd).
    In the essay you refer to these items by using the author's name and the date from your list of references, as the next example indicates.

It may be assumed that every one likes to be remembered, as A. Non has argued (see Non [2001] p. 4), although the contrary has been claimed by A. Nonentity (see Nonentity [2003] p. 61).

As this example illustrates, you can usually replace footnotes with bracketed references. The alternative is:

It may be assumed that every one likes to be remembered, as
A. Non has argued (see Non [2001] p. 4), although the contrary has
been claimed by A. Nonentity.[4]

A better use for a footnote is a long yet relevant digression, thus (and yes,
I should really supply a reference to the Marx quote):

> Careers, once lost, can seldom be restarted, unless it be in the sense
> that Marx had in mind when he said of Napoleon III that when history
> repeats itself the first time is tragedy and the second farce.[5] There are
> in fact few examples of mathematicians falling from grace only to rise
> again.

The famous quotation comes from Marx's essay *The eighteenth Brumaire of
Louis Napoleon*, available in many editions, e.g. Marx [1852], and runs, "Hegel
remarks somewhere that all great world-historic facts and personages appear,
so to speak, twice. He forgot to add: the first time as tragedy, the second time
as farce." It then continues rather pertinently: "Men make their own history,
but they do not make it as they please; they do not make it under self-selected
circumstances, but under circumstances existing already, given and transmitted
from the past. The tradition of all dead generations weighs like an Alp on the
brains of the living."

A long remark which is not a digression merits a paragraph in the text.

Every substantial piece of information should be referenced, especially if
it is a quotation. This may seem pedantic, but it is the only way to keep
misinformation from creeping in. No one's memory is completely reliable.

Always check that every book or article you refer to is in the list of references, and that everything in the list of references is referred to in the text
(including footnotes).

It is acceptable, and sometimes preferable, to refer to major works that are
known by their titles by giving author and title, thus Gauss, *Disquisitiones
arithmeticae*, rather than Gauss [1801]. As this example illustrates, an abbre-
viated title is acceptable if in general use, but give the full title in the list of
references.

12.2.1 Appendices

In essays of 500 to 2000 words, try to avoid them. They are the proper place
for substantial new information.

---

[4] See Nonentity [2003] p. 61.

[5] See Marx, *Eighteenth Brumaire*, and, in this connection, the comments in Nonentity [2003].

12.2.2 Names

English usage is idiosyncratic. You will find Gauss referred to as Gauss or, sometimes, as C. F. Gauss, Klein often as Felix Klein, Henri Poincaré almost always as Poincaré, Potter always as Harry Potter. I advocate, but cannot enforce, the rule that last names are enough. American (US) English has moved strongly to giving the full name on first mention: Carl Friedrich Gauss, Christian Felix Klein, and so on. I find this suggests an untoward degree of intimacy. Another common habit in books is to add the dates on first mention, as in sentences like "C. F. Gauss (1777–1855) was commonly regarded as the greatest mathematician of his day". I dislike this, if only because it gives an irritating sensation that Gauss was bound to die in 1855 and knew so all along.

12.2.3 Your essays

Your instructors may wish to ask for two copies of each piece of work, for their records and to deal with questions arising from the grading process.

12.3 An assignment on the first 12 chapters

General advice on tackling such an assignment is given below the question.

Question 12.6 (Essay Question 1)

Write an essay of not more than 500 words that answers either (a) or (b).

(a) Describe how Poncelet came to rediscover projective geometry. Give an example of his work at its most rigorous, and indicate one way in which what he wrote was controversial. How important was his work by 1840?

(b) Describe briefly the discoveries Bolyai and Lobachevskii made concerning non-Euclidean geometry. Why was their work so poorly received in their lifetimes?

12.3.1 Advice

When you are asked for an opinion or judgement (as in "How important was ..." or "Why was ...") give a brief argument in support of your opinion.

Try to give arguments on both sides of the question and at least consider stating an opinion only to disagree with it.

Distinguish between contemporary criticisms of someone's work and your own judgements (but don't be afraid of offering your own).

You will probably find it helpful to break your essay into parts corresponding to the various parts of the question.

You may find it helpful to write notes for the essay first. Try to use not more than 150 words. For each answer, observe the following rules.

Each note should be one sentence long, and should contain exactly one idea.

The sentences should be organised in groups according to topic.

The sentences, and the topics, should be arranged in a sensible order.

When you have finished, you should be confident that you can write an essay of the required length in which the topics come in this order.

13

Across the Rhine – Möbius's Algebraic Version of Projective Geometry

Figure 13.1 Möbius

13.1 Möbius's *Barycentric calculus*

In 1827 Möbius published his *Der barycentrische Calcul* [165] or *The barycentric calculus.*[1] The word "barycentre" means centre of gravity, but the book is not about mechanics but geometry. It is most concerned with, and is best

---

[1] The best single source on Möbius is *Möbius and his band*, Fauvel et al. (eds.) [73]. This chapter closely follows my account there, pp. 78–103.

remembered for, introducing a new system of coordinates, the barycentric co-ordinates.

Möbius built up to this new idea slowly, as was his way. He began by raising and solving this problem: given a line segment AB, a direction (other than AB, of course) along which lines AA' and BB' point, and coefficients a and b, find positions for A' and B' and a point P on AB such that

$$a \cdot AA' + b \cdot BB' = (a + b) \cdot PP',$$

where PP' is parallel to AA' and BB'. This dry, even dull, and very algebraic problem is typical of his style. The very dryness should not, however, obscure what is already a novelty in his presentation; the idea of directed or vectorial quantities, here symbolised by expressions like AA', but shortly to be repre-sented by a single letter. (However, the idea did not catch on. Grassmann's subsequent presentation also failed; only when three-dimensional vectors came as three-quarters of Hamilton's quaternions in 1843, did slow steady progress begin that reached a turning point around 1900 – but that's another story).[2]

Möbius solved his simple problem by first locating the point P which divides AB in the ratio $b : a$ (so $AP : PB = b : a$). Then any line through P will meet lines through A and B in points A' and B' such that

$$a \cdot AA' + b \cdot BB' = 0,$$

because the triangles PAA' and PBB' are similar. He then located A'' on AA' and B'' on BB' respectively, such that $A'A'' = B'B'' = PP'$, say, so

$$a \cdot A'A'' = a \cdot PP' \quad \text{and} \quad b \cdot B'B'' = b \cdot PP',$$

whence addition yields

$$a \cdot A'A'' + b \cdot B'B'' = (a + b) \cdot PP',$$

as required.

13.1.1 Barycentric coordinates

Then came the analogy that motivated the title of the book. By the law of the lever, the point P above lies at the centre of gravity of a weightless rod with a weight of a at A and b at B. You can think of the direction AA' as the direction of the force of gravity. Now the ratio $b : a$ forms the barycentric coordinates of the point P.

---

[2] For which see Crowe [47].

Möbius next generalised his simple argument to obtain the barycentric coordinates of points in the plane. Given a triangle ABC, any other point, D, generates three triangles: ABD, BCD and CAD. If you consider their areas, then it is easy enough to show, as Möbius did, that the ratio of the areas can be taken as the coordinates of the point D. In terms of weights and centres of gravity, if the triangle ABC is weightless, and weights a, b and c hang from its vertices, then the centre of gravity of the weights – their barycentre in old terminology – is at the point with coordinates $a : b : c$.

To get a feeling for these new coordinates we study them in the simplest cases first, which is how Möbius himself came to grips with them, as we know from his notebooks. Suppose the weight c is zero. It then has no effect, so the centre of gravity of the three weights is somewhere along the line joining the other two. By the opening argument, if the weight at A is 1 and the weight at B is 3 then the centre of gravity is 3/4 of the way along AB, because that is the point where a lever with the weights 1 at A and 3 at B would balance. In general, if weights a and b hang from the vertices A and B respectively, the centre of gravity and so the point of balance divides the segment AB at P, where $a \cdot AP = b \cdot BP$.

If now we fix the weights at A and B (say, 1 and 3 again) and hang a positive weight from C, the centre of gravity will be pulled off the line AB and into the triangle ABC, in the direction of PC. The triangle will now balance on a knife-edge placed along the line PC. If we start this argument again but look first at the weights at B and C and then introduce a weight at A we again find the triangle will balance along the line PA, and similarly it balances along PB too. These lines meet at P, by definition of the barycentre. The whole triangle would balance on a pin placed underneath the centre of gravity. If you now imagine that the weight at A, say, balances the triangle PBC you can deduce that the coordinates are proportional to the areas of the individual triangles PBC, PCA, PAB.

Points outside the triangle ABC also have barycentric coordinates. Return to the line AB, and set the weight at C to 0 again. A centre of gravity beyond B, at Q say, corresponds to the lever ABQ balanced at Q with certain weights at A and B, and now one of the weights must be negative; think of it as a balloon pulling the lever upwards. Either a or b may be negative, because only their ratio matters. In same way, points in the region outside the triangle are given coordinates with a negative entry.

Möbius showed that if two sets of weights are used, $[a, b, c]$ and $[a', b', c']$ for which the ratios are different so $a : b : c \neq a' : b' : c'$, then the corresponding barycentres are at different points. Conversely, each point in the plane is specified by a unique set of ratios given by three coefficients, the weights, for which the centre of gravity lies at the given point. So Möbius proposed that those coefficients be taken as what he called the barycentric coordinates of the point. It may seem odd that three numbers should be needed to specify a point in a plane, but actually only the ratios of the weights matter; weights of 2, 3 and 7 grams and weights of 2, 3 and 7 tons have their centres of gravity in the same place. So the barycentric coordinates of a point are given by their two independent ratios. To distinguish them from the more usual Cartesian coordinates, barycentric coordinates will be written in square brackets thus: $[a, b, c]$. Notice also that one combination of weights is meaningless: weights of zero at each vertex of the triangle. That combination is the only one that fails to have a unique centre of gravity.

What is better about barycentric coordinates than the usual Cartesian ones? To see their advantage, we need to compare the barycentric coordinates and the Cartesian coordinates of a point in the plane. To make life easier for ourselves, we choose the triangle ABC to be right-angled and isosceles, put its right angle, C, at the Cartesian origin, and let A and B lie at $(1, 0)$ and $(0, 1)$ respectively.

We consider the point P with Cartesian coordinates (p, q) and barycentric coordinates $[a, b, c]$. We know from the barycentric coordinates that the knife-edge through A and P meets the line CB at the point with Cartesian coordinates $(0, \frac{b}{b+c})$, so this line has equation $bx + (b + c)y = b$. Similarly, the knife-edge through B and P meets the line CA at the point with Cartesian coordinates $(\frac{a}{a+c}, 0)$, so this line has equation $(a + c)x + ay = a$. A little work shows what is easy to check, that these lines meet at the point with Cartesian coordinates $(\frac{a}{a+b+c}, \frac{b}{a+b+c})$. So the Cartesian coordinates of the point with barycentric coordinates $[a, b, c]$ are

$$ p = \frac{a}{a + b + c} \quad \text{and} \quad q = \frac{b}{a + b + c}. $$

Conversely, the barycentric coordinates of a point whose Cartesian coordinates are (p, q) are found by noticing that $p + q = \frac{a+b}{a+b+c}$ so $1 - (p + q) = \frac{c}{a+b+c}$. Therefore the required barycentric coordinates are

$$ [a, b, c] = [p, q, 1 - (p + q)]. $$

Nothing too bad there – or too interesting, until one notices that the above argument breaks down when $a + b + c = 0$, as it certainly may. When $a + b + c = 0$, there is no point in the Cartesian plane that corresponds to the point with barycentric coordinates $[a, b, c]$. If we call a point defined by giving its Cartesian coordinates a Cartesian point, and one defined by giving its barycentric coordinates a barycentric point, then we may say that the plane of barycentric points has more points than the plane of Cartesian ones. Möbius spoke of the points for which $a + b + c = 0$ as lying at infinity. Indeed, if you place such weights at the vertices A, B and C, then you find that the corresponding knife-edges are parallel.

13.1.2 Projective transformations

So we seem to have more points using barycentric coordinates than using Cartesian coordinates. Either this shows that the whole system is fatally flawed or
else it's exciting – and happily it's the second of these alternatives. To see why,
we consider simple shadow projection from a point source of light L. This casts
pictures drawn on one translucent screen onto a second screen. The image of a
point is generally a point, of a line is a line. But the image of two intersecting
lines, while usually of two intersecting lines, need not always be so. If the lines
meet at the point N, say, then the image of N is found by following the line LN
until it meets the second screen. If it happens, however, that this screen is parallel to the line LN, then the line never meets the screen, the point N has
no image, and the images of the lines on the first screen are lines that do not
meet; they appear instead, to be parallel. The moral of this tale is that the
seemingly simple transformation of the plane that involves shining light from a
point source in the fashion described is not, strictly speaking, a transformation
at all, because it loses points. Conversely, if you run the light backwards you
see that the image of two parallel lines can be cast as two intersecting lines; a
point appears to have been conjured up out of nowhere. The traditional form of
words used on such occasions, and which Möbius himself employed, are these:
one says that the point of intersection of the parallel lines has been projected
to infinity.

These "transformations" cannot be followed algebraically using Cartesian
coordinates to find where the intersection point has gone, because the argument breaks down. But they can be followed in barycentric coordinates. We
can give barycentric coordinates to the image of the point of intersection, and
there are no missing points. To do so, we put the light source, L, at the point
whose usual, three-dimensional Cartesian coordinates are $(0, -1, 1)$. We are going to shine light through the (x, z)-plane onto the (x, y)-plane, and investigate
the shadow of a triangle. In the (x, z)-plane we choose a point with Cartesian coordinates (p, q); meaning that its x-coordinate is p and its z-coordinate
is q (since its y-coordinate must be zero, we shall forget about it). So in the
(x, z)-plane we have a point with Cartesian coordinates (p, q) and therefore
with barycentric coordinates $[p, q, 1 - (p + q)]$. It is sent to the point on the
line joining it to L that lies in the (x, y)-plane. That point turns out to have
Cartesian coordinates $(\frac{p}{1-q}, \frac{q}{1-q})$ for which the corresponding barycentric coordinates are $[\frac{p}{1-q}, \frac{q}{1-q}, \frac{1-p-2q}{1-q}]$, which simplifies to $[p, q, 1 - p - 2q]$. So, in
barycentric coordinates, the projection sends the point $[p, q, 1 - (p + q)]$ to the
point $[p, q, 1 - p - 2q]$.

Now, the points that fail to have an image are exactly those that lie in
the horizontal plane through the light source L, namely those with Cartesian

coordinates $(p, 1)$ and barycentric coordinates $[p, 1, -p]$. They are sent, as you see (algebraically!) to the points $[p, 1, -1 - p]$, precisely the barycentric points that have no Cartesian equivalents. This means that the study of projective transformations is much easier using barycentric coordinates than using Cartesian coordinates because you can always see the symbols. It's a balance between the higher price we pay to get started and the ease we have later on.

Only now do we get some geometry. The fact that all non-degenerate conics are projectively equivalent then motivated Möbius to study projective transformations in terms of their effect on barycentric coordinates. He soon found that a further algebraic simplification helped. He transformed the coordinates $[x, y, x + y + z]$ to coordinates $[x, y, z]$. This further simplified the algebra, as we shall see, but lost the connection to centres of gravity. The new coordinates are simply called projective coordinates. Note that projective coordinates are also homogeneous: $[a, b, c]$ and $[ka, kb, kc]$ represent the same point for all nonzero k. Moreover, the triple $[0, 0, 0]$ stands for no point at all.

To see how projective coordinates are related to the usual Cartesian coordinates, we need first to find the equation of a line in barycentric coordinates. Consider a line in the Cartesian plane. It consists of points with Cartesian coordinates (x, y) such that $ax + by + c = 0$, for some constants a, b and c. The points with barycentric coordinates $[x, y, z]$ and Cartesian coordinates $\left(\frac{x}{x+y+z}, \frac{y}{x+y+z} \right)$ lie on this line, therefore, if and only if

$$a \frac{x}{x + y + z} + b \frac{y}{x + y + z} + c = 0,$$

which is the same equation as $ax + by + c(x + y + z) = 0$. So the corresponding equation of the line in barycentric coordinates is

$$ax + by + c(x + y + z) = 0.$$

In Möbius's new projective coordinates we replace $x + y + z$ by z, so the corresponding line in projective space has the equation $ax + by + cz = 0$. The point with Cartesian coordinates (x, y) has barycentric coordinates $[x, y, 1 - (x + y)]$ and projective coordinates $[x, y, 1]$. Conversely, the point with projective coordinates $[x, y, z] = \left[\frac{x}{z}, \frac{y}{z}, 1 \right]$ has Cartesian coordinates $\left(\frac{x}{z}, \frac{y}{z} \right)$. This means that an equation in Cartesian coordinates of the form $ax + by + c = 0$ has this projective equation: $ax + by + cz = 0$. Moreover, a curve with equation $ax^2 + bxy + cy^2 + dx + ey + f = 0$ in Cartesian coordinates has this projective equation: $ax^2 + bxy + cy^2 + dxz + eyz + fz^2 = 0$, obtained by replacing x with x/z, y with y/z and clearing of fractions. Nothing could be simpler.

A projective transformation sends lines to lines. It follows that it sends equations of the form $ax + by + cz = 0$ to equations of the same form, and this means that each of x, y and z must be transformed into linear combinations of x, y and z, something of the form $\alpha x + \beta y + \gamma z$. So a projective transformation emerges as a 3×3 matrix, with the proviso that it too is homogeneous (matrices A and kA have the same effect). As a result, Möbius had a novel, entirely algebraic, description of conic sections and their projective transformations.

13.1.3 Duality

But he went on to do something virtually new. As he was finishing writing the book, he tells us in its eventual preface, he heard that in French geometrical work it was proving exciting to associate to a point in the plane a line and vice versa (to every line a point): the phenomenon of duality.

Möbius had a reasonably slick algebraic way to do all this. From the equation of a line in projective coordinates, $ax + by + cz = 0$, he extracted the coefficients, $\{a, b, c\}$, introducing parentheses to stress the fact that this is a line we are talking about. Möbius now took two planes, and said that to the point $[u, v, w]$ in the one plane there corresponded the line defined by the triple $\{u, v, w\}$ in the other plane, and to the line defined by the triple $\{p, q, r\}$ in the one plane there corresponded the point defined by the triple $[p, q, r]$ in the other plane. This automatically ensured that collinear points went to concurrent lines and concurrent lines to collinear points, and also ensured that if you dualised again you returned to your starting point.

Later writers observed that one may think of a line as having coordinates. Indeed, if the line has equation $ax + by + cz = 0$, let its coordinates be $\{a, b, c\}$. Now the duality is as easy as it can be. To the point with coordinates $[a, b, c]$ associate the line with coordinates $\{a, b, c\}$. Conversely, to the line with coordinates $\{a, b, c\}$, associate the point with coordinates $[a, b, c]$. This gives you everything you require of a duality, and the algebra couldn't be easier.

A word about Möbius's way of introducing duality in the planar case in his *Barycentric calculus* [165]. The equation $x^T A x = 0$ represents a conic for which the duality $(a) \mapsto \{Aa\}$ sends each point (a) to its polar line, and each polar line back to its pole. Möbius's implicit choice of conic was evidently $A = I$, which gives the conic $x^T x = 0$, but what is this locus? In projective coordinates, $x^2 + y^2 + z^2 = 0$, which has no points, or at least no real points. Remember that the projective coordinates $[0, 0, 0]$ make no sense. So there is a duality but the sense Möbius had of it was surely algebraic, not geometric. Much later, in the 1850s, he was to investigate the way in which thinking of this as a locus with complex points can help, but that would take us too far afield.

It is also worth noting that the algebraic tests for collinearity and concurrence behave as they should under duality. Indeed, consider the three points $A = [a, b, c]$, $A' = [a', b', c']$ and $P = [p, q, r]$. They lie on the line $\{\alpha, \beta, \gamma\}$ if and only if

$$\begin{pmatrix} a & b & c \\ a' & b' & c' \\ p & q & r \end{pmatrix} \begin{pmatrix} \alpha \\ \beta \\ \gamma \end{pmatrix} = \begin{pmatrix} 0 \\ 0 \\ 0 \end{pmatrix}.$$

Now consider the three lines $\{a, b, c\}$, $\{a', b', c'\}$ and $\{p, q, r\}$. They meet in the point $[\alpha, \beta, \gamma]$ if and only if

$$\begin{pmatrix} a & b & c \\ a' & b' & c' \\ p & q & r \end{pmatrix} \begin{pmatrix} \alpha \\ \beta \\ \gamma \end{pmatrix} = \begin{pmatrix} 0 \\ 0 \\ 0 \end{pmatrix}.$$

Since not all of α, β and γ are zero, the test that three points lie on a line is the same as the test that their duals meet in a point, and vice versa, namely that

$$\det \begin{pmatrix} a & b & c \\ a' & b' & c' \\ p & q & r \end{pmatrix} = 0.$$

That concludes this account of Möbius's work on geometry in 1827. It was well received in its day. Baltzer, the author of the biography of him that accompanies his *Collected works* (in four volumes) [167] claimed that Möbius's influence was unmistakable in Steiner's *Systematische Entwicklung* (1832) [231] and Chasles' *Aperçu* (1837) [31], but it could equally well be argued that it was overtaken by the work of Plücker in the 1830s. Möbius's later discoveries, to which we shall presently turn, led mathematicians like Clebsch and Klein back to them, and they acquired a new lease of life, which they continue to enjoy to this day.

Baltzer also quoted the views of Cauchy and Gauss. Cauchy, reviewing Möbius's [165] in Férussac's *Bulletin* in 1828, found that everything was new in it, the ideas, the notation, the terms, so the whole basis for geometry was less simple than before, and it would only be after a difficult process of study that one could decide if the advantages compensated for the difficulties. He never embarked on such a programme. But Gauss, who only found a copy of the book many years after it came out, wrote to his friend Schumacher in 1843 that although he had been doubtful about buying it, he soon found it highly satisfying, because it brought the quintessence of the theory of conics into the light, and that it did what novel accounts should: it replaced the need for genius, upon which no one can depend, with the ability to solve problems mechanically.

13.1.4 Central projection from one plane to another

We map the plane $y = 0$ onto the plane $z = 0$ by central projection from the point (a, b, c). This joins the point $(x, 0, z)$ to the point $\lambda (a, b, c) + (1 - \lambda)(x, 0, z) = (?, ?, 0)$. This implies that $\lambda = \frac{z}{z-c}$ and $1 - \lambda = \frac{-c}{z-c}$, so we have, after a little work,

$$\lambda (a, b, c) + (1 - \lambda)(x, 0, z) = \left(\frac{az - cx}{z - c}, \frac{bz}{z - c}, 0 \right).$$

We regard this as a map

$$(x, z) \mapsto \left(\frac{az - cx}{z - c}, \frac{bz}{z - c} \right),$$

and put both sides into barycentric coordinates, thus:

$$[x, z, 1 - x - z] \mapsto \left[\frac{az - cx}{z - c}, \frac{bz}{z - c}, 1 - \frac{az - cx}{z - c} - \frac{bz}{z - c} \right].$$

To find what this does to a typical point $[x, z, w]$, we write $[x, z, w] = \left[\frac{x}{s}, \frac{z}{s}, \frac{w}{s} \right]$ and observe that this point goes to

$$\left[\frac{az/s - cx/s}{z/s - c}, \frac{bz/s}{z/s - c}, 1 - \frac{az/s - cx/s}{z/s - c} - \frac{bz/s}{z/s - c} \right],$$

which simplifies after some more work to

$$[az - cx, bz, z - az - bz - cz - cw].$$

So we may write the map as

$$\begin{pmatrix} x \\ z \\ w \end{pmatrix} = \begin{pmatrix} -c & a & 0 \\ 0 & b & 0 \\ 0 & 1 - (a + b + c) & -c \end{pmatrix} \begin{pmatrix} x \\ z \\ w \end{pmatrix}.$$

13.2 A note on duality

Duality in n-dimensional space associates to each point a hyperplane and vice versa. Let the point have projective coordinates $[\mathbf{a}]$. A plane has equation $\mathbf{b}^T \mathbf{x} = 0$, so we can speak of a plane having coordinates $\{\mathbf{b}\}$. The duality is linear, so the plane $\{\mathbf{b}\}$ corresponding to the point $[\mathbf{a}]$ because of this duality must be of the form $\{\mathbf{b}\} = A[\mathbf{a}]$ for some matrix A representing a projective transformation. If the point $[\mathbf{a}]$ is to lie on the dual plane, then $(A[\mathbf{a}])^T [\mathbf{a}] = 0$, that is, $[\mathbf{a}]^T A^T [\mathbf{a}] = 0$. If this is to be true for all points $[\mathbf{a}]$ then the matrices A and A^T must represent the same projective transformation, as this little argument shows: write $\mathbf{a} = \mathbf{u} + \mathbf{v}$, and deduce

$$\mathbf{u}^T A^T \mathbf{v} + \mathbf{v}^T A^T \mathbf{u} = 0,$$

so

$$\mathbf{u}^T A^T \mathbf{v} = -\mathbf{v}^T A^T \mathbf{u} = -\left(\mathbf{v}^T A^T \mathbf{u}\right)^T = -\mathbf{u}^T A \mathbf{v}$$

whence A^T and A satisfy $A^T = A$. But we are thinking of these transformations as projective transformations, so $A = kA$ for any matrix A and (non-zero) scalar k. So our equation becomes $A^T = kA$.

What can k be? Möbius was interested in projective three-space, where points are described by homogeneous quadruples, and a projective transformation is represented by a 4×4 matrix, so kI is also a 4×4 matrix. So if we take determinants on both sides of the equation $A^T = kA$, we obtain the equation $\det\left(A^T\right) = k^4 \det(A)$. But $\det\left(A^T\right) = \det(A)$, so $k^4 = 1$, whence $k = \pm 1$. So we find that A may be either a symmetric matrix ($A = A^T$) or else an anti-symmetric matrix ($A = -A^T$).

The symmetric case – in any number of dimensions – is the one associated to a conic, or to a quadric surface in higher dimensions. Indeed, the projective equation of a conic can be written $\mathbf{x}^T A \mathbf{x} = 0$, where A is a symmetric matrix. Given a point $[\mathbf{a}]$, its polar line with respect to this conic is the line with line coordinates $\{A\mathbf{a}\}$, and the equation $\mathbf{a}^T A \mathbf{a} = 0$ says that a point lies on its polar if and only if the point lies on the conic and its polar is the tangent to the conic at that point. Exactly analogous statements hold in three or higher dimensions when the equation $\mathbf{x}^T A \mathbf{x} = 0$ describes a quadric and the duality is between point and plane or point and hyperplane.

However, the anti-symmetric case is new, and is one of Möbius's finest discoveries. When $k = -1$, $A^T = -A$, so A is now an anti-symmetric matrix. There are no non-singular anti-symmetric matrices of odd dimension, but there are such matrices of even dimension. The discovery that in odd-dimensional spaces there is a new kind of duality, not associated with conics, is due to Möbius. It arose from his study of geometrical mechanics (but that's another story[3]) and it resolved the dispute between Gergonne and Poncelet about the nature of duality in plane projective geometry in a way neither of them could have guessed.

---

[3] Told in Gray's essay in Fauvel et al. [73, pp. 78–103].

13.3 Möbius's introduction of projective coordinates

Möbius took four arbitrary points, A, B, C, D, no three of which lay on a line, as a quadrangle of reference, and observed that by joining up the sides in pairs (AB and CD, AC and BD, and AD and BC) one obtained three more points, E, F and G in that order. Now, the line through say F and G meets AB at the fourth harmonic point of A, B and E. Recall that if E is at infinity, the fourth harmonic point is the midpoint of AB, so this construction is the projective analogue of dividing a square into four equal pieces. By choosing from among the points, he thus obtained other sets of four non-collinear points, and repeating the construction indefinitely, Möbius obtained what he called a net of points and lines. By keeping careful track of how sets of four collinear points are generated in the net, Möbius could calculate their cross-ratios, which are always rational numbers. Now, as he showed, a projective transformation (the most general continuous transformation sending lines to lines) preserves the cross-ratio of four collinear points. He deduced (by an implicit limiting argument) that his net allowed any set of four collinear points to be assigned a number (its cross-ratio) that was projectively invariant. So the net played the role in projective geometry that pairs of points a rational distance apart on lines with rational slopes would play in Euclidean geometry. Moreover, coordinates could now be imposed on the projective plane in a way that was projectively invariant. It would be enough to join the point to two of the three diagonal points of the quadrangle of reference and to consider the cross-ratios on two of the opposite sides. These cross-ratios would serve as the coordinates.

Plücker, Hesse, Higher Plane Curves, and the Resolution of the Duality Paradox

14.1 Higher plane curves

An algebraic plane curve is, by definition, the locus in the plane corresponding to a polynomial equation of some degree, n: $f(x, y) = 0$. For example $x^3 y + y^3 + x = 0$ represents a quartic. The equation may be written in homogeneous coordinates $[x, y, z]$ by setting $x = x'/z'$, $y = y'/z'$ and multiplying through by the lowest power of z' that produces a polynomial equation (if you wish you may then remove the primes) when the curve is considered to lie in the projective plane. The example above becomes $x^3 y + y^3 z + z^3 x = 0$ in homogeneous form.

The study of higher plane curves, as the curves were called when $n > 2$, goes back at least as far as Newton, who in 1667–1668 made a thorough study of cubics[1], but for present purposes a start can be made with Plücker, who, in his *System der analytischen Geometrie* of 1835 [194, p. 264], showed that every curve of degree n in the projective plane has in general $3n(n-2)$ inflection points. An inflection point is one where the tangent has at least three-fold contact with the curve. By "in general" he meant that the curve has no multiple points or cusps. Plücker (and here we modernise his notation, but not the essentials of his argument) took a curve, let us call it C, with an equation of the form $F(x, y) = 0$, and noted that it met a line with equation $y = ax + b$ where these two equations have a common root, which is where x is a root of

---

[1] See Newton's *Mathematical papers* [176, vol. II, pp. 10–89]

J. Gray, *Worlds Out of Nothing*,
Springer Undergraduate Mathematics Series,
DOI 10.1007/978-0-85729-060-1_14, © Springer-Verlag London Limited 2011

Figure 14.1 Plücker

the equation

$$F(x, ax + b) = \Xi(x) = 0. \tag{14.1}$$

The line is a tangent to the curve when the equations $\Xi = 0$ and

$$\frac{d\Xi}{dx} = 0 \tag{14.2}$$

have a root in common. For later use he also observed that the tangent is an inflection tangent if Equations 14.1 and 14.2 have a root in common that is also a root of the equation

$$\frac{d^2\Xi}{dx^2} = 0. \tag{14.3}$$

Another form for Equation 14.2 is

$$\frac{\partial F}{\partial x} + a\frac{\partial F}{\partial y} = 0. \tag{14.4}$$

Now if $F(x, y)$ is of degree n, then Equation 14.4 is an equation of degree $n-1$, and the curve it defines, call it C', therefore meets the curve C in $n(n-1)$ points. So he concluded that through a given point there are $n(n-1)$ tangents to a given curve.

Plücker considered various special cases. If the chosen point actually lies on the curve then the tangent to the curve at that point must be counted twice. An inflectional tangent must be counted thrice and an osculating asymptote is a tangent at a point at infinity on the curve.

The most interesting part of the discussion was Plücker's interpretation of what a line meeting a curve in a double point could mean, in his *System* [194, §282]. As he noticed, the condition does not mean that the line is a tangent. If the curve has a double point at a point P, then any line through P will meet the curve in a double point there. But among all the lines through P, two remain tangents if they are moved slightly and meet the curve in a point near, but distinct from the point P. This can happen in three ways. It might be that these lines have real slopes, which is what happens if the curve has two real branches crossing at the double point. Or it might be that the lines have conjugate imaginary slopes, which is what happens if the curve has an isolated real point at P. Or the lines might themselves coincide, in which case the curve has a cusp there.

14.1.1 Cubic curves

All these different phenomena can be illustrated with cubic curves. The curve with equation $y^2 = x\,(x + a)\,(x - b)$ crosses the x-axis at the points $x = 0$, $x = -a$, $x = b$. If we suppose $0 < a, b$, then the curve looks like an oval between $-a$ and 0 and then an open part in the range $b \leq x$. To obtain a curve with a real double point at the origin, let $b = 0$, so the equation becomes $y^2 = x^2\,(x + a)$. To obtain a curve with a real isolated double point at the origin, suppose instead that $a = 0$, so the equation of the curve becomes $y^2 = x^2\,(x - b)$. Finally, to obtain a curve with a cusp, let $a = 0 = b$, so the equation becomes $y^2 = x^3$. In the last case, the tangent to the curve at the cusp is the x-axis, and it is clear that it is the limiting case of a line which is a tangent to the curve at a point near the cusp and which meets the curve again somewhere else.

14.2 Plücker's resolution of the duality paradox

Plücker now argued that a line through a double point is not a "true" tangent ("true" is our word), and indeed miscounts the number of "true" tangents by 2, and a line through a cusp misrepresents the number of tangents there by 3. So the formula for the number of tangents from a point to a curve of degree n with d double points and c cusps must be reduced by 2 for each double point and by 3 for each cusp, so the number of tangents in this case is $n\,(n - 1) - 2d - 3c$. The reason that a double point reduces the degree of the dual by 2, he explained, comes from looking at the tangents to the curve from an external point. If the curve is changed slightly, so that it no longer has a double point, there are two tangents from the given external point which, when the double point is restored, collapse into the line joining the external point to the double point. A similar argument deals with the cusp.

Plücker now attempted a count of the number of inflection points on a curve. The inflection points satisfy Equation 14.3 above which says that the direction of the tangent there is stationary, or, equivalently, as he noted, the equation says that at these points on the curve the radius of curvature is infinite. This equation is of degree $3(n-2)$, and so he concluded that the number of inflection points on a curve of degree n is $3n(n-2)$. So a cubic has in general 9 inflection points, no more than 3 of which, he then proved, can be real. This confirmed results first obtained in the 18th century by the Scottish mathematician Colin MacLaurin. However, he added, this number can be less if there are repetitions.

Plücker then put this computation to good use, both at the very end of his *System der analytischen Geometrie* (1835) [194] and more systematically in his *Theorie der algebraischen Curven* (1839) [195, ch. 4]. He sketched a resolution of the duality paradox, which he called "a vital question for the general theory of curves" [194, §329]. He argued that a cubic curve with no double points or cusps has a dual of degree 6. But such a cubic curve has 9 inflection points, and they give rise to 9 cusps on the dual curve. As noted, he had already shown earlier in the book that the effect of a cusp is to lower the degree of the dual curve by 3, so 9 cusps lower the degree of the dual of the dual by 27. This reduces the degree of the dual of the dual from 30 to 3, which is necessary because, of course, the dual of the dual must be the original cubic curve.

He went on to argue that a curve of degree n with no double points or cusps has a dual of degree $n(n-1)$ and $3n(n-2)$ inflection points, which give rise to $3n(n-2)$ cusps on the dual curve. These therefore lower the degree of the dual of the dual by $9n(n-2)$. This reduces the degree of the dual of the dual from $n(n-1)(n(n-1)-1)$, but does not reduce it to n as is required. The discrepancy is

$$n(n-1)(n(n-1)-1) - 9n(n-2) - n = n(n-2)(n^2-9).$$

However, a bitangent (a line which touches the curve in two distinct places) to the original curve produces a double point on the dual, and he had already shown that each double point reduces the degree of the dual by 2. He therefore conjectured (or rather, simply asserted) that a curve of degree n with no double points or cusps must have $\frac{1}{2}n(n-2)(n^2-9)$ bitangents, thus resolving the duality paradox.

Exercise 14.1

Confirm Plücker's argument about cusps in a special case, by showing that the dual of the cubic curve with a cusp is another cubic curve (so the degree of its dual has indeed been reduced by 3).

The solution to this exercise is as follows. Let the curve have equation $y^2 = x^3$, or, in projective coordinates $y^2 z = x^3$. With respect to affine coordinates the cusp is at the origin. A typical point on the curve and near the cusp has the parametric representation $(t^2, t^3, 1)$. The cusp is at the point with parameter $t = 0$. The equation of the tangent to a curve at a point $p = (a, b, c)$ is

$$x F_x(p) + y F_y(p) + z F_z(p),$$

where $F(x, y, z) = y^2 z - x^3$. In this case the tangent is $x\left(-3t^4\right) + y\left(2t^3\right) + z\left(t^6\right) = 0$, which reduces to $3tx - 2y - t^3 z = 0$. So the tangent has line coordinates $\{3t, -2, -t^3\}$, and the corresponding dual point has coordinates $\left(3t, -2, -t^3\right) = \left(-\frac{3}{2}t, 1, \frac{1}{2}t^3\right)$. By inspection, we see that this point lies on the curve with equation $4x^3 + 27zy^2 = 0$, which is another cubic, having an inflection point at the point where $t = 0$. In affine coordinates, near that point the curve has equation $4x^3 + 27z = 0$, which is the standard elementary form of the curve with an inflection point; the inflection tangent is $z = 0$.

14.3 Confirmation by others

Plücker's arguments were rather sketchy, but all that was known about the topic in the 1830s was Poncelet's suggestion in his article of 1832 [203] that a curve of degree n could have only finitely many bitangents. In 1848 Hesse proved that there are 28 bitangents to a quartic curve [114, §3]. This was difficult work. Hesse wrote to Jacobi (8 May 1845, see *Werke* [114, p. 717]) that he had devoted the greatest part of his efforts to the task of finding canonical forms for, and systematic properties of, curves of contact, when everything seemed a jumble of properties and now "In the solution of my problem of the bitangents, I advance very slowly for every step forward requires the greatest effort." In June of that year he added: "I hope to reach the end by the same means through which Newton is believed to have discovered his law of gravity, namely, that he always believed in it."

Plücker's estimate of the number of bitangents to a curve of degree n was first proved by Jacobi [125], who confirmed that the number is in general $\frac{1}{2}n(n-2)(n^2 - 9)$. Jacobi, who had a truly remarkable ability with formulae, proved Plücker's estimate directly by using the condition that the equation which a line must satisfy in order to be a tangent (to the given curve) has a repeated root when the line is a bitangent. He found that this condition yields an equation of degree $(n-2)\left(n^2-9\right)$ and hence a curve meeting the original curve in $n\left(n-2\right)\left(n^2-9\right)$ points, the points of contact of $\frac{1}{2}n\left(n-2\right)\left(n^2-9\right)$ bitangents.

14.4 Plücker

Plücker's biography tells us something about reputation and influence in the first half of the 19th century. He was born in Elberfeld and went to school in Düsseldorf, where he took up mathematics because he had an inspiring teacher. He studied mathematics at the universities of Bonn, Heidelberg and Berlin – such travelling around was usual in 19th-century Germany and even encouraged – and in 1824 he spent some time in Paris. There he submitted his first article for publication, to Gergonne, who duly published it in his *Annales* [89]. But Gergonne, as was apparently his custom, rewrote the article before allowing it to appear, and did not even send proof pages to Plücker for correction. As a result, the paper appeared with errors, introduced by the editor, and drew some not very flattering criticisms from Poncelet. It was some years before Plücker could get the corrections published, and then only in another journal (Férussac's *Bulletin*).

In 1825 Plücker took the post-doctoral qualification from Bonn that qualified him to be a university teacher, and by 1835 he was a professor of mathematics at Halle. By then he had published a number of articles on geometry, and also the two-volume *Analytisch-geometrische Entwicklungen* (*Analytic-geometric developments*) of 1828 and 1831 [192]. He averred in the preface to volume 2 (written in Bonn in autumn 1830) that he was ignorant of Poncelet's *Traité* while writing volume 1, which he had concluded in September 1827. He had been drawn, however, into the dispute between Poncelet and Gergonne over the discovery of duality – the implication is that he had been drawn in unwillingly. Nonetheless, the idea of duality, which he called reciprocity, impressed him strongly, and volume 2 ends with a section of some 50 pages about it. He did not use it to derive new results but to display it at work in the geometry of conics. However, he did make one prescient remark in volume 2 [192, p. 267]:

> From the above remarks it is clear, as it is quite natural, that the general theory of curves is to be developed with the idea of singular points and singular straight lines.

In his article of 1830, Plücker set out his theory of homogeneous coordinates. He said [193, p. 2] he was struck by the fact that on the one hand Poncelet's *Traité* was full of entirely new ideas, and on the other it agreed so completely with the methods of analytic geometry that these latter methods could seem like a mere rewriting, even plagiarism, of Poncelet's. Plücker found part of the explanation for this in the idea of points on a line at infinity, which emerged so naturally in his formulation of analytic geometry in homogeneous coordinates. This coincidence helps to explain why most of the mathematical results in the paper concern families of curves of the second degree.[2] Towards the end of the paper he took up the theory of reciprocity "one of the most beautiful and significant extensions to geometry in recent times" [193, p. 26]. The honour for discovering this theory he divided between Gergonne and Poncelet, noting [193, p. 27] that he himself had come to exactly similar conclusions by a direct analytic route somewhat later that made no reference to a conic. But again Plücker contented himself with setting out the simplest parts of the theory of tangents, and when he came to write the major books of the 1830s he withdrew not only from the use of homogeneous coordinates but also the geometric insights they bring about how to handle the line at infinity.

In 1837 Plücker transferred back to the University of Bonn, but as a professor of physics, which he remained for many years. The reason for this shift, apart from the intrinsic interest that physics, and questions of spectroscopy in particular, had for Plücker was that he was dissatisfied with the reception his work was getting. In Germany it was judged to be unfruitful by comparison with the more synthetic approach of Poncelet and Steiner. This was ironic, because by then Poncelet was drawing similarly bleak conclusions about the impact of his own work. But relations between Plücker and Steiner were particularly bad, and Steiner was in the growing University of Berlin with ready access to Leopold Crelle, the successful editor of the first journal in Germany devoted exclusively to mathematics. Steiner identified strongly with synthetic geometry, and let it be known that he would not be willing to write for Crelle's journal if it continued to carry articles by Plücker. Plücker felt completely denigrated in Berlin, and switched fields.

In a further irony, he thereupon became much more famous abroad than in his native country. Not only did he eventually win the Copley Medal of the Royal Society of London for his work on spectroscopy – he is one of the discoverers of cathode rays – but his reputation as a profound geometer was to flourish there too. At a meeting of the British Association for the Advancement of Science held in Swansea in 1848, Sylvester hailed him as "the master" of the

---

[2] Oddly, and without referring to the work of Möbius, which he may not have known, he predicted that the greatest service his new coordinate system would provide for mathematics was in mechanics [193, p. 3].

English mathematicians, and said that when it came to the relation of geometry to analysis there was none between Plücker and Descartes. Cayley, who with his friend Sylvester dominated British pure mathematics in the second half of the 19th century, hailed Plücker's work on the singular points of curves as "the most important beyond all comparison in the entire subject of modern geometry" [30].

14.5 Hesse

Figure 14.2 Hesse

However, while Plücker was in self-imposed exile in the domain of experimental physics, another German mathematician, Otto Hesse, was sharpening up his arguments. His work improved on Plücker's by the growing use of homogeneous coordinates. So, for example, to obtain the inflection points on a curve, Hesse argued, as Plücker had, that one expects "adjacent" normals to a curve to meet at the appropriate centre of curvature, and therefore at a point of inflection the adjacent normals will be parallel, and the corresponding radius of curvature infinite. Its reciprocal is therefore zero, but in homogeneous coordinates the reciprocal of the radius of curvature of a curve with equation

$$f(x, y, z) = 0 \text{ at the point } (x, y, z) \text{ is } \begin{vmatrix} f_{xx} & f_{xy} & f_{xz} \\ f_{yx} & f_{yy} & f_{yz} \\ f_{zx} & f_{zy} & f_{zz} \end{vmatrix}.$$

(Proof deferred. Take this result on trust; it goes back to Newton according to some historians and certainly to the early 18th century; its history, and its validity, need not concern us.) This determinant has come to be called the Hessian of f, in recognition of Hesse's accomplishment. The Hessian of the curve C with equation $f(x, y, z) = 0$ is the curve, sometimes denoted $H(C)$ with equation $H(f) = 0$. Since f has degree n and each element of the determinant $H(f)$ has degree $n - 2$, it is clear that $H(f)$ has degree $3(n - 2)$, so the Hessian of a cubic is another cubic. The curve $H(C)$ has the property of passing through the inflection points of the curve C and also through any triple points on the curve, and any cusps, as you can see by finding the Hessian of $f(x, y, z) = y^2 - x^3 = 0$, which turns out to be xy^2. So for a non-singular curve of degree n, the Hessian shows that the curve has $3n(n - 2)$ and, in particular, the non-singular cubic has 9, inflection points.

Otto Hesse was born in Königsberg in 1811 and from 1832 to 1837 studied mathematics and physics at the university there under Jacobi, Bessel, Neumann and Richelot. This parade of names does much to show that Königsberg, although small, was perhaps the most high-powered university in Germany at the time. Jacobi was a prodigious mathematician, and in 1832 had just won a prize from the Academy of Sciences in Paris for his part in the discovery of elliptic functions. Bessel, after whom Bessel functions are named (although he was not the first to discover them) was a leading astronomer, a colleague of Gauss, and an expert, among other things at the determination of stellar parallax. Franz Neumann was a physicist with a particular interest in optics, and Richelot was a mathematician and one of Jacobi's best students. Moreover, Königsberg was the university that pioneered the student seminar; mathematicians adopted it from the linguists in 1834 and in this way promoted the study of mathematics at the research level, and it was the university that led the way in proper, systematic education in laboratory work and seminar discussions and therefore in the professional education of physicists.[3] Hesse could hardly have gone to a better place than the university in his native town.

After graduation Hesse taught for a while in local schools, travelled round Germany and on to Italy in the summer of 1838 mostly walking, because he came from a poor background, and on his return taught physics and chemistry at the newly founded trade school until he passed his Habilitation at Königsberg University. This qualification is a necessary and sufficient one for becoming a university professor (but not sufficient to be paid). It was obtained by writing an advanced thesis and giving a public defence of a topic, chosen from a short list presented by the candidate. Hesse then embarked as a lecturer, covering most branches of mathematics. He became a so-called extraordinary

---

[3] See Olesko [179].

professor (a junior position) at Königsberg in 1845 and, after Jacobi left to go to Berlin and Bessel died, he was the sole teacher of geometry, analysis and mechanics.

The influence of Jacobi on Hesse was marked. Jacobi, Hesse later wrote, had a tendency to go beyond the textbooks and lecture on the depths of the subject, specifically on those parts where he had worked himself. "Because he exerted himself to determine the fundamental guiding ideas of each theory, he worked on problems with his listeners with simplicity so that they may similarly hope to do the same."[4] Hesse took from Jacobi the algebraic instruments with which to clarify and master the new geometric ideas of Poncelet, Bobillier, Chasles, Steiner and Plücker. He also acquired the ambition, and the ability, to bring his listeners at every level to the stage where they could work and study independently. He was rewarded with a string of distinguished pupils who did much to spread the good name of Königsberg and the quality of its teaching methods across Germany: in 1843/1844 Kirchhoff, Aronhold and Durège; in 1849/1850 Lipschitz, Carl Neumann (Franz's son) and Schroeter; and in 1850 Rudolf Clebsch, the man he regarded as his best and closest student and with whom he became good friends until Clebsch's early death in 1872. But neither these achievements, nor the quality of his research, impressed the university sufficiently to promote him. Sylvester called him (*Phil Trans*, 1853, vol. 143) "The worthy pupil of his illustrious master Jacobi, but who, to the scandal of the mathematical world, remains still without a chair in the university which he adorns with his presence and his name."[5] Finally, in 1856 Hesse left his native city for the southern-German University of Heidelberg, where he had as colleagues no less than Kirchhoff, Bunsen (the chemist, after whom the burner is named) and Helmholtz, the leading German scientist of the century.

In Heidelberg he was popular with and much respected by his students, in whom he took a great deal of interest. In this period he worked on his textbook on analytic geometry, which Clebsch praised as "the first sign of life that modern algebra shows outside the narrow circle of journals"[6] – which shows how closely algebra and geometry were connected in the minds of Hesse and Clebsch.

A number of honours came his way: membership of the Göttingen Society in 1856 and corresponding membership of the Berlin Academy of Sciences in 1859, but despite this it seems that the education ministry in Heidelberg no more appreciated him than the ministry had in Königsberg, and in 1868 Hesse moved again, this time to Munich and the newly founded polytechnic there. He was

---

[4] See Hesse's *Lebenslauf*, in Hesse, *Gesammelte Werke* [114, p. 713].
[5] Quoted in Hesse [115, p. 718, n. 1].
[6] Quoted in Hesse [115, p. 719]

now interested in mathematics education in the schools; descriptive geometry was taught at the polytechnic but by someone else, and to an audience of 200. Hesse was awarded Berlin's Steiner prize for geometry in 1872, but by then a long illness had claimed him, and he died in 1874 at the age of 63.

15
The Plücker Formulae

This chapter provides a summary of the basic facts about what are called singular points on a curve, and an account of how they resolve the duality paradox. The next chapter gives the mathematical theory behind these facts.

15.1 Singular points

A curve given by a polynomial equation of degree n in two variables x and y is said to be a curve of degree n, and will be denoted by and referred to as a C_n.

A double point on a curve is a point through which two branches of the curve pass. For example, the origin on $x^3 + y^3 = 3xy$.

A cusp is another example of a double point on a curve. For example, the origin on $y^2 = x^3$.

A curve without double points, cusps (or worse) is said to be non-singular. Curves with double points, cusps (or worse) are singular.

Plücker's resolution of the duality paradox was as follows.

A non-singular curve of degree n has a dual curve of degree $n(n-1)$, but a singular curve of degree n with d double points and c cusps has degree $n(n-1) - 2d - 3c$.

The dual of a curve with a double point is a curve with a bitangent (a line tangent to the curve at two distinct points) and vice versa, and the dual of a curve with a cusp is a curve with an inflection point and vice versa.

J. Gray, *Worlds Out of Nothing*,
Springer Undergraduate Mathematics Series,
DOI 10.1007/978-0-85729-060-1_15, © Springer-Verlag London Limited 2011

So the dual of a non-singular curve of degree n with b bitangents and e inflection points is a curve of degree $n(n-1)$ with b double points and e cusps, and the degree of the dual of the dual therefore has degree

$$n(n-1)(n(n-1)-1) - 2b - 3e.$$

This will resolve the duality paradox if and only if

$$n(n-1)(n(n-1)-1) - 2b - 3e = n.$$

Since he had already proved that $e = 3n(n-2)$, Plücker could propose that

$$b = \tfrac{1}{2}n(n-2)(n^2-9).$$

15.1.1 The non-singular cubic curve in the plane

This is a C_3 with $d = 0$, $c = 0$, $b = 0$, and e is to be determined. The dual curve is a C_6 with e cusps, and the dual of the dual is of degree 3, so $6 \cdot 5 - 3e = 3$ and therefore $e = 9$. This once again recaptures an 18th-century result that a non-singular cubic curve has 9 inflection points. It turns out that no more than 3 of them can be real.

15.1.2 The non-singular quartic curve in the plane

This is a C_4 with $d = 0$, $b = $?, $c = 0$, $e = $?.

Its dual is a C_{12} with b double points, e cusps, and no bitangents and no inflection points. The dual of this C_{12} is therefore a curve of degree $132 - 2b - 3e = 4$, and you can check that this works with $b = 28$ and $e = 24$.

Singular curves behave differently, and in this way properties of singular curves were discovered with almost no calculation. For example, consider cubic curves. Necessarily $b = 0$ (be sure you know why). If a cubic curve has a double point and a cusp, its dual will be of degree 1, and its dual will be a point, which is not the original curve, so a singular cubic curve has either a double point or a cusp, but not both.

Case 1: $d = 1$, $c = 0$

	Curve	Dual curve	Dual of dual
Degree	3	$6 - 2 \cdot 1 - 3 \cdot 0 = 4$	$12 - 3e = 3$
Double points $= d$	1	0	1
Bitangents $= b$	0	1	0
Cusps $= c$	0	e	0
Inflection points $= e$	e	0	e

so $e = 3$.

Case 2: $d = 0$, $c = 1$

	Curve	Dual curve	Dual of dual
Degree	3	$6 - 2\cdot0 - 3\cdot1 = 3$	$6 - 3e = 3$
Double points $= d$	0	0	0
Bitangents $= b$	0	0	0
Cusps $= c$	1	e	1
Inflection points $= e$	e	1	e

so $e = 1$.

Singular quartic curves yield to the same sort of treatment, as follows.

	Curve	Dual curve	Dual of dual
Degree	4	$12 - 2d - 3c > 2$	?
Double points $= d$	d	b	d
Bitangents $= b$	b	d	b
Cusps $= c$	c	e	c
Inflection points $= e$	e	c	e

Only a limited number of cases can occur:

1. $12 - 2d - 3c = 3$, so either $c = 1$ and $d = 3$, or $c = 3$ and $d = 0$.

2. $12 - 2d - 3c = 4$, so either $c = 0$ and $d = 4$, or $c = 2$ and $d = 1$.

3. $12 - 2d - 3c = 5$, so $c = 1$ and $d = 2$.

4. $12 - 2d - 3c = 6$, so either $c = 0$ and $d = 3$, or $c = 2$ and $d = 0$.

5. $12 - 2d - 3c = 7$, so $c = 1$ and $d = 1$.

6. $12 - 2d - 3c = 8$, so $c = 0$ and $d = 2$.

7. $12 - 2d - 3c = 9$, so $c = 1$ and $d = 0$.

8. $12 - 2d - 3c = 10$, so $c = 0$ and $d = 1$.

9. $12 - 2d - 3c = 12$, so $c = 0$ and $d = 0$.

In each case the next stage, finding the dual of the dual, yields a certain number of possible values for b and e, and this gives a complete description of all possible quartic curves, each of which one would then have to construct directly in some other way.

In this connection, another Plücker formula must be introduced, along with some more of his observations about the number of double points a curve may possess. Plücker advanced the investigation of the number of points one curve can have in common with another by taking double points into consideration. Consider, for example, a quartic curve Q with four double points, the maximum it is allowed to have by the Plücker formulae. Let C be the conic passing through these four points and an arbitrary but fixed fifth point, so the curves Q and C meet in nine points, because the first four points are double. Bézout's theorem says that a quartic curve and a conic may only meet in $2 \cdot 4 = 8$ distinct points, unless the quartic curve is reducible and contains the conic as a component.

More generally, Plücker argued that an irreducible curve C_n of degree n can have at most $d_n = \frac{1}{2}(n-1)(n-2)$ double points. For, suppose it had $d+1$ double points. A curve of degree n is determined by $d_{n+3}-1 = \frac{1}{2}(n+1)(n+2)-1$ points, so, in particular, a curve C_{n-2} of degree $n-2$ can be found passing through these double points, because $\frac{1}{2}(n-1)(n)-1 \geq \frac{1}{2}(n-1)(n-2)+1$ whenever $n \geq 3$. The C_{n-2} meets the C_n in a further $\frac{1}{2}n(n-1)-1-(\frac{1}{2}(n-1)(n-2)+1)$ points, that is, in $n-3$ more points. So the C_{n-2} meets the C_n in $(n-1)(n-2)+2+n-3 = (n-1)^2$ points, which exceeds by 1 the number of common points permitted by Bézout's theorem for irreducible curves (which is of course $n(n-2)$).

The simple Plücker formula $m = n(n-1)-2d-3c$ that relates the degree n of a curve with the degree m of its dual, when the curve has d double points and c cusps, already sets bounds on what can happen when a curve is dualised, and is enough to resolve the duality paradox. However, Plücker found a further constraint, which is best explained by considering the Hessian of a curve. If the curve is non-singular it meets its Hessian in $3n(n-2)$ points, which, as we have seen, are the inflection points of the original curve. So in this case we have the formula $e = 3n(n-2)$. But if the original curve has a double point, the Hessian can be shown to pass through it, and indeed to have six-fold contact with it, because it has tangents in common with the original curve (as the example of the folium of Descartes shows). It follows that the number of inflection points is reduced by 6, and so for a curve with d double points $e = 3n(n-2)-6d$. The situation with a cusp is similar, but the contact is even stronger, indeed eight-fold. Hence, for a curve with c cusps, $e = 3n(n-2)-8c$. For a curve with d double points and c cusps, the formula is $e = 3n(n-2)-6d-8c$.

Some years later, Clebsch noted that, as a result, the quantity which is the maximum number of double points it can have minus the number of double points and cusps it does possess is an invariant, being the same for a curve and its dual. This number $D = \frac{1}{2}(n-1)(n-2)-d-c$ was called the deficiency for a time, for example by Salmon, and thought of as measuring the extent to which a curve failed to have the maximum number of double points it could have. It was given a much more fundamental meaning by Riemann (as a topological invariant of the curve) and renamed the genus.

With these formulae, Plücker claimed to have calculated all possibilities for curves of degree n up to $n, m \leq 10$, where m is the degree of the dual curve. But it should be noted that this claim holds up only if the class of singularities is restricted to double points and cusps, and, dually, to inflection points and bitangents. Such singularities are called Plückerian singularities by

some writers.[1] So, for example, Plücker's analysis entirely omits the quartic
curve consisting of four lines through a point. Because the dual is constrained
if the curve is to fit Plücker's approach, the analysis also omits all reducible
curves with components that are lines, such as the quartic composed of four
lines. There is also a number of curves of higher degree with more complicated
singularities that cannot be dealt with in this way, and to that extent Plücker's
resolution of the duality paradox was left dependent on the belief – later a
theorem – that any singularity can be regarded as being composed of Plückerian
singularities. Work was started on this idea by Cayley, continued by Smith and
Brill, and arguably consummated by Charlotte Scott. Yet again, this is another
story, to which I hope to return.

15.1.3 28 real bitangents

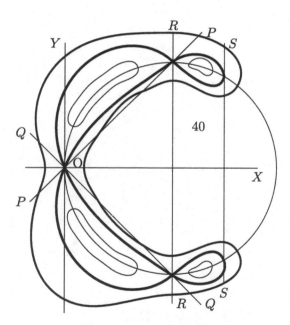

Figure 15.1 Plücker's quartic

Plücker's energy also went into the construction of a plane quartic curve
with all its 28 bitangents real, thus showing that such a curve does exist, and
that he was more of a geometer than Möbius, for example, because he was

---

[1] For example, by Fischer [76], which should be consulted for an accessible but
rigorous modern account of the mathematics described here.

sensitive to such questions. In fact, the 28 bitangents have a rich subsequent history, even if that is another story. The 28 bitangents arise as follows: one for each bean-shaped piece of the curve, four for each pair, and there are 6 pairs. (See Figure 15.1.)

Plücker had made a study of the bitangents to a quartic in his book of 1839 [195, ch. V]. He correctly established that all 28 bitangents may be real [195, §115] by considering deformations of the curve[2]

$$F\left(x,y,z\right) = \left(y^2 - x^2\right)\left(x - 1\right)\left(x - \tfrac{3}{2}\right) - 2\left(y^2 + x\left(x - 2\right)\right)^2 = 0$$

to $F\left(x,y,z\right) = \kappa$.

The curve made up of the four meniscas has 28 real bitangents. He also showed that quartics may be found having 16 or 8 real bitangents but no other values (greater than 8) [195, §122].

---

[2] See Plücker, *Theorie* [195, p. 247, fig. 40].

16
The Mathematical Theory of Plane Curves

We are interested in the local behaviour of a curve, that is to say what small pieces of them look like. Either they will have one branch passing through a point, or more than one. If only one, then the curve will have a tangent there. To study the nature of a curve near a particular point we may always suppose that we have moved the point to the origin.

Let the equation of the curve be $f(x, y) = 0$ and write the expression for f in this form:

$$f(x, y) = a_{00} + a_{10}x + a_{01}y + a_{20}x^2 + a_{11}xy + a_{02}y^2 + \cdots.$$

We write f_k for the part of degree k, namely $a_{k0}x^k + a_{k-1,1}x^{k-1}y + \cdots + a_{0k}y^k$.

If the highest degree term is of degree n we say the curve is of degree n. Sometimes the equation $f_1 = 0$ is called the linear part of the curve, and the equation $f_2 = 0$ is said to define the conic or quadratic part.

16.1 Non-singular points and tangents

Now, if we assume the origin is a point on the curve, we must have $a_{00} = 0$. If the origin is a point at which the first-degree terms do not vanish identically, the point is called a smooth or non-singular point. When a curve has no singular points, it is said to be a smooth or non-singular curve, otherwise it is singular.

J. Gray, *Worlds Out of Nothing*,
Springer Undergraduate Mathematics Series,
DOI 10.1007/978-0-85729-060-1_16, © Springer-Verlag London Limited 2011

It is reasonably clear that if the term of degree 1 does not vanish identically at the origin, then it is a very good approximation to the curve for small values of x and y, which suggests that $a_{10}x + a_{01}y = 0$ should be the tangent to the curve at the origin. This is indeed true.

So we have the following result. If the origin is a non-singular point of the curve defined by the above equation, $f(x, y) = 0$, then the equation of the tangent to the curve at the origin is

$$a_{10}x + a_{01}y = 0. \tag{16.1}$$

To verify this, consider the line $a_{10}x + a_{01}y = 0$. We assume $a_{01} \neq 0$ (if it is, make a linear change of variable that, for example, rotates the figure). This line meets the curve where

$$a_{20}x^2 + a_{11}\left(-\frac{a_{10}}{a_{01}}\right)x^2 + a_{02}\left(-\frac{a_{10}}{a_{01}}\right)^2 x^2 + \cdots = 0. \tag{16.2}$$

Now this equation has a repeated root at $x = 0$, showing that the line is a tangent.

On the other hand, any other line (which we may take without loss of generality to have equation $y = mx$) meets the curve at the points where $a_{10}x + a_{01}mx + a_{20}x^2 + a_{11}mx^2 + a_{02}m^2x^2 + \cdots$.

We may write this as

$$x\left(a_{10} + a_{01}m\right) + x^2\left(a_{20} + a_{11}m + a_{02}m^2\right) + \cdots$$
$$= x\left(\left(a_{10} + a_{01}m\right) + x\left(a_{20} + a_{11}m + a_{02}m^2\right)\right) + \cdots,$$

for which $x = 0$ is another (i.e. repeated) root if and only if $a_{10} + a_{01}m = 0$ which is the case if and only if the line we are considering is indeed $a_{10}x + a_{01}y = 0$.

16.2 Double points

Suppose now that the first-degree terms also vanish at the origin. The equation of the curve is then

$$f(x, y) = a_{20}x^2 + a_{11}xy + a_{02}y^2 + \cdots.$$

Suppose that the terms of degree two do not vanish identically, so $a_{20}x^2 + a_{11}xy + a_{02}y^2 \neq 0$. The curve is now said to have a double point at the origin. Double points come in three kinds, as we can see by looking at the equation

$$f_2(x, y) = a_{20}x^2 + a_{11}xy + a_{02}y^2 = 0,$$

which is a very good approximation to the curve for small values of x and y. This equation is the equation of a line pair.

Case 1 There are real points on this line pair if and only if $a_{11}^2 - 4a_{20}a_{02} > 0$. We say the curve has two real branches at the double point.

Case 2 If $a_{11}^2 - 4a_{20}a_{02} = 0$ the line pair is the line $2a_{02}y + a_1x = 0$ taken twice.

Case 3 If $a_{11}^2 - 4a_{20}a_{02} < 0$ then the line pair is in the form of a pair of complex conjugate lines whose only real point is the origin.

As noted above, all this is illustrated by the family of cubic curves with equations of the form $y^2 = (x + a)\, x\, (x - b)$, where we assume $a, b \geq 0$. If we look for real points on the curve, we find that there are real points in the ranges $-a \leq x \leq 0$ and $b \leq x$. So if $a = 0 \neq b$ the origin is a real point on the curve, and there are real points for whenever $b \leq x$. If $a \neq 0 = b$ there are real points whenever $-a \leq x$, and the curve has a double point at the origin. If $a = 0 = b$ the equation of the curve becomes $y^2 = x^3$, which is well known to be a cusp.

Exercise 16.1

Sketch the cubic curves under discussion.

The equation of the curves can be rewritten as

$$abx - (a - b)\, x^2 + y^2 - x^3 = 0.$$

The non-singular case:

If $ab \neq 0$ then the first-degree term defines the tangent at the origin: $x = 0$.

The three singular cases:

If $a \neq 0 = b$ then the first-degree term vanishes and the second-degree term is $-ax^2 + y^2$, which equated to zero gives the equation of the two lines $y = \pm\sqrt{a}\, x$.

If $a = 0 = b$ then the first-degree term vanishes and the second-degree term is y^2, which equated to zero gives the equation of the line $y = 0$, taken twice.

If $a = 0 \neq b$ then the first-degree term vanishes and the second-degree term is $bx^2 + y^2$, which equated to zero gives the equation of the two complex conjugate lines $y = \pm\sqrt{-b}\, x$.

Vigorous partial differentiation, which amounts to re-proving the Taylor expansion theorem for f, shows that (evaluating the partial derivatives at the origin) $f_x = a_{10}$, $f_y = a_{01}$, $f_{xx} = 2a_{20}$, $f_{xy} = a_{11}$, $f_{yy} = 2a_{02}$ and, generally, $\frac{\partial^{m+k} f}{\partial^m x \partial^k y} = m!k!a_{m,k}$.

This gives another expression for the tangent at the origin. It is $xf_x(0,0) + yf_y(0,0) = 0$.

16.3 Homogeneous coordinates

In homogeneous coordinates, the curve has equation

$$F(X,Y,Z) = a_{00}Z^n + a_{10}XZ^{n-1} + a_{01}YZ^{n-1} + a_{20}X^2Z^{n-2}$$
$$+ a_{11}XYZ^{n-2} + a_{02}Y^2Z^{n-2} + \cdots = 0.$$

It passes through the point $(0,0,1)$ if and only if $a_{00} = 0$. The equation of the tangent to the curve at the origin, by what was said above, is $a_{10}X + a_{01}Y = 0$, provided this makes sense, which is does provided at least one of $a_{10}, a_{01} \neq 0$, i.e. provided, as you can easily check, at least one of $\frac{\partial F}{\partial X}(0,0,1)$, $\frac{\partial F}{\partial Y}(0,0,1) \neq 0$.

To investigate tangents at a general point on the curve, assume, without significant loss of generality, that we join the points $[X,Y,Z]$ and $[X',Y',Z']$ by a line. This line consists of the points $[X+tX', Y+tY', Z+tZ']$. It meets the curve in a number of points, which are the roots of an equation in t.

The coefficient of t^n in this equation will be $F(X',Y',Z')$ and the coefficient of t^{n-1} will be $XF_X + YF_Y + ZF_Z$ evaluated at $[X',Y',Z']$. The coefficient of t is $X'F_X + Y'F_Y + Z'F_Z$ evaluated at $[X,Y,Z]$, and the constant term is $F(X,Y,Z)$. Following Salmon, we write the expression $X\frac{\partial}{\partial X'} + Y\frac{\partial}{\partial Y'} + Z\frac{\partial}{\partial Z'}$ as $p \cdot \Delta$, where $p = [X,Y,Z]$, so the coefficient of t^{n-1} can be written as $p \cdot \Delta F(X',Y',Z')$ and the coefficient of t can be written as $p' \cdot \Delta' F(X,Y,Z)$, where the expression $\Delta' F$ denotes $X'\frac{\partial}{\partial X} + Y'\frac{\partial}{\partial Y} + Z'\frac{\partial}{\partial Z}$ and $p' = [X',Y',Z']$.

Exercise 16.2

Show by direct calculation that the coefficient of t^{n-2} is

$$\Delta^2 F = \left(X\frac{\partial}{\partial X'} + Y\frac{\partial}{\partial Y'} + Z\frac{\partial}{\partial Z'} \right) \left(X\frac{\partial F}{\partial X'} + Y\frac{\partial F}{\partial Y'} + Z\frac{\partial F}{\partial Z'} \right)$$

evaluated at $[X',Y',Z']$. It is a quadratic expression in X,Y,Z.

The values of t for which the line meets the curve are given by the roots of the equation

$$F(X,Y,Z)t^n + \Delta F(X,Y,Z)t^{n-1} + \cdots = 0.$$

We suppose that $[X,Y,Z]$ is a point on the curve, and ask that the line be a tangent there. For this we require that the equation has $t = 0$ as a repeated root, and since the constant term is now zero this condition is simply that

$$X'F_X(X,Y,Z) + Y'F_Y(X,Y,Z) + Z'F_Z(X,Y,Z) = 0,$$

or

$$\Delta' F(X,Y,Z) = 0.$$

The equation of the tangent at the point $[X_0, Y_0, Z_0]$ on the curve is now easy to find. It is

$$X F_X (X_0, Y_0, Z_0) + Y F_Y (X_0, Y_0, Z_0) + Z F_Z (X_0, Y_0, Z_0) = 0, \qquad (16.3)$$

obtained simply by substituting into the equation above. You can check that this reduces to the local, affine, expression given earlier. This equation makes sense as long as at least one of the first partial derivatives is non-zero, so the singular points of the curve $F = 0$ are at the common points, if any, of the curves $F_X (X, Y, Z) = 0$, $F_Y (X, Y, Z) = 0$, $F_Z (X, Y, Z) = 0$.

16.4 First and subsequent polars

The curve $\Delta' F$ with equation

$$X' F_X (X, Y, Z) + Y' F_Y (X, Y, Z) + Z' F_Z (X, Y, Z) = 0$$

meets the curve F at the points where the tangents from $[X', Y', Z']$ touch the curve. The curve $\Delta' F$ is of degree n and is called the first polar of F with respect to $[X', Y', Z']$. It is of degree $n - 1$ in X, Y, Z so we immediately deduce that the number of tangents that may be drawn to a curve of degree n through a point not on the curve is $n(n - 1)$.

Or rather, we would if the curve was non-singular. If however the curve has a double point, then the repeated root condition is satisfied automatically, but the line is not really a tangent there. So the two intersections the first polar has with the curve at the double point spuriously raise the number of tangents counted by the formula $n(n - 1)$ by 2.

Question 16.1

What happens if the first polar is tangent to one of the branches of the curve at a double point? Try this example: $F (X, Y, Z) = X^3 + Y^3 - 3XYZ = 0$ and take the first polar with respect to the point $(1, 0, 1)$.

16.4.1 The first polars of a circle

Let the circle, without loss of generality, have equation $X^2 + Y^2 - Z^2 = 0$, so the first partial derivatives are $F_X = 2X$, $F_Y = 2Y$, $F_Z = -2Z$. Take the point $[a, b, c]$, then the first polar of the circle with respect to this point has the equation $aX + bY - cZ = 0$. If the point $[a, b, c]$ lies on the circle, this is the equation of the tangent at that point, so the first polar of a point on the circle is the tangent there. The first polar of the centre of the circle is the line at infinity, $Z = 0$.

16.4.2 Inflection points

The tangents from $p = [a, b, c]$ to the curve $F(X, Y, Z) = 0$ meet the curve at the common points of $F = 0$ and $p.\Delta F = 0$. Applying this result to the curve $p.\Delta F = 0$, we see that the tangents from $p = [a, b, c]$ to the curve $p.\Delta F(X, Y, Z) = 0$ meet the curve at the common points of $p.\Delta F = 0$ and $p.\Delta [p.\Delta F] = 0$. So at the points (if any) where $p.\Delta F = 0$ touches (i.e. has a common tangent with) $F = 0$, the tangent from the point $p = [a, b, c]$ to the curve is an inflection tangent.

We have already seen that the tangents from $p = [a', b', c']$ to the curve $F[X, Y, Z] = 0$ meet the curve at the common points of $F = 0$ and $p.\Delta F = 0$. Let q be one of these common points, and suppose that the line pq is an inflectional tangent to the curve $F = 0$ at the point q. Let us choose coordinates, as we may without loss of generality, such that $q = (0, 0, 1)$ and the line pq has equation $y = 0$. In local coordinates (x, y) the curve

$$F = 0 = a_0 + a_{10}x + a_{01}y + a_{20}x^2 + a_{11}xy + a_{02}y^2 + \text{higher terms}$$

has the following properties.

It passes through the origin, so $a_0 = 0$.

It is non-singular there and has the line $y = 0$ as a tangent, so on evaluating $\frac{dy}{dx}$ at the origin we find $a_{10} = 0$.

It is has the line $y = 0$ as an inflectional tangent, so on evaluating $\frac{d^2 y}{dx^2}$ at the origin and equating it to zero, we find $a_{20} = 0$.

We are interested in $q.\Delta F = F_Z$. In local coordinates (x, y) the curve has equation

$$F_Z = 0 = a_{01}y + a_{11}xy + a_{02}y^2 + \text{higher terms},$$

and we notice that what may be called the "conic part" of this expression $a_{01}y + a_{11}xy + a_{02}y^2$ is reducible. Indeed, $a_{01}y + a_{11}xy + a_{02}y^2 = y(a_{01} + a_{11}x + a_{02}y)$, and one of the lines in this line pair is the inflectional tangent.

In general, without transferring the point $q = (a, b, c)$ to the origin, if the point q is one at which the curve has an inflectional tangent, this means that the conic part of $aF_X + bF_Y + cF_Z = 0$ is reducible. Now, the conic part is this conic in a, b, c:

$$\left(a\frac{\partial}{\partial X} + b\frac{\partial}{\partial Y} + c\frac{\partial}{\partial Z} \right) \left(a\frac{\partial F}{\partial X} + b\frac{\partial F}{\partial Y} + c\frac{\partial F}{\partial Z} \right) = 0$$

which is
$$G(a, b, c) := a^2 F_{XX} + 2ab F_{XY} + \cdots + c^2 F_{ZZ} = 0.$$

This conic is reducible if the equations
$$\frac{\partial G}{\partial a} = 0, \qquad \frac{\partial G}{\partial b} = 0, \qquad \frac{\partial G}{\partial c} = 0$$

have a common solution. Now,
$$\frac{\partial G}{\partial a} = 0 = 2a F_{XX} + 2b F_{XY} + 2c F_{XZ},$$

with similar expressions for the other terms, so the equations that have to have a common solution can be written in matrix form as
$$\begin{pmatrix} F_{XX} & F_{XY} & F_{XZ} \\ F_{XY} & F_{YY} & F_{YZ} \\ F_{XZ} & F_{YZ} & F_{ZZ} \end{pmatrix} \begin{pmatrix} a \\ b \\ c \end{pmatrix} = \begin{pmatrix} 0 \\ 0 \\ 0 \end{pmatrix}.$$

Since the solution cannot be all zeros, the determinant of the matrix must vanish. This determinant is the Hessian, so we see that
$$H(F) := \begin{vmatrix} F_{XX} & F_{XY} & F_{XZ} \\ F_{XY} & F_{YY} & F_{YZ} \\ F_{XZ} & F_{YZ} & F_{ZZ} \end{vmatrix} = 0.$$

So an inflection point on the curve $F = 0$ lies on the intersection of the curve with the curve with equation $H(F) = 0$.

16.5 Hessians

Hessians of cubic curves are fun to compute, and instructive. Given the equation of the cubic, you first put it in homogeneous form $F(X, Y, Z) = 0$, then you calculate the first partial derivatives F_X, F_Y, F_Z, then the second partial derivatives F_{XX}, F_{XY}, F_{XZ}, F_{YY}, F_{YZ}, F_{ZZ} and then the Hessian.

For example, starting with the cubic curve $y = x^3$, which has the homogeneous form $YZ^2 - X^3 = 0$, it is easy to see that the Hessian is
$$\begin{vmatrix} -6X & 0 & 0 \\ 0 & 0 & 2Z \\ 0 & 2Z & 2Y \end{vmatrix} = 24XZ^2.$$ The Hessian curve, $H(F) = 0 = 24XZ^2$ does indeed pass through the inflection point, because the line $X = 0$ passes through the point $(0, 0, 1)$. But what of the other component of the Hessian curve, with equation $Z^2 = 0$? It passes through the cusp that the curve $y = x^3$ (or, better

$YZ^2 - X^3 = 0$) has at the point $(0,1,0)$ and indeed the line $Z = 0$ is tangent there (with a three-fold point of contact). So the double line $Z^2 = 0$ has a six-fold point of intersection there, as required by the second of Plücker's formulae.

We can now check out Bézout's theorem in this case. The curves $YZ^2 - X^3 = 0$ and $24XZ^2 = 0$ should meet at nine points, and indeed they do: one at the origin (counted once) and one at the point $(0,1,0)$ (counted eight times – the six-fold intersection just mentioned, plus the double intersection of the line $X = 0$ with the cusp).

Example 16.2

An example of a cubic curve, $Y^2Z - (X + Z)X(X - Z)$, which in non-homogeneous coordinates is $y^2 - (x + 1)x(x - 1)$ and its Hessian (check that the Hessian is $3X^2Z - 3XY^2 + Z^3$ which in non-homogeneous coordinates is $3x^2 - 3xy^2 + 1$) is shown in Figure 16.1. The Hessian, shown dashed, picks out the inflection points of the cubic.

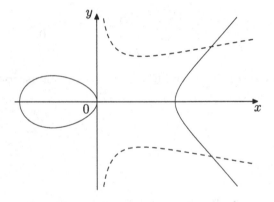

Figure 16.1   Inflection points of a cubic picked out by its Hessian

Example 16.3

Another good example to take – see Figure 16.2 – is the folium of Descartes, $X^3 + Y^3 - 3XYZ = 0$. Its Hessian curve works out to be $X^3 + Y^3 + XYZ = 0$. Work out what this tells you about the origin on the folium and check that Bézout's theorem holds. Check that the line $X + Y + Z = 0$ is an inflectional tangent to the curve at the point $[1, -1, 0]$.

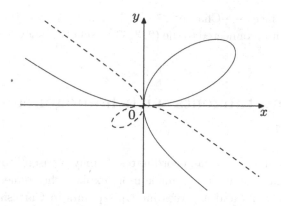

Figure 16.2 The folium of Descartes and its Hessian

Example 16.4

Although it is a vast digression, I cannot resist one more example, which is of a curve much studied by Klein in the late 1870s. In homogeneous coordinates its equation is $f_4(X, Y, Z) = X^3Y + Y^3Z + Z^3X = 0$ which is highly symmetric, and indeed Klein was much interested in its symmetries. We can also find its Hessian, which is $5X^2Y^2Z^2 - XY^5 - YZ^5 - ZX^5$. If we change coordinates by replacing Z with $Z - (X + Y)$ and then consider the part of the plane where $Z = 1$, we can plot the graph of the transformed equation that results and we can plot the Hessian of the transformed equation (this Hessian has an equation that is too large for comfort and is omitted). The curves come out looking like Figure 16.3 (the transform of f_4 is shown dark, and its Hessian dashed).

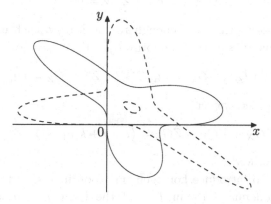

Figure 16.3 Klein's quartic curve and its Hessian

The Riemann surface (see Chapter 17) of the curve f_4 was studied by Klein, who showed that it is connected to the $(2, 3, 7)$ tessellation shown in Chapter 25.

16.6 Finding tangents with homogeneous coordinates

We start with familiar Cartesian coordinates. It may be useful to go over some of the previous material again and make more explicit the connection with the way one finds tangents and analyses multiple points in Cartesian coordinate geometry.

Let the curve have equation $f(x, y) = 0$. Differentiation gives us $f_x dx + f_y dy = 0$, where f_x is the partial derivative of f with respect to x: $\frac{\partial f}{\partial x}$, and f_y is the partial derivative of f with respect to y: $\frac{\partial f}{\partial y}$. So, if (x_0, y_0) is a point on the curve, the slope of the tangent at that point is $\frac{dy}{dx} = \frac{-f_x}{f_y}$ evaluated at that point. The equation of the tangent at (x_0, y_0) is therefore

$$(y - y_0)(f_y)_0 = -(x - x_0)(f_x)_0,$$

where $(f_y)_0$ is the value of f_y at (x_0, y_0), and similarly, $(f_x)_0$ is the value of f_x at (x_0, y_0). This equation can be rewritten as

$$x(f_x)_0 + y(f_y)_0 - (x_0(f_x)_0 + y_0(f_y)_0) = 0.$$

Now we introduce Euler's formula for homogeneous expressions. Let $F(X, Y, Z)$ be a homogeneous polynomial in the given variables of degree n. Then

$$XF_X + YF_Y + ZF_Z = nF.$$

To prove this, observe that it is enough to check it on each monomial in F. Any such monomial, G say, is of the form $a_{ij}X^iY^jZ^k$, where $i + j + k = n$. So

$$G_X = ia_{ij}X^{i-1}Y^jZ^k, \quad G_Y = ja_{ij}X^iY^{j-1}Z^k, \quad G_Z = ka_{ij}X^iY^jZ^{k-1}.$$

Adding these together, we get

$$XG_X + YG_Y + ZG_Z = (i + j + k)a_{ij}X^iY^jZ^k$$

and the result is proved.

Now we start to introduce homogeneous coordinates, setting $x = \frac{X}{Z}$ and $y = \frac{Y}{Z}$. Any monomial term in f is of the form $g = a_{ij}x^iy^j$. So $g_x = ia_{ij}x^{i-1}y^j = ia_{ij}\left(\frac{X}{Z}\right)^{i-1}\left(\frac{Y}{Z}\right)^j = \frac{1}{Z^{i+j-1}}ia_{ij}X^{i-1}Y^j = \frac{1}{Z^{n-1}}ia_{ij}X^{i-1}Y^jZ^k$. On the other hand, if we homogenise g we get $G = a_{ij}X^iY^jZ^k$ where

$i + j + k = n$. It follows that $G_X = ia_{ij}X^{i-1}Y^jZ^k$, so we deduce that $Z^{n-1}g_x = G_X$, with a similar expression for the partial derivatives with respect to y and Y.

If we now homogenise the equation for the tangent, we obtain

$$\frac{X}{Z}\left(\frac{F_X}{Z^{n-1}}\right)_0 + \frac{Y}{Z}\left(\frac{F_Y}{Z^{n-1}}\right)_0 - \left\{\frac{X_0}{Z_0}\left(\frac{F_X}{Z^{n-1}}\right)_0 + \frac{Y_0}{Z_0}\left(\frac{F_Y}{Z^{n-1}}\right)_0\right\} = 0,$$

which simplifies to

$$X(F_X)_0 + Y(F_Y)_0 - \frac{Z}{Z_0}(X_0(F_X)_0 + Y_0(F_Y)_0) = 0.$$

This expression simplifies, using Euler's formula, to

$$X(F_X)_0 + Y(F_Y)_0 + \frac{Z}{Z_0}Z_0(F_Z)_0 = 0,$$

and thence to the expression for the tangent that we seek:

$$X(F_X)_0 + Y(F_Y)_0 + Z(F_Z)_0 = 0.$$

This expression is memorable, unlike the Cartesian one, and very easy to use.

Consider now the very similar expression

$$X_1(F_X) + Y_1(F_Y) + Z_1(F_Z) = 0.$$

This defines a curve, of degree $n - 1$, called the first polar of F with respect to the point $[X_1, Y_1, Z_1]$. Now let $[X_0, Y_0, Z_0]$ be a point on both the original curve $F = 0$ and the first polar. We have

$$X_1(F_X)_0 + Y_1(F_Y)_0 + Z_1(F_Z)_0 = 0,$$

which says that the point $[X_1, Y_1, Z_1]$ lies on the tangent at $[X_0, Y_0, Z_0]$. So the first polar with respect to the point $[X_1, Y_1, Z_1]$ meets the curve where the tangents from $[X_1, Y_1, Z_1]$ touch the curve.

It is immediate that there are in general $n(n - 1)$ tangents from a point to a curve of degree n.

You should now check the following results on simple special cases.

1. The first polar of point $[X_1, Y_1, Z_1]$ with respect to the "circle" $X^2 + Y^2 - Z^2 = 0$ is indeed the polar line discussed earlier. Try, for example, the point $[1, 1, 1]$.

2. If a curve has a double point (say at $[0, 0, 1]$) then the first polar (with respect to an arbitrary point) passes through it. Notice that the first polar meets the curve at a point counted twice (one for each branch of the curve) so the number of true tangents is overestimated by 2, as Plücker argued.

3. If the first polar of a point touches a curve at a point, that point is an inflection point of the curve.

Exercise 16.3

It is stated above that the Hessian of a (non-singular) curve passes through the inflection points of the curve. Confirm this argument in simple cases, e.g. the curve $YZ^2 = X^3$ and the point $[1, 0, 1]$.

16.7 References

No book on the history or the mathematics of algebraic curves should go without mentioning the rich and beautiful book by Brieskorn and Knörrer, *Plane algebraic curves* [28]. This remarkable volume not only moves through a variety of topics with historical sensitivity, it goes a long way into the rigorous modern theory.

17
Complex Curves

17.1 Complex by necessity

In all this welter of original work on the geometry of curves, one matter has remained stubbornly unclear, although to a modern mathematician the need for it is painfully apparent. Is the subject the algebraic geometry of real curves in the real projective plane, or has everything migrated to the complex projective plane? In a sense, the question is anachronistic. It does not arise for Poncelet because, as we saw, he had his own way of talking about imaginary points and resisted very strongly Cauchy's suggestion of making everything algebraic. But Plücker and Hesse were avowedly algebraic, yet they do not seem to have openly confronted the issue. At one stage, Plücker even outlined a way in which complex coordinates could be written out of the theory in favour of certain symmetry considerations (a topic there is not room to explore here). Generally speaking, he and Hesse seem to have pursued a policy of quiet acceptance: intersection points of one curve with another, tangents to a curve, and all manner of objects may be complex if the algebra forces them to be so – but one will not explore too closely. This is a curious position. A cubic curve, let us say, was thought of as a real object in the real plane with nine inflection points, six of which at least were necessarily imaginary. A cubic and a quartic curve meet in 12 points, of which in any given case quite a number might be complex. But a curve was not made up of complex points – points with complex coordinates – or if it was one did not enquire too closely how this could be so.

J. Gray, *Worlds Out of Nothing*,
Springer Undergraduate Mathematics Series,
DOI 10.1007/978-0-85729-060-1_17, © Springer-Verlag London Limited 2011

This unsatisfactory state of affairs points to a more interesting historical observation. The situation was unsatisfactory, it could not last indefinitely, but it was not urgent. Lack of clarity on the issue was not impeding work, there was, if not a remedy, at least a patch to hand. It could wait. This, it turns out, is quite often the situation in research mathematics. A body of ideas has been stretched too far, old concepts are no longer adequate, but a lot can be done without new concepts being invented. In most walks of life one can work around a problem or tackle it directly, and mathematics is no different. The historical point is not to reproach the old mathematicians for their confusion, still less is it to slip them our resolution of it, but to notice the confusion and to see how, if at all, it is finally tackled.

So intimate has the link between algebra and geometry become, that to a modern mathematician the equation $f(x, y) = 0$ in an algebraic-geometrical context almost automatically suggests some kind of algebraic dependence of one variable, y, on another, x. A geometer will naturally think of a curve depicting this dependence in the Cartesian plane with coordinates x and y.

One potent effect of algebra, and a reason for its success, is that polynomial equations have as many solutions as their degree requires. Of course, part of the price is that the solutions are generally complex. But as long as the exact solution of such an equation is not required and knowledge of the mere existence of the solutions suffices, there is a sense in which to turn a problem into algebra is to solve it. But one can still ask: what was the geometric meaning of points whose coordinates were complex? In the opening years of the 19th century a controversy smouldered about the nature of $\sqrt{-1}$ which was only gradually resolved, so there was no simple unproblematic solution available to geometers in the 1830s, when the concept of magnitude ceased to be taken as the basis of mathematics. Naturally, confusion at this point caused problems for geometers.

Yet despite these firmly algebraic origins, today the situation is that to a modern mathematician the equation $f(z, w) = 0$ suggests more or less all that the equation $f(x, y) = 0$ does, except that the choice of z and w evokes the idea that the variables are complex, and the geometrical realisation of the curve is not quite so intuitive or elementary. So we can ask: when and how did this consensus about the complex case begin to emerge? Who gave us complex curves and complex space? To sharpen this question, let me observe that Cayley, when explaining in 1878 how a curve can be thought of as a set of points in $\mathbb{C} \times \mathbb{C}$, explicitly remarked that "I was under the impression that the theory was a known one; but I have not found it anywhere set out in detail." [30]

17.1.1 Complex numbers in geometry

Euler was among the first to define a curve via formulae, and he saw early on that algebra promotes the introduction of complex numbers into geometry. In this setting, the natural claim (what we call Bézout's theorem) is that a curve defined by an equation of degree k meets another, defined by an equation of degree m, in exactly km points, but it requires proof. This is difficult, as Euler found, because it is necessary to add points at infinity (the axis of a parabola meets the parabola only once), to find the multiplicity of an intersection (tangents are counted as meeting the curve twice) and to count intersections that seemingly do not exist (a line may fail to meet a conic at all when the points of intersection are then complex).

17.1.2 The introduction of complex curves

This was an innovation of Bernhard Riemann in his doctoral dissertation (1851), following the lead of mathematicians working on the big subject of elliptic functions (beyond this book, but see below). For those who know some complex function theory, Riemann's great idea was that an algebraic curve with equation $f(s, z) = 0$ of degree n in s and m in z, where s and z are complex variables, can be thought of as an n-fold covering of the complex z-plane (or, better, z-sphere). The resulting object rapidly became known as the Riemann surface of the function f.

Riemann surfaces are a good example of Riemann's philosophy of magnitudes and quantities at work. From Riemann's standpoint there was no need for a real interpretation of complex points on a curve. For the first time, a curve may simply have complex points on it.

Riemann's ideas took time to get across, for all sorts of reasons. For example, as several mathematicians observed, given an equation for a complex curve, it was difficult to see and get to grips with the associated Riemann surface.

The person who most energetically took up Riemann's ideas, so far as algebraic geometry is concerned, was R. F. A. Clebsch, who took them up quite deliberately in 1862. From then until his death 10 years later at the age of 39, Clebsch was the leader of a growing group of German geometers, and he and his colleague founded the *Mathematische Annalen* in 1869 as a forum for their views. Clebsch's most important paper on the subject came out in 1863. There he generalised and extended what Riemann had done in lectures, and dealt with a wide variety of topics in the theory of curves by means of Riemann's theory. In particular, he pushed the theory of the bitangents to a plane quartic further, and re-derived results of Hesse's.

Even though he died young, Clebsch had left a group of mathematicians behind able to study curves as consisting of complex points and, by the 1870s, the new view was successfully installed.

17.2 The introduction of complex points – the example of elliptic functions

The modern solution comes from an entirely different branch of mathematics, elliptic function theory. The integral $y(x) = \int_0^x \frac{dt}{\sqrt{1-t^2}}$ defines the function $x = x(y) = \sin y$ by the process known as inverting the integral. The sine function is a much more tractable function than $y(x) = \arcsin(x)$; it is periodic, whereas the arcsin function is infinitely many valued (and for that reason not strictly a function at all in the modern sense of the term). Abel and Jacobi independently in the 1820s showed that the integral $\int_0^v \frac{dt}{\sqrt{1-t^4}} = u(v)$ is more tractable when regarded as defining not $u = u(v)$ but, inversely, $v = v(u)$, and it is seen even more clearly when v is allowed to be complex. Gauss had dropped a cryptic hint in his *Disquisitiones arithmeticae* of 1801 [83, §335] about the principles of his theory of cyclotomy: "Not only can they be applied to the theory of circular functions [the familiar trigonometric functions], but also to many other transcendental functions, e.g. those which depend on the integral $\int \frac{dt}{\sqrt{1-t^4}}$." He never gave his promised treatment of this and related topics, but in the 1820s Abel took up the lemniscate, whose arc length is given by the simplest of what are called elliptic integrals: $s = \int_0^x \frac{dt}{\sqrt{1-t^4}}$. He wrote to Crelle that he had found that the lemniscatic arc can be divided by ruler and compass into n equal arcs for exactly the same values of n as the circular arc, as Gauss had hinted in the *Disquisitiones arithmeticae*. To another friend he said that he now saw as clear as day how Gauss had come to discover his results.[1]

His work, and that of Jacobi, produced a vast and stimulating theory of the new complex functions, which came to be called elliptic functions. Their merits, and the problems they posed, were a major stimulus to the introduction of complex functions, and therefore of complex numbers, in mathematics.

---

[1] Quoted in Ore [181, pp. 154–155].

18
Riemann: Geometry and Physics

Figure 18.1 Riemann

18.1 Riemann

Riemann was the archetype of the shy mathematician, not much drawn to topics other than mathematics, physics and philosophy, devout in his religion, conventional in his tastes, close to his family and awkward outside them.[1] As a child, he was taught by his father, a pastor, and then for some years at school before going to Göttingen University. There he had initially intended to study theology, in accordance with his father's wishes – Göttingen was the only university in Riemann's native Hanover with strong links to the Hanover church – but his remarkable ability at mathematics led him to switch subjects. He was

---

[1] For a biography of Riemann, see Laugwitz, *Bernhard Riemann* [148].

J. Gray, *Worlds Out of Nothing*,
Springer Undergraduate Mathematics Series,
DOI 10.1007/978-0-85729-060-1_18, © Springer-Verlag London Limited 2011

always inclined to the conceptual side of things, rather than the computational or algorithmic. His written German is scholarly and old-fashioned – Victorian, one might say – and his Latin (required for academic purposes) is no easier.

However, the level of mathematical education in Göttingen was not particularly high, and in accordance with the German freedom to study anywhere Riemann left Göttingen for Berlin in 1847, where he spent two years learning from Dirichlet about potential theory and partial differential equations, number theory and theory of integration. He also attended Jacobi's lectures on analytical mechanics and higher algebra. In particular, the influence of Jacobi's lectures might well have been to stimulate Riemann's tendency to think in an abstract and sophisticated way about the relation of mathematics to physics and the real world.

But Berlin was a bustling capital, and in other ways Göttingen was the ideal choice for Riemann. His brief involvement in the revolutionary events of 1848 in Berlin also seem to have been a factor in his decision to return, as the subsequent reactionary crackdown sought to punish those who had been involved.[2]

He returned home in 1848, then went back to Berlin, and then came back to Göttingen in 1849. Here he attended Wilhelm Weber's lectures on mathematical physics and for a while he devoted himself to studies in physics and Naturphilosophie. This influenced his way of dealing with geometry. There he quickly met Richard Dedekind, who was five years younger than him, and who was also a truly conceptual thinker, but unlike Riemann, one who was to move slowly and steadily throughout a long life. In 1855 matters improved greatly for Riemann. Gauss had for some years been a distant but real presence in Göttingen, but after his death in 1855 Dirichlet was called as the senior professor. Socially there was something of a strain between the sociable and highly musical Dedekind and Dirichlet (who was married to a sister of Mendelssohn-Bartholdy) on the one hand, and the gauche Riemann. But mathematically, Dirichlet, who may be regarded as the man who brought rigorous mathematics to Germany, was the ideal abstract thinker to guide Riemann, and the young man remained very grateful to him. His influence is visible in Riemann's work, not just the published papers but the lectures he began to give once he was working for his Habilitation.

However, Dirichlet died in 1859 and Riemann was very quickly appointed his successor (the more charismatic figure of Dedekind being by then a professor in Zürich). He still attracted only a few students, and one wonders how matters might have developed had he not contracted pleurisy in 1862. This led to a permanent weakening of his lungs, and he was advised to spend as much time as possible in the south. He accordingly spent as much time as he could in Italy,

---

[2] See the letters in Neuenschwander [175, pp. 85–131].

but his health deteriorated and on 20 July 1866 he died near Lake Maggiore, where he is buried. He left behind a number of published papers, several more in a good enough state to be published, and yet more that could be edited and printed in the first edition of his *Werke* (edited by Dedekind and Heinrich Weber in 1876).

It was not only the difficult and boldly original nature of Riemann's ideas that hindered their reception. He had had a few good students, but several of them, Hankel, Hattendorff and Roch, died young. After 1866 the Göttingen tradition then lapsed into the hands of Schering, who was not a profound mathematician. More might have come from Clebsch's discovery of Riemann's ideas, had he also not died, aged 39, in 1872. Thereafter, the German scene was increasingly dominated by the school around Weierstrass and Kronecker, which placed much more emphasis on algebra than geometry, let alone topology.

18.2 Riemann's publications

The first of Riemann's publications was the privately distributed doctoral thesis (1851) on the foundations of a theory of functions of a complex variable, where the idea of a "Riemann surface" was presented for the first time. The thesis was accepted and his defence successfully conducted in December 1851. In the German university system in the 19th century, and indeed until recently, there was a crucial post-doctoral qualification called, in untranslatable German, the Habilitation. In 1854 he successfully presented his Habilitationsschrift (essay – on the theory of functions) and gave his Habilitationsvortrag (lecture – on geometry).

Riemann is a prime example, with Galois, of the turn towards conceptual thinking and away from prodigious calculation that characterises the 19th century, and it was his conceptual novelties that made his work hard to accept. He was deeply involved for a time in studying the philosophy of Herbart, and worked enthusiastically in experimental physics, where Wilhelm Weber exerted a life-long influence upon him, even though Riemann's ideas in physics on action at a distance became distinctly unWeberian. In particular, Riemann speculated on how distortions in space might enable forces such as electricity, magnetism and gravity to travel. His idea was that these distortions of space would somehow enable influence to travel from one object to another, thus explaining the otherwise mysterious action at a distance of these fundamental forces, and mathematically they would show up as variations in the underlying metrical relations, thus altering the geometry of space.[3]

---

[3] For more details, see Bottazzini and Tazzioli [25].

His involvement in both philosophy and physics enabled him to reformulate the concept of a mathematical quantity. He came to argue that mathematics is about "n-fold extended quantities" ("n-tuples of real numbers" is the equally unattractive modern term) to which is added some appropriate extra structure. In particular, complex numbers are pairs of real numbers; some extra structure (given by the Cauchy–Riemann equations) enables one to define analytic functions.

This new metaphysical basis for mathematics marks a break with the old theory of magnitudes that still preoccupied Gauss. In the context of differential geometry it freed the mind to propose many different descriptions of physical space (any set of n-fold extended quantities with a metric will do) and to regard Euclidean geometry as just one possibility among many. It also fits very well to a view of physics as being about variable quantities that one can measure and whose variations obey some physical "laws".

18.3 Riemann on geometry

In 1854 Riemann submitted three topics for his Habilitation, and contrary to his hopes, as he later wrote, Gauss chose the one "On the hypotheses which lie at the foundations of geometry" that therefore formed his lecture for the Habilitation (the *Habilitationsvortrag* (1854)).[4] The lecture, of which lengthy extracts will be found below, is not an easy read. It lacks the formulae which would help mathematicians understand it, because the lecture was given to the philosophy faculty, of which mathematics formed a department. Moreover, the philosopher Lotze was on the panel, and Riemann may have known what Klein was to find out to his pain some years later (1870) that Lotze was very dismissive of non-Euclidean geometry. Gauss, on the other hand, knew that his health was very poor and was eager to hear what this bright young man had to say on a subject that was among his lifelong interests. According to Dedekind, the lecture exceeded all his expectations, he sat through it in the greatest astonishment and spoke to his friend Wilhelm Weber afterwards in a rare state of excitement about the profundity of Riemann's ideas.[5]

In it Riemann did not explicitly mention non-Euclidean geometry, and he never mentioned the names Bolyai and Lobachevskii. He referred only to Legendre when describing the darkness which he said has covered the foundations since the time of Euclid, and later in [209, II, §5 and III, §1] he described two homogeneous geometries in which the angle sum of all triangles is deter-

---

[4] Riemann's *Lebenslauf* in Riemann, *Gesammelte Werke*, 3rd edn. [210, p. 579].
[5] Riemann's *Lebenslauf* in Riemann, *Gesammelte Werke*, 3rd edn. [210, p. 581].

mined once it is known for one triangle, as occurring on surfaces of constant non-zero curvature. This oblique glance at the characterisation of three homogeneous geometries is typical of the formulation of the "problem of parallels" since Saccheri; it can be found in many editions of Legendre's *Éléments de géométrie*, e.g. [149], and would have alerted any mathematician to the implications for non-Euclidean geometry without its being mentioned by name. On the other hand, not naming it explicitly would avoid philosophical misapprehensions about what he had to say, so one may perhaps ascribe the omission to a Gaussian prudence.

It is less certain, however, that Riemann had read Lobachevskii, and very unlikely he had read Bolyai. Only Gauss appreciated them in Göttingen at that time, and it is not known if he discussed these matters with Riemann. There is, however, the tantalising fact that a major paper of Dirichlet's on the theory of Fourier series, which we can be certain Riemann read, was published in the same volume of Crelle's *Journal* as Lobachevskii's "Géométrie imaginaire" of 1837 [152, vol. 17]. It is hard to believe that Riemann's eyes would not have caught that title and, thus drawn, read the article itself.[6] That said, the crucial idea in Riemann's paper is his presentation of geometry as intrinsic, grounded in the free mobility of infinitesimal measuring rods, and to be expressed mathematically in terms of curvature. This is an immense generalisation of Gauss's idea of the intrinsic curvature of a surface [84], itself a profound novelty.

18.3.1 Surfaces

Applied to surfaces, Riemann's proposals went like this. A surface is a two-dimensional set of points (something that is swept out by a moving curve, just as a curve is swept out by a moving point). Given a path in the surface, its length is measured approximately by using a finite ruler, and smaller and smaller rulers give better and better approximations. The length is measured correctly only in the limit and, by means of an infinitesimal ruler, Riemann gave a precise description in formulae. Accordingly, computing length is done by integration, and the concept of length is infinitesimal and calculus based, hence the name differential geometry for this branch of mathematics. Once you have a concept of length you can find the geodesic between two points and compute the curvature of the surface as Gauss had shown. If the surface is a plane, a geodesic is a straight line, and on a sphere it is a great circle.

---

[6] The page numbers are 35–56 for Dirichlet and 295–320 for Lobachevskii [152, vol. 17].

But almost all Riemann said about non-Euclidean geometry was that the angle sum of a triangle on a surface of constant negative curvature is always less than π. He never mentioned the subject by name, or any investigator of it except Legendre. (See the extract below, taken from the first English translation of the hypotheses, produced, oddly unclearly, by his leading English contemporary, W. K. Clifford [39].) He did, however, drop one remark, one formula, which shows just how deep his insight was into the topic.

Riemann wished to discuss surfaces without reference to any ambient space. His inspiration here was Gauss's discovery of curvature, which Gauss showed was something that could be determined from quantities measured in the surface alone. Indeed, Riemann's whole insight into geometry may be summarised by saying that geometry is first of all the study of the intrinsic properties of n-dimensional manifolds, and that the study of how a manifold inherits properties from a larger ambient space should be reformulated accordingly. Now, when one studies a surface intrinsically, one has only coordinates. For example, the familiar latitude and longitude coordinates on the earth (regarded as a perfect sphere) may be thought of as specifying points in a plane, and indeed this is exactly how a map of the surface may be constructed and depicted in the pages of an atlas. In fact, maps of the earth may be constructed in many ways. Stereographic projection, for example, maps the sphere of radius r onto the tangent plane to the sphere at the north pole by joining a point on the sphere with a straight line from the south pole to a point P in the plane. If the point P has coordinates (x, y) and distance is measured according to the formula

$$ds = \frac{\sqrt{dx^2 + dy^2}}{1 + (1/4r^2)\left(x^2 + y^2\right)}$$

the intrinsic geometry of the plane is identical with spherical geometry, because this formula is what the usual metric on the sphere looks like after stereographic projection. A rather complicated calculation done once and for all in courses on differential geometry allows mathematicians to compute the curvature of a surface from its metric, and if one starts with the plane with the metric just given, then, happily, the space being described is indeed a space of constant curvature r^{-2}.

This little formula conceals a remarkable statement. If one sets r^{-2} negative, say, for simplicity, $r^{-2} = -4$, the formula for distance becomes $ds = \frac{\sqrt{dx^2 + dy^2}}{1 - (x^2 + y^2)}$, which makes sense only when $x^2 + y^2 < 1$. But inside this region, which is the interior of the circle of radius 1, the formula for distance describes a two-dimensional space of constant negative curvature (-4, to be precise). There can be no doubt that Riemann knew this perfectly well, even though he did not draw attention to it, because he had been talking about negative curvature just a page before.

18.4 Riemannian geometry

The vital point about Riemann's lecture is that Riemann said that all geometry is based on intrinsic measurement. Gauss had made only the much more limited claim that it is possible to do geometry on a surface without referring to the surrounding Euclidean (three-dimensional) space. Following Gauss, Riemann showed that any two surfaces have different geometries, that is, different mathematical theorems are true for them if they have different curvatures anywhere; but he did much more. Riemann gave a wholly novel answer to the question: what is geometry? To him, geometry was to do with concepts like length and angle which could be intrinsically defined on a surface or space of some sort. It follows that there are many geometries, one for each kind of surface and each definition of distance: a geometry arises from anything in which it makes sense to talk of a distance between two points, and this geometry will have a set of theorems associated with it.

This is a radical step: we go from having only one true geometry, to having infinitely many different geometries, none of which has any special status. There is not even a universal ambient space which endows all subspaces with their metrics. To see how radical this is, if we have a three-dimensional set of points which possess some sense of distance, then it carries a geometry. As an example, consider physical space. Do Riemann's ideas mean that space is Euclidean? Not at all, since we have not mentioned Euclidean geometry. Far from being the origin of geometrical properties, Euclidean space becomes just one candidate for physical space. To discover whether it is the correct one it would be necessary to make measurements and calculate the three-dimensional analogue of the curvature.

A point upon which Riemann insisted, indeed he opened the lecture with it, was that Euclid's postulates are completely subverted: no longer can they be regarded as unproblematically true assumptions about physical space. Instead, Riemann argued, all geometry is based on specific metrical considerations, and Euclid's geometry cannot occupy a paramount position as the geometry of space and the source of geometrical concepts which are induced onto embedded surfaces. Riemann's starting point was so radical that discussion of any axiom system for Euclidean geometry would miss the point, and this also contributes to what can seem to be the "case of the missing geometry". Non-Euclidean geometry is not centre stage in Riemann's lecture because axioms for (Euclidean) geometry are not fundamental, or even interesting, for Riemann.

Riemann made some other remarkable observations. He distinguished between a space (spaces?) which is unbounded and that which is infinite. He pointed out that there was no evidence that physical space was bounded, in the sense that it had a boundary beyond which one could not go. But this did not mean that it was infinite. It could simply be a (three-dimensional) sphere, for example. This idea has become commonplace since the days of Einstein and the modern theories of cosmology which go from the Big Bang to a finite universe, but it is hard to underestimate how shocking it must have seemed to everyone accustomed to thinking of space as infinite (whether for physical or theological reasons). He also had provocative, and remarkably modern, ideas about how geometry might break down in very very small regions.

Riemann did not seek to publish these ideas. They were somewhat further developed in his Paris prize entry of 1861, also unpublished until 1876, and first appeared in 1867, after his death. By then Beltrami had independently discovered the import of Gauss's ideas for non-Euclidean geometry.

18.5 From Riemann's *Habilitationsvortrag*

The extracts below are taken from Clifford's translation of Riemann's *Habilitationsvortrag* [209], originally published in *Nature* in 1873. William Kingdon Clifford was by common consent the best English mathematician of his generation, with a particular ability in geometry, and by this translation he signalled his appreciation of Riemann's ideas. He went on to do work on the difficult theory of Riemann surfaces before succumbing to tuberculosis. He died in 1879 at the age of 33.

The translation [39] has more than its share of obscure Victorian phraseology, which may be attributed to Riemann's dense German style and the fact that mathematical terminology was being created and complexities uncovered as Riemann and Clifford were writing. Clifford usually wrote more clearly than this, and was regarded as a gifted lecturer. Some of these are easy enough to penetrate. Riemann's German word "Mannigfaltigkeit" was translated by Clifford as "manifoldness" and sometimes as "extent" and can be taken to mean "manifold" (with some of the same lofty overtones) provided it is remembered that here it is an informal concept and today it is a precise one.

ON THE HYPOTHESES WHICH LIE AT THE BASES OF GEOMETRY
Plan of the Investigation.
    It is known that geometry assumes, as things given, both the notion of space and the first principles of constructions in space. She gives definitions of them which are merely nominal, while the true determinations appear in the form of axioms. The relation of these assumptions remains consequently in darkness; we neither perceive whether and how far their connection is necessary, nor, *a priori*, whether it is possible.

From Euclid to Legendre (to name the most famous of modern reforming geometers) this darkness was cleared up neither by mathematicians nor by such philosophers as concerned themselves with it. The reason of this is doubtless that the general notion of multiply extended magnitudes (in which space-magnitudes are included) remained entirely unworked. I have in the first place, therefore, set myself the task of constructing the notion of a multiply extended magnitude out of general notions of magnitude. It will follow from this that a multiply extended magnitude is capable of different measure-relations, and consequently that space is only a particular case of a triply extended magnitude. But hence flows as a necessary consequence that the propositions of geometry cannot be derived from general notions of magnitude, but that the properties which distinguish space from other conceivable triply extended magnitudes are only to be deduced from experience. Thus arises the problem, to discover the simplest matters of fact from which the measure-relations of space may be determined; a problem which from the nature of the case is not completely determinate, since there may be several systems of matters of fact which suffice to determine the measure-relations of space – the most important system for our present purpose being that which Euclid has laid down as a foundation. These matters of fact are – like all matters of fact – not necessary, but only of empirical certainty; they are hypotheses. We may therefore investigate their probability, which within the limits of observation is of course very great, and inquire about the justice of their extension beyond the limits of observation, on the side both of the infinitely great and of the infinitely small.

[...]

II. *Measure-relations* ...

§1. Measure-determinations require that quantity should be independent of position, which may happen in various ways. The hypothesis which first presents itself, and which I shall here develop, is that according to which the length of lines is independent of their position, and consequently every line is measurable by means of every other. Position-fixing being reduced to quantity-fixings, and the position of a point in the n-dimensioned manifoldness being consequently expressed by means of n variables x_1, x_2, $x_3, \ldots, x_n$ the determination of a line comes to the giving of these quantities as functions of one variable. The problem consists then in establishing a mathematical expression for the length of a line, and to this end we must consider the quantities x as expressible in terms of certain units. I shall treat this problem only under certain restrictions, and I shall confine myself in the first place to lines in which the ratios of the increments dx of the respective variables vary continuously. We may then conceive these lines broken up into elements, within which the ratios of the quantities dx may be regarded as constant; and the problem is then reduced to establishing for each point a general expression for the linear element ds starting from that point, an expression which will thus contain the quantities x and the quantities dx. I shall suppose, secondly, that the length of the linear element, to the first order, is unaltered when all the points of this element undergo the same infinitesimal displacement, which implies at the same time that if all the quantities dx are increased in the same ratio, the linear element will vary also in the same ratio.

On these suppositions, the linear element may be any homogeneous function of the first degree of the quantities dx, which is unchanged when we change the signs of all the dx, and in which the arbitrary constants are continuous functions of the quantities x. To find the simplest cases, I shall seek first an expression for manifoldnesses of $n-1$ dimensions which are everywhere equidistant from the origin of the linear element; that is, I shall seek a continuous function of position whose values distinguish them from one another. In going outwards from the origin, this must either increase in all directions or decrease in all directions; I assume that it increases in all directions, and therefore has a minimum at that point. If, then, the first and second differential coefficients of this function are finite, its first differential must vanish, and the second differential cannot become negative; I assume that it is always positive. This differential expression, then, of the second order remains constant when ds remains constant, and increases in the duplicate ratio[7] when the dx, and therefore also ds, increase in the same ratio; it must therefore be ds^2 multiplied by a constant, and consequently ds is the square root of an always positive integral homogeneous function of the second order of the quantities dx, in which the coefficients are continuous functions of the quantities x. For space, when the position of points is expressed by rectilinear co-ordinates, $ds = \sqrt{\Sigma(dx)^2}$; space is therefore included in this simplest case. The next case in simplicity includes those manifoldnesses in which the line-element may be expressed as the fourth root of a quartic differential expression. The investigation of this more general kind would require no really different principles, but would take considerable time and throw little new light on the theory of space, especially as the results cannot be geometrically

---

[7] That is, as the square. This terminology was obsolete when Clifford used it; even Riemann's German is more direct.

expressed; I restrict myself, therefore, to those manifoldnesses in which the line-element is expressed as the square root of a quadric differential expression. Such an expression we can transform into another similar one if we substitute for the n independent variables functions of n new independent variables. In this way, however, we cannot transform any expression into any other; since the expression contains $n(n+1)$ coefficients which are arbitrary functions of the independent variables; now by the introduction of new variables we can only satisfy n conditions, and therefore make no more than n of the coefficients equal to given quantities. The remaining $n(n-1)$ are then entirely determined by the nature of the continuum to be represented, and consequently $n(n-1)$ functions of positions are required for the determination of its measure-relations. Manifoldnesses in which, as in the Plane and in space, the line-element may be reduced to the form $\sqrt{\Sigma dx^2}$ are therefore only a particular case of the manifoldnesses to be here investigated; they require a special name, and therefore these manifoldnesses in which the square of the line-element may be expressed as the sum of the squares of complete differentials I will call flat. In order now to review the true varieties of all the continua which may be represented in the assumed form, it is necessary to get rid of difficulties arising from the mode of representation, which is accomplished by choosing the variables in accordance with a certain principle.

[...]

§3.   In the idea of surfaces, together with the intrinsic measure-relations in which only the length of lines on the surfaces is considered, there is always mixed up the position of points lying out of the surface. We may, however, abstract from external relations if we consider such deformations as leave unaltered the length of lines – i.e., if we regard the surface as bent in any way without stretching, and treat all surfaces so related to each other as equivalent. Thus, for example, any cylindrical or conical surface counts as equivalent to a plane, since it may be made out of one by mere bending, in which the intrinsic measure-relations remain, and all theorems about a plane – therefore the whole of planimetry – retain their validity. On the other hand they count as essentially different from the sphere, which cannot be changed into a plane without stretching. According to our previous investigation the intrinsic measure-relations of a two-fold extent in which the line-element may be expressed as the square root of a quadric differential, which is the case with surfaces, are characterised by the total curvature. Now this quantity in the case of surfaces is capable of a visible interpretation, viz., it is the product of the two curvatures of the surface, or multiplied by the area of a small geodesic triangle, it is equal to the spherical excess of the same. The first definition assumes the proposition that the product of the two radii of curvature is unaltered by mere bending; the second, that in the same place the area of a small triangle is proportional to its spherical excess. To give an intelligible meaning to the curvature of an n-fold extent at a given point and in a given surface-direction through it, we must start from the fact that a geodesic proceeding from a point is entirely determined when its initial direction is given. According to this we obtain a determinate surface if we prolong all the geodesics proceeding from the given point and lying initially in the given surface-direction; this surface has at the given point a definite curvature, which is also the curvature of the n-fold continuum at the given point in the given surface-direction.

§4.   Before we make the application to space, some considerations about flat manifoldnesses in general are necessary; i.e., about those in which the square of the line-element is expressible as a sum of squares of complete differentials. In a flat n-fold extent the total curvature is zero at all points in every direction; it is sufficient, however (according to the preceding investigation), for the determination of measure-relations, to know that at each point the curvature is zero in $\frac{1}{2}n(n-1)$ independent surface-directions. Manifoldnesses whose curvature is constantly zero may be treated as a special case of those whose curvature is constant. The common character of these continua whose curvature is constant may be also expressed thus, that figures may be moved in them without stretching. For clearly figures could not be arbitrarily shifted and turned round in them if the curvature at each point were not the same in all directions. On the other hand, however, the measure-relations of the manifoldness are entirely determined by the curvature; they are therefore exactly the same in all directions at one point as at another, and consequently the same constructions can be made from it: whence it follows that in aggregates with constant curvature figures may have an arbitrary position given them. The measure-relations of these manifoldnesses depend only on the value of the curvature, and in relation to the analytic expression it may be remarked that if this value is denoted by α, the expression for the line-element may be written

$$\frac{1}{1+\frac{1}{4}\alpha\varSigma x^2}\sqrt{\varSigma dx^2}.$$

[...]

III. Application to Space

§1.   By means of these inquiries into the determination of measure-relations of an n-fold extent the conditions may be declared which are necessary and sufficient to determine the metric properties of space, if we assume the independence of line-length from position and express-ibility of the line-element as the square root of a quadric differential, that is to say, flatness in the smallest parts.

First, they may be expressed thus: that the curvature at each point is zero in three surface-directions; and thence the metric properties of space are determined if the sum of the angles of a triangle is always equal to two right angles.

Secondly, if we assume with Euclid not merely an existence of lines independent of position, but of bodies also, it follows that the curvature is everywhere constant; and then the sum of the angles is determined in all triangles when it is known in one.

Thirdly, one might, instead of taking the length of lines to be in-dependent of position and direction, assume also an independence of their length and direction from position. According to this conception changes or differences of position are complex magnitudes expressible in three independent units.

§2. In the course of our previous inquiries, we first distinguished between the relations of extension or partition and the relations of measure, and found that with the same extensive properties, differ-ent measure-relations were conceivable; we then investigated the sys-tem of simple size-fixings by which the measure-relations of space are completely determined, and of which all propositions about them are a necessary consequence; it remains to discuss the question how, in what degree, and to what extent these assumptions are borne out by experience. In this respect there is a real distinction between mere extensive relations and measure-relations; in so far as in the former, where the possible cases form a discrete manifoldness, the declara-tions of experience are indeed not quite certain, but still not inac-curate; while in the latter, where the possible cases form a contin-uous manifoldness, every determination from experience remains al-ways inaccurate: be the probability ever so great that it is nearly exact. This consideration becomes important in the extensions of these empirical determinations beyond the limits of observation to the infinitely great and infinitely small; a latter may clearly become more inaccurate beyond the limits of observation, but not the for-mer.

In the extension of space-construction to the infinitely great, we
must distinguish between *unboundedness* and *infinite extent*, the former belongs to the extent relations, the latter to the measure-relations.
That space is an unbounded three-fold manifoldness, is an assumption
which is developed by every conception of the outer world according
to which every instant the region of real perception is completed and
the possible positions of a sought object are constructed, and which by
these applications is for ever confirming itself. The unboundedness of
space possesses in this way a greater empirical certainty than any external experience. But its infinite extent by no means follows from this;
on the other hand if we assume independence of bodies from position,
and therefore ascribe to space constant curvature, it must necessarily
be finite provided this curvature has ever so small a positive value.
If we prolong all the geodesics starting in a given surface-element, we
should obtain an unbounded surface of constant curvature, i.e., a surface which in a flat manifoldness of three dimensions would take the
form of a sphere, and consequently be finite.

[. . .]

§3.   The questions about the infinitely great are for the interpretation of nature useless questions. But this is not the case with the questions about the infinitely small. It is upon the exactness with which we
follow phenomena into the infinitely small that our knowledge of their
causal relations essentially depends. The progress of recent centuries in
the knowledge of mechanics depends almost entirely on the exactness
of the construction which has become possible through the invention of
the infinitesimal calculus, and through the simple principles discovered
by Archimedes, Galileo, and Newton, and used by modern physics. But
in the natural sciences which are still in want of simple principles for
such constructions, we seek to discover the causal relations by following
the phenomena into great minuteness, so far as the microscope permits.
Questions about the measure-relations of space in the infinitely small
are not therefore superfluous questions.

If we suppose that bodies exist independently of position, the curvature is everywhere constant, and it then results from astronomical
measurements that it cannot be different from zero; or at any rate its
reciprocal must be an area in comparison with which the range of our
telescopes may be neglected. But if this independence of bodies from
position does not exist, we cannot draw conclusions from metric relations of the great, to those of the infinitely small; in that case the
curvature at each point may have an arbitrary value in three directions, provided that the total curvature of every measurable portion of

space does not differ sensibly from zero. Still more complicated relations may exist if we no longer suppose the linear element expressible as the square root of a quadric differential.

Now it seems that the empirical notions on which the metrical determinations of space are founded, the notion of a solid body and of a ray of light, cease to be valid for the infinitely small. We are therefore quite at liberty to suppose that the metric relations of space in the infinitely small do not conform to the hypotheses of geometry; and we ought in fact to suppose it, if we can thereby obtain a simpler explanation of phenomena.

The question of the validity of the hypotheses of geometry in the infinitely small is bound up with the question of the ground of the metric relations of space. In this last question, which we may still regard as belonging to the doctrine of space, is found the application of the remark made above; that in a discrete manifoldness, the ground of its metric relations is given in the notion of it, while in a continuous manifoldness, this ground must come from outside. Either therefore the reality which underlies space must form a discrete manifoldness, or we must seek the ground of its metric relations outside it, in binding forces which act upon it.

The answer to these questions can only be got by starting from the conception of phenomena which has hitherto been justified by experience, and which Newton assumed as a foundation, and by making in this conception the successive changes required by facts which it cannot explain. Researches starting from general notions, like the investigation we have just made, can only be useful in preventing this work from being hampered in too narrow views, and progress in knowledge of the interdependence of things from being checked by traditional prejudices. This leads us into the domain of another science, of physics, into which the object of this work does not allow us to go to-day.

19
Differential Geometry of Surfaces

19.1 Basic techniques

Let us consider a surface in space, given by an equation of the form $F(x, y, z) = 0$. We shall assume we have a map from a region of the plane, with coordinates (u, v) onto part of the surface, and that we can differentiate this map as often as we like. We assume that at each point of the surface the directions u-increasing and v-increasing are distinct.

For example, on the sphere of radius R, we may take u as the longitude and v as the latitude, so a typical point on the sphere has coordinates $R(\cos u \cos v, \sin u \cos v, \sin v)$.

Let $\mathbf{r}(u, v)$ be a point on an arbitrary surface. Then when u and v are functions of t, we have $\frac{d\mathbf{r}}{dt} = \frac{\partial \mathbf{r}}{\partial u}\frac{du}{dt} + \frac{\partial \mathbf{r}}{\partial v}\frac{dv}{dt}$, and so the square of the speed of r as a function of t is

$$\left(\frac{d\mathbf{r}}{dt} \cdot \frac{d\mathbf{r}}{dt}\right) = \frac{\partial \mathbf{r}}{\partial u} \cdot \frac{\partial \mathbf{r}}{\partial u}\left(\frac{du}{dt}\right)^2 + 2\frac{\partial \mathbf{r}}{\partial u} \cdot \frac{\partial \mathbf{r}}{\partial v}\frac{du}{dt}\frac{dv}{dt} + \frac{\partial \mathbf{r}}{\partial v} \cdot \frac{\partial \mathbf{r}}{\partial v}\left(\frac{dv}{dt}\right)^2.$$

This looks less intimidating if we write $\mathbf{r}_u$ for $\frac{\partial \mathbf{r}}{\partial u}$, $\mathbf{r}'$ for $\frac{d\mathbf{r}}{dt}$ and drop the dt. Then it becomes

$$\|\mathbf{r}'\|^2 = \mathbf{r}_u \cdot \mathbf{r}_u du^2 + 2\mathbf{r}_u \cdot \mathbf{r}_v du\, dv + \mathbf{r}_v \cdot \mathbf{r}_v dv^2.$$

By tradition, the quantities $\mathbf{r}_u \cdot \mathbf{r}_u$, $\mathbf{r}_u \cdot \mathbf{r}_v$, $\mathbf{r}_v \cdot \mathbf{r}_v$ are denoted E, F, G, and the expression $Edu^2 + 2Fdu\, dv + Gdv^2$ is called the First Fundamental Form.

J. Gray, *Worlds Out of Nothing*,
Springer Undergraduate Mathematics Series,
DOI 10.1007/978-0-85729-060-1_19, © Springer-Verlag London Limited 2011

By analogy with Pythagoras's theorem, it measures the distance between $\mathbf{r}(u, v)$ and $\mathbf{r}(u + du, v + dv)$, so we write

$$ds^2 = E\,du^2 + 2F\,du\,dv + G\,dv^2.$$

For the sphere with the above coordinates, the First Fundamental Form is $ds^2 = R^2 \left(\cos^2 v\,du^2 + dv^2\right)$.

This makes sense when either u or v is constant. In the case $u = \text{const.}$, so $du = 0$, we deduce that $ds = R\,dv$, and when $v = \text{const.}$, so $dv = 0$, we deduce that $ds = R \cos v\,du$.

From elementary plane geometry, we know that the angle between two vectors $\mathbf{a}$ and $\mathbf{b}$ is given by $\frac{\mathbf{a}.\mathbf{b}}{\|\mathbf{a}\|\,\|\mathbf{b}\|}$, so the angle between the u-curves and the v-curves is $\frac{F}{\sqrt{EG}}$, so the curves are orthogonal if and only if $F = 0$. The angle between $a_1\mathbf{x}_u + a_2\mathbf{x}_v$ and $b_1\mathbf{x}_u + b_2\mathbf{x}_v$ is found from the same formula, and works out to be ϕ, where

$$\cos\phi = \frac{a_1 b_1 E + (a_1 b_2 + a_2 b_1)\,F + a_2 b_2 G}{\left(a_1^2 E + 2a_1 a_2 F + a_2^2 G\right)^{1/2} \left(b_1^2 E + 2b_1 b_2 F + b_2^2 G\right)^{1/2}}.$$

The element of area is $\sqrt{EG - F^2}$.

With this information, we can do all the elementary geometry of the surface, but the formulae are often very complicated. It is therefore desirable to find coordinates in which various quantities are constant or even vanish. For example, one may ask that the u- and v-curves be parameterised by arc length, in which case $E = 1 = G$. Or one can ask that they always meet orthogonally, in which case $F = 0$.

A geodesic is a curve of shortest length between its points. Longitudes are geodesics on the sphere but latitudes are not (except for the equator). A useful coordinate system is obtained by taking a geodesic, and drawing all the geodesics orthogonal to it. The longitude and latitude coordinates on the sphere are an example of this. Another is obtained by mimicking polar coordinates: draw all the geodesics from a fixed point, and measure the angle between an arbitrary one and a fixed one.

19.1.1 Geodetic projection

This is a map from the sphere (strictly speaking, a hemisphere) onto a plane. We start by considering the inverse map, from the plane to the sphere. Consider the map from the horizontal plane of height a to the sphere (centre the origin and of radius R) given by projection along a radius: $(u, v, w) \mapsto (\lambda u, \lambda v, \lambda w)$, where $w = a$ (the domain is a plane) and $(\lambda u)^2 + (\lambda v)^2 + (\lambda a)^2 = R^2$ (the codomain is the sphere). (See Figure 19.1.)

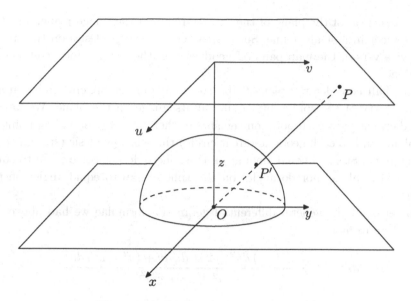

Figure 19.1 Geodetic projection (1)

This implies that $\lambda^2 \left(u^2 + v^2 + a^2\right) = R^2$, so $\lambda = \frac{R}{\left(u^2+v^2+a^2\right)^{\frac{1}{2}}}$, and so the map is

$$(u, v, a) \mapsto \left(\frac{uR}{\left(u^2 + v^2 + a^2\right)^{\frac{1}{2}}}, \frac{vR}{\left(u^2 + v^2 + a^2\right)^{\frac{1}{2}}}, \frac{aR}{\left(u^2 + v^2 + a^2\right)^{\frac{1}{2}}}\right).$$

On the sphere, we have that $ds^2 = dx^2 + dy^2 + dz^2$. After a long calculation, which I omit, this becomes

$$ds^2 = R^2 \frac{\left(a^2 + v^2\right) du^2 - 2uv \, du \, dv + \left(a^2 + u^2\right) dv^2}{\left(a^2 + u^2 + v^2\right)^2}.$$

Let us set $u = 0$, so the map becomes

$$(0, v, a) \mapsto \left(0, \frac{vR}{\left(v^2 + a^2\right)^{\frac{1}{2}}}, \frac{aR}{\left(v^2 + a^2\right)^{\frac{1}{2}}}\right)$$

and the metric becomes $ds = R\frac{a \, dv}{\left(a^2+v^2\right)}$. This says that, as v increases, successive equal steps of dv make smaller and smaller increments in ds, or, equivalently, as v increases, equal increments in ds come about from successively larger increments in dv. This is what one expects from the picture.

The inverse of this map has the property that it maps a great circle on the sphere to a straight line on the plane. For, a great circle is cut out by a plane through the centre of the sphere, and the image of the great circle must be the intersection of the plane of the great circle and the image plane, and two planes meet in a straight line. Since great circles are geodesics on the sphere (and vice versa), the map pairs off geodesics on the surface and geodesics on the plane.

We shall consider the plane to be the image of the sphere, and examine the geometry of the sphere by looking at its image on the plane. We have a little dictionary: to each point on the sphere there is a corresponding point on the plane, and to each geodesic (great circle) there is a geodesic (straight line) on the plane. Steps of equal amount on the sphere do not correspond to equal steps on the plane, nor do angles on the sphere map to equal angles on the plane.

In terms of the general differential geometric formulae we had above, we can deduce from

$$ds^2 = R^2 \frac{\left(a^2 + v^2\right) du^2 - 2uv \, du \, dv + \left(a^2 + u^2\right) dv^2}{\left(a^2 + u^2 + v^2\right)^2}$$

that

$$E = R^2 \frac{\left(a^2 + v^2\right)}{\left(a^2 + u^2 + v^2\right)^2}, \quad F = R^2 \frac{-uv}{\left(a^2 + u^2 + v^2\right)^2}, \quad G = R^2 \frac{\left(a^2 + u^2\right)}{\left(a^2 + u^2 + v^2\right)^2}.$$

So the angle θ between the u- and v-curves is given by $\cos \theta = \frac{uv}{\sqrt{(a^2+v^2)(a^2+u^2)}}$, which is, as we would expect, not zero. But all the curves $u = $ const. are orthogonal to the geodesic $v = 0$, and all the curves $v = $ const. are orthogonal to the geodesic $u = 0$.

A geodesic through the (u, v) origin can be represented parametrically by $u = r \cos \mu$, $v = r \sin \mu$, where r is the usual arc-length parameter – but not the one for the metric induced from the sphere. From this we deduce that $du = dr \cos \mu$ and $dv = dr \sin \mu$, and putting this in the formula we find (after some work) that $ds = \frac{aR}{a^2 + r^2} dr$. The substitution $r = a \tan \theta$ reduces this to $ds = R \, d\theta$, so $s = R\theta = R \arctan (r/a)$, as we expect from Figure 19.2.

A circle with centre the (u, v) origin can be represented parametrically by $u = r \cos \mu$, $v = r \sin \mu$, where r is the usual arc-length parameter but not the one for the metric induced from the sphere. This circle corresponds to one on the sphere with radius $s = R\theta = R \arctan (r/a)$, which implies $r = a \tan (s/R)$. From this we deduce that $du = -r \sin \mu \, d\mu$ and $dv = r \cos \mu \, d\mu$, and putting this in the formula we find (after some work) that $ds = \frac{Rr}{\sqrt{(a^2 + r^2)}} d\mu$. So the circumference of a circle on the sphere with radius s is $R \sin (s/R)$. This checks with what we can compute directly, but it has been deduced by working on the plane with the metric induced from the sphere.

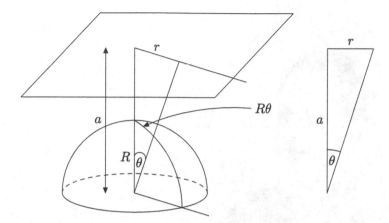

Figure 19.2 Geodetic projection (2)

19.2 Introducing Beltrami's *Saggio*

The extract or, rather, paraphrase from Beltrami's *Saggio* [13] given below shows how Beltrami took this account of how the geometry of the sphere can be described on a plane, and modified it in two ways to obtain his account of non-Euclidean geometry. First, he changed the metric, the formula expressing the distance between two points whose coordinates are known, so that the figure in the plane describes non-Euclidean geometry. The new description has the consequence that the entire figure is contained within a disc of a certain radius. The second thing he did is reverse the direction of the argument. We naturally start with the sphere, the earth, and proceed to make a map of it on a plane. We can do this in many ways and fill whole atlases accordingly. Beltrami began with the atlas, the picture in the disc, and deduced the existence of that which the picture in the disc described, the two-dimensional space in which non-Euclidean geometry is the intrinsic geometry. He did this without also producing a surface in three-dimensional Euclidean space that is somehow depicted in the map – which is just as well, because it was later shown that (under very modest requirements that Beltrami tacitly assumed) there is no such surface.

Beltrami's achievement is the disc and its interpretation. In many accounts this is confused with the pseudosphere, a confusion to which, unfortunately, Beltrami contributed. Beltrami called the surface of constant negative curvature a pseudosphere. Minding's surface, the surface of revolution obtained by rotating the tractrix about its axis, he referred to as simply a surface of revolution. However, current usage mostly, but not always, calls Minding's surface the pseudosphere. To avoid ambiguity in this book, let it be decreed that Minding's

Figure 19.3 Beltrami

surface shall also be called the pseudosphere, and that the surface of constant negative curvature shall be called either the surface of constant negative curvature or, once that point is properly established, two-dimensional non-Euclidean space. Beltrami was very clear that the pseudosphere is not the same as the surface of constant negative curvature. In fact, they differ in much the same way that the cylinder differs from the plane, but all Beltrami did was hint that the pseudosphere must be cut open before there is any chance of a map between it and the disc. It seems that it was Klein who first saw clearly how to draw a pseudosphere in the disc: first cut it open, then place it with the cusp end (where the tractrix "meets" the axis) on the boundary of the disc – happily, that point is infinitely far away in the non-Euclidean metric on the disc.

Beltrami had read widely among his predecessors in the investigation of non-Euclidean geometry, a sign perhaps of how anxious his path to his discovery was. He cites both Minding and another Italian mathematician, Codazzi, who, in 1857, had shown that triangles on the pseudosphere are described by the hyperbolic trigonometric formulae we know belong to non-Euclidean geometry. But Codazzi for whatever reason had not made that connection, which was therefore left to Beltrami. Codazzi, however, did not cite Minding. He drew his information about the pseudosphere from Joseph Liouville's account, in the fourth of a number of lengthy additional notes Liouville made to the re-edition in 1850 of Monge's *Applications de l'analyse à la géométrie* [169]. Liouville was then at the height of his influence on French mathematics, and this fourth note is the route by which Gauss's ideas of differential geometry, and in particular the idea of intrinsic (Gaussian) curvature, entered France (see Lützen [158]). Liouville explained what it was and showed that there are surfaces of constant curvature: constant positive curvature on a sphere, zero curvature on a plane and constant negative curvature on the pseudosphere. He did not mention Minding, but nor did he write as if this surface was his own discovery. Perhaps it had just somehow established itself.

A rich source of information on how Beltrami came to work on non-Euclidean geometry, and much else besides, is provided by Voelke's book [244]. It seems that it was Richard Baltzer who first picked up the significance of the correspondence between Gauss and Gerling on the subject, and he encouraged Houël to translate Lobachevskii's *Geometrische Untersuchungen* into French. This he did in 1866, following it with a translation of Bolyai's *Appendix* in 1867. Beltrami picked up the idea of non-Euclidean geometry from Houël's translation of Lobachevskii, but for a time thought that it could only be a geometry on a surface in Euclidean three-dimensional space, and that the idea of three-dimensional non-Euclidean geometry was a "geometric hallucination", as he put it in a letter to Houël.[1] Meanwhile, Baltzer had put favourable remarks about Bolyai and Lobachevskii in the second volume of his textbook [7] and these had been picked up by Cremona. Beltrami and Cremona discussed the topic of non-Euclidean geometry, but remained confused until the publication of Riemann's *Habilitationsvortrag*, which seems to have removed any residual doubts in their minds about the distinction between intrinsic and embedded geometry.

19.2.1 Beltrami's *Teoria* of 1868

In 1868 Beltrami published his second paper on spaces of constant curvature [14]. He was by then fully familiar with Riemann's Habilitation paper, which had just been published, and his treatment is much more confident. Not only is it firmly n-dimensional, it compares Beltrami's and Riemann's metrics for non-Euclidean geometry inside the n-dimensional ball and shows how they are related by a simple geometric transformation.

An English translation of this paper will be found in Stillwell [232], where the editor makes the generous suggestion that Beltrami's name should be introduced into the usual names for models of non-Euclidean geometry. In particular, the Poincaré disc model should be called the Riemann–Beltrami model.

---

[1] Beltrami to Houël, 18 November 1868, quoted in [244, p. 163].

I did not feel able to adopt that suggestion here, but it has a lot to com-
mend it.[2] Stillwell also suggests [232, p. 36] that "perhaps Beltrami is the
missing link between Riemann and Poincaré." It seems unlikely that this was
the case in practice, because Beltrami's name is missing from Poincaré's writings in the crucial period when he was first putting non-Euclidean geometry to use, and Poincaré was usually open about his sources. In particular
it is missing from those occasions when Poincaré first explained the new geometry to his readers, precisely when a reference to it would surely have
helped.

This is also the point at which to bring out a disagreement with the usually
very helpful University of St Andrews' website. In their article on non-Euclidean
geometry [178] they write:

> In fact Beltrami's model was incomplete but it certainly gave a
> final decision on the fifth postulate of Euclid since the model pro-
> vided a setting in which Euclid's first four postulates held but the fifth
> did not hold. It reduced the problem of consistency of the axioms of
> non-Euclidean geometry to that of the consistency of the axioms of
> Euclidean geometry.
>
> Beltrami's work on a model of Bolyai–Lobachevsky's non-Euclidean
> geometry was completed by Klein in 1871. Klein went further than this
> and gave models of other non-Euclidean geometries such as Riemann's
> spherical geometry.

It is not clear to me in what sense Beltrami's work was "incomplete", other
than in the productive sense that more could be done with it by other people
with other questions. Not only did Beltrami do much of what it is suggested
Klein did by way of completing what he had begun, Klein's more fundamental
contribution, as we shall see below (§20.5), was to give Beltrami's work a setting
in projective geometry. Beltrami's contribution seems to me to have been a
rather complete study of metrics for spaces with constant negative curvature
and their connection to non-Euclidean geometry. And while it is true that it
reduced the consistency problem to the consistency of Euclidean geometry, no
one saw it that way. Beltrami's descriptions showed that the new geometry
made rigorous sense by the standards of the day, because at that time no one
contested the consistency of Euclidean geometry. It was to be another 30 years
before a renewed sophistication about axiom systems put that matter in doubt
(see below, §30.3).

---

[2] I am indebted to Bob Osserman for reminding me of this paper by Beltrami, and
Stillwell's discussion of it.

19.3 The *Saggio*

An English translation of Beltrami's *Saggio*, and several other papers, including Beltrami's *Teoria*, Klein's account of non-Euclidean geometry in 1871, and several pieces by Poincaré, will be found in a very useful volume with good introductions to the essays: John Stillwell, *Sources of hyperbolic geometry* [232]. Notice, however, that formula (1) for Beltrami's non-Euclidean metric in the disc is given incorrectly there: the denominator should read: $\left(a^2 - u^2 - v^2\right)^2$. What follows is a drive-through account, slightly paraphrased.

19.3.1 From Beltrami's *Saggio* (Essay) of 1868 [13]

The general setting

The formula

$$ds^2 = R^2 \frac{\left(a^2 - v^2\right) du^2 + 2uv\, du\, dv + \left(a^2 - u^2\right) dv^2}{\left(a^2 - u^2 - v^2\right)^2} \qquad (1)$$

represents the square of the line element on a surface whose spherical curvature is constant, negative, and equal to $-\frac{1}{R^2}$. The form of this expression ... has the particular advantage (from our point of view) that a linear equation in u, v represents a geodesic and, conversely, any geodesic is representable by a linear equation in these variables.

Beltrami here referred his readers to a note at the end of the paper.

In particular, the two systems of coordinate lines $u = $ const., $v = $ const. consist of geodesics, whose mutual positions are easily discerned. In fact, if we let θ denote the angle between the two coordinate curves at the point (u, v) we have

$$\cos\theta = \frac{uv}{\sqrt{\left(a^2 - u^2\right)\left(a^2 - v^2\right)}}, \quad \sin\theta = \frac{a\sqrt{\left(a^2 - u^2 - v^2\right)}}{\sqrt{\left(a^2 - u^2\right)\left(a^2 - v^2\right)}}, \qquad (2)$$

so for $u = 0$ or $v = 0$ we have $\theta = 90°$. Thus the geodesic components of the system $u = $ const. are all orthogonal to the geodesic $v = 0$ of the other system, and the geodesics of the system $v = $ const. are all orthogonal to the geodesic $u = 0$ of the first system. In other words, the point $(u = v = 0)$ is the intersection of the orthogonal geodesics $u = 0$, $v = 0$ which we take as fundamental, and any point of the surface is determined as the intersection of geodesics perpendicular to the fundamental ones; this is an obvious generalisation of the ordinary Cartesian method.

The formula (2) admits values of the variables u, v subject to

$$u^2 + v^2 \leq a^2. \tag{3}$$

Within these limits the functions E, F, G are real, single-valued, continuous and finite, and E, G, $EG - F^2$ are also positive and nonzero. Then, as we have shown from first principles [Beltrami referred to another of his papers here.] in the region considered, each pair of real values u, v satisfying (3) corresponds to a unique real point and, conversely, each point corresponds to a single pair u, v of real values satisfying the above condition.

Thus, if x, y denote the rectangular coordinates of an auxiliary plane, the equations $x = u$, $y = v$ establish a map of the region in question, a map for which each point of the region corresponds to a unique point of the plane and conversely. The whole region in question is mapped onto the inside of a circle of radius a with centre at the origin, the boundary being called the *limit circle*. In this map the geodesics of the surface are represented by chords of the limit circle and, in particular, the coordinate geodesics are represented by chords parallel to the coordinate axes.

We now see how this limits the region on the surface to which the preceding considerations apply.

A geodesic issuing from the point $(u = 0, v = 0)$ can be represented by the equations [Equation (4) has been omitted here, it is the equation of the limit circle, $u^2 + v^2 = a^2$.]

$$u = r \cos \mu, \ v = r \sin \mu, \tag{5}$$

where r and μ are the ordinary polar coordinates of the point (u, v) on the straight line representing the geodesic in question. Then, since μ is constant, it follows from (1) that

$$d\rho = R \frac{a\, dr}{a^2 - r^2}, \text{ whence } \rho = \frac{R}{2} \log \frac{a + r}{a - r},$$

where ρ is the arc length of the geodesic containing the origin, the point $(u = v = 0)$.

[...] [Equation (6) omitted here.]

This value is zero for $r = 0$ and it grows indefinitely as r or $\sqrt{u^2 + v^2}$ increases from 0 to a, becoming infinite for $r = a$ and hence for all values of u, v satisfying (4), and imaginary when $r > a$. Thus it is clear that the curve defined by equation (4), representing the limit circle in the Euclidean plane, is none other than the locus of infinite points of the surface, a locus which can be regarded as a geodesic circle with centre at the point ($u = v = 0$) and infinite (geodesic) radius. This geodesic circle of infinite radius does not exist except in the imaginary or ideal region of the surface, since the region considered a moment ago extends indefinitely far in all directions to embrace all real points of the surface. In this way the limit circle contains a representation of the whole real part of our surface, since all lines are determined by their points at infinity on the limit circle, just as they are determined by their points on concentric internal circles, which are geodesic circles on the surface itself with centre at point ($u = v = 0$).

If r is taken as a constant in equation (5), and μ as a variable, then the equations represent a geodesic circle, and formula (1) gives

$$\sigma = \frac{Rr\mu}{\sqrt{a^2 - r^2}} \tag{7}$$

where σ is the arc length on the geodesic circle, represented in the auxiliary plane by the circular arc of radius r and angle μ. Since σ is proportional to μ for any r, we easily see that the geodesics, at their common origin, make the same angles as their counterparts in the auxiliary plane; and that the infinitesimal part of the surface around the point ($u = v = 0$) is similar to its image on the plane, a property which does not hold for any other point.

It follows from (6)[3] that the semi-perimeter of a geodesic circle of radius ρ is given by

$$\pi R \sinh \frac{\rho}{R}. \tag{8}$$

It follows from the above that the geodesics on the surface are represented in their totality by chords of the limit circle, while their prolongations past the circle are devoid of a real interpretation. On the other hand, two real points of the surface are represented by two, likewise real, points inside the limit circle, which determine one chord of the circle itself. We therefore see that two real points of the surface, however chosen, are always connected by a *unique geodesic*, which is represented on the auxiliary plane by the chord through the corresponding two points.

---

[3] After a little work, here omitted.

Non-Euclidean geometry

Thus the surfaces of constant negative curvature [are such that] the theorems of non-Euclidean planimetry apply to them fully. These theorems, instead of lacking a concrete interpretation, refer rather to such surfaces, as we now proceed to demonstrate at length. To avoid circumlocution, we call the surface of constant negative curvature pseudospherical, and we retain the term radius for the constant R on which its curvature depends.

Beltrami soon found the following rules:

As a result, we can formulate the following rules.

   I. Two distinct chords which intersect inside the limit circle correspond to geodesics which intersect at a point of finite distance and at angle different from 0° or 180°.

  II. Two distinct chords which intersect on the limit circle periphery correspond to two geodesics which converge to a single point at infinity and which are inclined at a zero angle to each other.

III. Finally, two distinct chords which intersect beyond the limit circle, or which are parallel, correspond to two geodesics with no common point in their full extensions on the (real) surface.

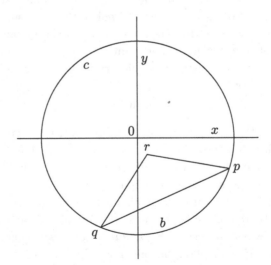

Figure 19.4 The Beltrami disc

Now let pq be any chord of the limit circle, and let r be a point inside the circle but not on the chord. [See Figure 19.4.]

This chord corresponds to a geodesic on the surface, $p'q'$, connecting the points p', q' at infinity (which correspond to p, q); the point r corresponds to a point r' at a finite distance from the geodesic $p'q'$. From this point extend infinitely many geodesics, some of which meet $p'q'$, and some of which do not. The former are represented by the lines from the point r to the arc pbq ($< 180°$), the latter by the lines from this point to the points of the arc pcq ($> 180°$). Two special geodesics divide one of these classes from the other: they are represented by the lines rp, rq, namely, the geodesics issuing from r' which meet $p'q'$ at infinity, one on one side, the other on the other. Since the rectilinear angles rqp, rpq have their vertices on the limit circle, it follows from (II) that the corresponding geodesic angles are zero, although at first sight they appear finite. On the other hand, since r is inside the limit circle and not on the chord pq, the angle prq is different from $0°$ or $180°$, so it follows from (I) that the corresponding geodesics $r'p'$, $r'q'$ make an angle at r' different from $0°$ or $180°$. Thus the geodesics $r'p'$, $r'q'$ can be called parallels to $p'q'$ in the sense that they separate the lines which intersect $p'q'$ from those which do not, and we can state the following result: through any point of the surface there are always two geodesic parallels to a given geodesic not passing through that point, and which meet each other at an angle different from $0°$ or $180°$. [...]

In consequence of this fact and the reasons already stated, the theorems of non-Euclidean planimetry for rectilinear figures necessarily become valid for the analogous geodesic figures on a pseudo-spherical surface. Examples of such are nos. 3–10, 16–24, 29–30 of Lobachevsky's *Theory of Parallels*.

We now consider the two geodesics from a given point that are parallel to a given geodesic. Let δ be the length of the geodesic normal from the point to the geodesic in question. This normal halves the angle between the parallels. ... The angle between either parallel and the normal is called the angle of parallelism and denoted by Δ. To calculate this angle, we make use of our usual analysis, putting the origin ($u = v = 0$) at the given point and drawing the fundamental geodesic $v = 0$ normal to the given geodesic. The latter then becomes represented by the equation $u = a \tan \frac{\delta}{R}$.

[Beltrami next deduced a formula which] agrees with that found by Battaglini. To compare with Lobachevsky, we rewrite it [after a little more work, including equation (9), as]

$$\tan \frac{\Delta}{2} = e^{-\delta/R},$$

which is exactly the formula of Lobachevsky (*Theory of Parallels*, no. 36), except for a difference in symbols due to choice of the unit.

Denoting the angle of parallelism relative to a normal distance z by $\Pi(z)$, as does Lobachevsky (no. 16), we obtain from (9) that

$$\cosh \frac{z}{R} = \frac{1}{\sinh \Pi(z)}, \quad \sinh \frac{z}{R} = \cot \Pi(z).$$

Now, by an observation of Minding (vol. XX of Crelle's journal), developed by Codazzi (in the *Annali* of Tortolini, 1857), the ordinary formulae for spherical triangles are converted into those for geodesic triangles on a surface of constant negative curvature by inserting the factor $\sqrt{-1}$ in the ratio of the side to radius and leaving the angles unaltered, which amounts to changing the circular functions involving the radius into hyperbolic functions.

Beltrami gave an example, and commented as follows.

The others can be obtained analogously. The inverse passage from these equations to those of spherical trigonometry was indicated by Lobachevsky, p. 45, but simply as an analytic fact.

The preceding results fully exhibit the correspondence between non-Euclidean planimetry and pseudo-spherical geometry. To verify the same fact from another point of view, we shall establish directly, by our analysis, the theorem about the sum of the three angles of a triangle.

Consider a right triangle formed by the fundamental geodesic $v = 0$, a perpendicular geodesic $u = \text{const.}$, and the geodesic issuing from the origin at angle μ, whose equation is $v = u \tan \mu$. [Beltrami then wrote down the element of area of a non-Euclidean figure, $\sqrt{EG - F^2} = R^2 \frac{du\,dv}{(a^2 - u^2 - v^2)^{3/2}}$, integrated it over the triangle with angles $\frac{\pi}{2}, \mu, \mu'$, and deduced that the area of the right-angled triangle is $R^2 \left(\frac{\pi}{2} - \mu - \mu'\right)$.] By dividing an arbitrary geodesic triangle ABC into right triangles by a geodesic normal from one vertex to the opposite side, one finds its area to be

$$R^2 (\pi - A - B - C).$$

This expression, since it must be positive, shows that the sum of the three angles of any geodesic triangle cannot exceed 180°. For it to equal 180° in any finite triangle, it is necessary that $R = \infty$, and then $A + B + C = \pi$ in any other finite triangle. But when $R = \infty$, (9) gives $\Delta = \frac{\pi}{2}$, whence the angle of parallelism is necessarily right; and conversely. These are also the conclusions of non-Euclidean geometry.

19.4 Legendre's error

The Beltrami disc model makes it very clear where Legendre's "best" argument falls into error. Recall that he needed a sequence of points $A, A_1, \ldots, A_n, \ldots$, and note that by their construction the distances AA_1, AA_2, and so on, roughly double each time. So they become arbitrarily large. Draw the picture in the Beltrami disc, placing the point A anywhere, and draw the arms of the angle ABC out to the boundary of the disc. Let B and C be the points on the boundary at the extremities of the arms of the angle, and draw the line BC. The sequence of points A_n can be shown to cross the line BC, but once it does it becomes impossible to draw a line from such a point A_n that crosses both AB and AC, and it is therefore impossible to iterate Legendre's construction indefinitely.

It is remarkable that, in the presence of the other axioms of Euclid's *Elements* [70], the innocuous looking statement "it is always possible to draw a line through a given point meeting both arms of a given angle" is equivalent to the parallel postulate – but it is. Legendre can surely be forgiven!

19.5 References

There are many places where one can learn non-Euclidean geometry from a differential geometric point of view. There is the book by Anderson, *Hyperbolic geometry* [2], and one to recommend if you want to go much further than is necessary here, is Beardon's *The geometry of discrete groups* [9]. Two works that can be particularly recommended for their treatment of differential geometry are Struik's *Lectures on classical differential geometry* [230] and Morgan's *Riemannian geometry: A beginner's guide* [171], the one closer to the classical material, the other to the modern theory of differentiable manifolds and differential geometry. Mention should also be made of the two-volume treatment by Berger, simply called *Geometry* [15, 16], for his coverage of the whole field.

Beltrami, Klein, and the Acceptance of Non-Euclidean Geometry

20.1 Beltrami's version

What Beltrami did in his *Saggio*, as the extracts given in Chapter 19 will have suggested, was to take the usual metric for a map of the sphere on the plane and modify it so that it was defined only within the unit disc, but the space mapped onto the interior of that disc had the following properties.

It was infinite.

A non-Euclidean parallel axiom could be defined.

The angle sum of a triangle was always less than π by an amount which depended on its area.

The trigonometric formulae for triangles were the ones obtained earlier by Codazzi and agreed with those given by Lobachevskii.

Geodesics in the space appeared as straight lines in the disc, although angles and distances were of course distorted (in a way described by his formulae).

This, coupled to a Riemannian idea of what a geometry is, meant that there could be no doubt about the meaningful nature of the geometry (and its higher-dimensional analogues that Beltrami went on to develop in his next paper). There simply was a space, because here in the *Saggio* is its (evidently consistent) description. Beltrami did not say that his new geometry was consistent if (or even if and only if) Euclidean geometry is, because at the time no one thought to question if Euclidean geometry made sense. Nor did the Riemannian philosophy of geometry invite one to reduce questions about a new geometry to questions about Euclidean geometry; Euclidean geometry was no longer the source of geometric ideas. Beltrami simply offered a new geometry in the Riemannian spirit. Doubts about the consistency of a description of Euclidean geometry were not to be raised until much later, when an axiomatic philosophy of geometry came in, and one could at least ask if a well-known, tried and tested geometry made sense.[1] Had anyone thought to debate the status of Euclidean geometry in 1868 one incredulous reply would have been "How can you doubt the truth of coordinate geometry?" There would have been others, but this one will do. To doubt the plausibility of a mathematics system one must have a platform from which to test it, and there was no such platform for scrutinising non-Euclidean geometry until the 1890s at the earliest.

The work of Riemann and Beltrami greatly assisted the reception of non-Euclidean geometry. For the first time, mathematicians could point to an account which was mathematically sound. To be sure, there was not a surface in Euclidean three-space analogous to the sphere but on which the induced geometry was non-Euclidean. No one had found one, at least, and it was shown by David Hilbert in 1902 that, under very modest restrictions, there could not be a surface of that kind. But Riemann had shown that it was not necessary to find such a surface. A cartographic map of a surface was just as good. Once Beltrami had published, in full detail, what Riemann's *Habilitationsvortrag* had hinted at in one formula, mathematicians could point to that account and the philosophy of geometry underlying it, and feel that they stood on solid ground. By and large, they did so. There were also other factors at work, of which the most forceful was the posthumous publication of Gauss's correspondence.

20.2 Gauss's posthumous contribution

After Gauss's death in 1855 the mathematicians in Göttingen discovered to their amazement that there was a prodigious amount of mathematics left behind in his notebooks that he had not bothered to publish. They chased up

---

[1] Although there is Riemann's passing remark at the start of his *Habilitationsvortrag* about the fundamental assumptions of geometry that "we neither perceive whether and how far their connection is necessary, nor, a priori, whether it is possible" [209]. Riemann does seem to have genuinely wondered if the infinitesimal structure of space had to be amenable to differential geometry at all, or that it might not be granular. He was joined in these speculations by Clifford, see [38], but it seems that a proper historical investigation of this idea has not yet been written.

his correspondents, and found more ideas that he had circulated among his friends but otherwise kept to himself. Prominent among them were his ever more strongly worded endorsements of non-Euclidean geometry and his support for Bolyai and Lobachevskii. If the prince of mathematicians had looked favourably on a geometry other than Euclid's, lesser mathematicians felt more willing to do so too. The publication of Gauss's letters to Schumacher in 1864 showed very clearly the extent of Gauss's belief in the possibility that space might be non-Euclidean; it also drew mathematicians' attention to the original publications of Bolyai and Lobachevskii. Inspired by the mention of these obscure names – it is quite possible that the name of Bolyai was completely unknown by the 1860s – the French mathematician Jules Hoüel produced translations of their work into French in the late 1860s, and for the first time it was read with sympathy. Beltrami, for example, read Lobachevskii's memoir of 1840 in Hoüel's French translation of 1866, having picked up the thread by reading some of what Gauss had written.

20.2.1 Kant?

A third factor is also worth mentioning, if only because the mathematician and historian Roberto Bonola gave it credence in his Italian book of 1906 which, in its handy English translation of 1912 [21] has dominated English accounts ever since. He attributed the change in attitude to non-Euclidean geometry to a decline in enthusiasm for Kantian philosophy. Kant had more or less regarded Euclidean geometry as a built-in feature of the mind, a necessary preliminary feature to any understanding or perception of space (I say "more or less" because, faced with the discovery of non-Euclidean geometry, a rearguard action by Kantians tried to argue that Kant had held a more nuanced position; this seems unlikely). It is true, for example, that Gauss tried to distance himself from Kant on this very issue. But in fact the fashion for Kant had died in the first half of the 19th century, when it was eclipsed by various competing forms of German idealism, and, just to complicate matters further, the philosopher of most influence on Riemann was Herbart, who, although more sympathetic to Kant than were the German idealists, was certainly critical of Kant's ideas about space and time – and it was just those ideas of Herbart's that were to prove stimulating to Riemann himself.[2] So it seems best to note merely that mathematicians could deal with non-Euclidean geometry without having to square it with a Kantian orthodoxy. As for the response of philosophers themselves to the new geometry, that is another story, but generally

---

[2] I write about this topic at greater length in my book, *Plato's ghost – the modernist transformation of mathematics.*

speaking they were not sympathetic, and tried to find reasons to dismiss it, contributing to a further estrangement between mathematics and philosophy at the time.

20.3 Felix Klein

Figure 20.1 Felix Klein

It was in this context that, in 1871, the next major mathematical contribution was made, and with it the subjects of projective and non-Euclidean geometry started to come together. (Christian) Felix Klein was born on 25 April 1849 in Düsseldorf. In 1857 he entered the eight-class humanistic Gymnasium, but he later wrote that "This purely philological type of education could offer me next to nothing for my early interest in scientific subjects, especially since the teaching of mathematics, which I could follow with great ease, had a strong formal character."[3] Instead he owed his education in mathematics to family friends and his own efforts, but his precocious skills enabled him to enter the University of Bonn in autumn 1865, when he was only sixteen and a half, in order to study mathematics and natural sciences. However,

---

[3] Klein's *Autobiography* [141]. I am indebted to David Rowe for providing me with a copy of this.

the entire instructional programme in mathematics and physics was at a low level in Bonn and had little to interest him. What was of decisive importance for Klein was that at Easter 1866, he became Plücker's assistant, setting up and carrying out demonstrations in his lectures on experimental physics. He also assisted Plücker with his mathematical researches, to which he had returned.

Klein had always intended to specialise in physics, but Plücker died unexpectedly in May 1868 and Klein, whose doctorate in December 1868 extended Plücker's results, was invited to take on the task of publishing Plücker's posthumous geometrical works. This brought him into contact with Clebsch in Göttingen. This was a remarkable early career, so fast indeed that at that time Klein had never heard lectures on the integral calculus! Göttingen swept Klein into a circle of algebraic geometers inspired by Clebsch. Physics took a back seat, although he was soon introduced to the house of Wilhelm Weber, because the work struck him as too finicky [140]. He travelled to Berlin in autumn 1869 to take part in the lively mathematical seminars of Kummer and Weierstrass, in which the members lectured on topics they had chosen themselves. While there he heard for the first time from Otto Stolz in Innsbruck of the existence of non-Euclidean geometry. He immediately conjectured that it was closely connected to the so-called Cayley metric but got the reply, most likely from Weierstrass, that these were two quite separate circles of ideas, and that the foundations of geometry had to be built on the property of straight lines that they were the shortest connection between any two points [140, p. 152].

Klein was not persuaded of this piece of conventional wisdom. In summer 1870 he went to Paris with his friend Sophus Lie, but plans to stay there for a while and then go to England were ended by the Franco-Prussian War. He tried to get the Ministry of Education and Cultural Affairs in Berlin to support the trip, but only got the lofty official answer: "We don't need any French or English mathematics." He then took part in the war, which nearly cost him his life when he got typhus, and then on his recovery he returned to Göttingen, where he habilitated in January 1871.

In the summer of 1871 he was back in Göttingen reunited with his friend Stolz. Stolz introduced him to the writings of Bolyai and Lobachevskii, which Klein had not yet then read, and what Klein later recalled as an endless debate ensued. Stolz was the sharp critic, insistent upon precise mathematical arguments; Klein was convinced that non-Euclidean geometry had to be part of projective geometry in Cayley's sense, and eventually he overcame his friend's persistent resistance [140, p. 152]. He then published the first of two accounts on non-Euclidean geometry [129]; we shall look at the longer of these below.

20.3.1 Klein at Erlangen ...

On Clebsch's recommendation, Klein was appointed as a full professor of mathematics in Erlangen in autumn 1872, at the astonishing age of 23. Erlangen was not in a good condition. The library was "antediluvian" (Klein's description, see [141]) and funds for the purchase of books, let alone models, to which he attached a great importance, were not available at first. Only two listeners came to his first lecture, and one disappeared at once, which imperilled the entire course. But then, in November, Clebsch died suddenly of diphtheria, and a number of Clebsch's special followers, who were often older than Klein, followed him to Erlangen.

As was the custom, on the occasion of his becoming a professor at Erlangen, Klein presented an inaugural address (a position paper). He presented a written text, which circulated at the time, in which he gave a uniform account of the existing directions of geometrical research and sorted them into a system. This famous "Erlangen Program" became in the 1890s a retrospective guideline for his research. He also gave a public address, which emphasised the importance of cultivating applied mathematics alongside pure mathematics to preserve the connection with related fields of knowledge, such as physics and technology; developing visual perception as an equal factor with logical abilities, and above all mathematical imagination and the spontaneity that flows from it. Therefore, he said, periodically repeated lectures must be given and alongside them special lectures for a smaller number of academically interested people; both have to rest on exercises and seminars. In addition, courses in descriptive geometry must take place, which emphasise the ability to draw. Furthermore there must be a reading room with a reference library, which enables students to study the relevant literature, while extensive collections of models ensure the education of the mathematical visual perception.

20.3.2 ... and beyond

He got somewhere with his ideas – the number attending his lectures once reached a total of ten! – but his career continued to rocket. He went next to the technical university in Munich at Easter 1875. There, the plan was to create a polytechnical school, such as existed in Paris and Zurich. In autumn 1880 he was called to the University of Leipzig, expressly to teach geometry, "since this discipline had been given only second-class treatment [there] until then" [133]. He started a cycle of lectures on geometry, which were intended to cover analytic, projective and differential geometry. He managed to create a collection of models and a public mathematical reading room, and mathematical drawing lessons were established. But he encountered considerable opposition, often from older colleagues, and in autumn 1882 his health collapsed completely. By then too he had come into contact with Poincaré.

20.4 Klein's Cayley metric

When Klein spoke of the Cayley metric, he had in mind what he published in two papers entitled "On the so-called non-Euclidean geometry", in 1871 [129] and 1873 [131]. Cayley had noticed a way in which one can start with projective geometry and smuggle in the idea of Euclidean distance, but had not realised the generality of the idea. It was Klein who saw that. It rests on the use of cross-ratio.

Consider a chord PQ of a circle passing through A and B (see Figure 20.2).

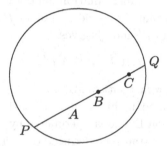

Figure 20.2 Klein's disc

Set $CR(P, A, Q, B) = \frac{PA}{PB} \cdot \frac{QB}{QA}$, the so-called cross-ratio of the four points P, A, Q, B. Then

$$CR(P, A, Q, B) \cdot CR(P, B, Q, C) = \frac{PA}{PB} \cdot \frac{QB}{QA} \cdot \frac{PB}{PC} \cdot \frac{QC}{QB}$$
$$= \frac{PA}{PC} \cdot \frac{QC}{QA}$$
$$= CR(P, A, Q, C).$$

So, defining $d(AB) = -\frac{1}{2} \log CR(P, A, Q, B)$, we have

$$d(AB) + d(BC) = d(AC),$$

whenever A, B and C lie on a line in that order. So it makes sense to define $d(AB)$ as the length from A to B (the minus sign is to make lengths positive).

(You can check this with coordinates. Let $P = -1, Q = 1, A = 0, B = b$ and $C = c$. Then $d(AB) = -\frac{1}{2} \log\left(\frac{1-b}{1+b}\right)$, and the Möbius transformation $z \mapsto \frac{z-b}{1-bz}$

which preserves cross-ratio (proof by computation, but to be taken on trust here!) sends b to 0 and c to $c' = \frac{c-b}{1-bc}$, and indeed $d(AB) + d(BC) = d(AC)$. As a check on the arithmetic, note that $\frac{1-c'}{1+c'} = \frac{(1+b)(1-c)}{(1+c)(1-b)}$.)

To explain the "fudge factor" of $-\frac{1}{2}$, note that $a' = -\frac{1}{2}\log\left(\frac{1-a}{1+a}\right) = \frac{1}{2}\log\left(\frac{1+a}{1-a}\right)$ implies that $a = \frac{e^{a'}-e^{-a'}}{e^{a'}+e^{-a'}} = \tanh a'$. So a point a Euclidean distance of a from the origin is a non-Euclidean distance of $\tanh^{-1} a$ from the origin. Note that $0 < a < 1$ is consistent with $0 < \tanh a' < 1$.

To define distance properly, we require that as a length is moved around, it does not alter and, in particular, it cannot return to only a part of its original position. Invariance is guaranteed in this case, because cross-ratio is invariant under projective transformations. If a projective transformation sends A, B, P, Q to A', B', P', Q' respectively, the key property of cross-ratio is that

$$CR(P, A, Q, B) = CR(P', A', Q', B').$$

It follows that a projective transformation mapping the circle to itself and mapping the line AB meeting the circle at points P and Q to the line $A'B'$ meeting the circle at points P' and Q' gives

$$d(AB) = d(A'B'),$$

so the above definition is in fact satisfactory.

In this way Klein showed in his article of 1871 [129] how to derive the fundamental formulae of non-Euclidean geometry by using projective ideas. In addition to the formula for distance he deduced one for angles, and in his article of 1873 [131] he showed how to extend these arguments from 2 to n dimensions. He also cited a formidable range of literature (Klein liked scholarship). He mentioned not only Lobachevskii, as Beltrami had, but Bolyai (in the original and in an Italian translation), he quoted both Beltrami's papers, and of course he mentioned Cayley. In the circumstances, it is interesting to note that the papers are called "On the *so-called* non-Euclidean geometry" (my emphasis): one presumes he did not want ill-informed criticism from non-mathematicians, among whom had to be counted the philosopher Lotze whom Klein later singled out in the biographical notes he added to his collected papers.

20.5 Klein's unification of geometry

Klein took great pleasure in bringing non-Euclidean geometry and projective geometry together, for it was his view that geometry had become too fragmented. There was not only Euclidean geometry and non-Euclidean geometry, there was projective geometry and Klein also knew that inversion in circles is geometrically interesting.[4] An inversion sends angles to equal but opposite angles, and maps lines and circles to lines and circles (but may confuse the two). This is a fourth geometry, and Klein discussed it in §6 of his Erlangen Program [130].

There was also another geometry, one introduced by Möbius in his *The barycentric calculus* but which has not so far been mentioned in this book: affine geometry. Affine geometry is the geometry you get when figures may be moved around continuously in such a way that lines go to lines (but angles do not necessarily go to equal angles). This could have given Klein a fifth geometry in addition to projective, Euclidean, non-Euclidean and inversive geometries, but as Klein admitted in his article in *Ges. Math. Abh.* vol. 1 [130, p. 497], he first learned of Möbius's work through the publication of Möbius's *Gesammelte Werke* in 1885–1887. To complicate matters still further, Möbius was one of the founders of inversive geometry, but Klein did not mention his name in that connection either.

Klein belonged to a tradition, common in mathematics and persisting to the present day, that dislikes too much diversity and seeks to keep it together by finding a suitable, possibly abstract, unifying standpoint. This desire remained with him all his life: by 1900 he was organising what became a 23-volume *Encyklopädie der mathematischen Wissenschaften* (*Encyclopaedia of the mathematical sciences*) [139] to provide a series of overviews, arguing that what Gauss alone had been able to do at the start of the century now required a team to accomplish, but was no less necessary.

For Klein, his Erlangen Program was an attempt to create an underlying unity for what had become the fragmented discipline of geometry. He did this through his innovative use of the group concept, which was not then widely known. Klein's presentation, as he himself noted when the work was reprinted in the 1890s, was less than perfect – the only property of a group that he insisted upon was closure – but the point Klein was making was important. Mathematicians were accustomed to using transformations of figures, say to replace a figure with an equivalent but simpler one, or to choose more convenient coordinate axes. Klein shifted attention from the figures to the transformations, and argued that henceforth geometry should be about groups as well as the properties of shapes. So a geometric property was one that was invariant under all the operations of the group associated to that geometry. He specifically employed the idea of one group being a subgroup of another. This enabled him to fix a space but vary the group, either to introduce a new geometry or to recognise a known one in an unexpected setting. So cross-ratio is a property of projective geometry because it is not altered by projective trans-

---

[4] Inversive geometry is outlined below in §25.4.

formations, and it does not need to be built up out of lengths, which are not, indeed, projective invariants. In this way the paradoxical nature of cross-ratio was resolved, although Klein wrote in 1921 that Cayley and some other mathematicians never got over their suspicion that this argument somehow contained a vicious circle.

Klein recognised that by selecting a figure in a space and considering the subgroup that maps that figure to itself was a fundamental way to inter-relate geometries and so to find a unifying principle that would encompass all of geometry. According to his own testimony, by December 1871 he had arrived at the following general formulation: "that there are as many different ways to treat the study of a manifold as there are ways to construct continuous groups of transformations, and that the Euclidean and non-Euclidean metrics are just as certainly subsumed by the projective treatment as their 'groups' are contained in the entire group of projective transformations by a suitable choice of coordinates" [130].

Klein ended his Erlangen Program with a series of seven notes of varying length and significance. One is worth picking out. Note 5 referred to what Klein cautiously continued to call the "so-called non-Euclidean geometry" in order to avoid debates with non-mathematicians. Non-Euclidean geometry was the subject of two important memoirs by Klein written on either side of the Erlangen Program [129, 131] which probably did more to convey the message of the Program than did his obscurely published pamphlet. (These are the papers cited in the extract from Klein's Erlangen Program, the papers were entitled (in German) "On the so-called non-Euclidean geometry" I and II.[5])

20.6 The Erlangen Program in the 1890s

There matters rested for Klein, and he proceeded to work on other topics, until in the 1890s he found himself again at Göttingen but now as a senior professor with a career to look back on, an empire to build (he most certainly was to succeed) and willing foreign mathematicians to translate his papers. The Erlangen Program was now put into English, Italian, French, Russian, Polish and Hungarian, and presented as a vision of a research programme uniting geometry and group theory, two subjects that had not only progressed considerably in the intervening 20 years but had indeed grown closer together. Most likely it had never been really intended as more than a philosophy, and it would be hard to argue that Klein himself had worked intensively on drawing out its implications, but the situation in the 1890s gave the Erlangen Program a prominence

---

[5] For a modern account of Klein's Erlangen Program, see Silvester [224].

that it has continued to enjoy. And inasmuch as it does argue, together with the related papers, that there is only one geometry, projective geometry, with various geometries such as non-Euclidean geometry as special cases, it is worthy of note. The use of group theory to classify geometries did indeed bring a unity to the subject that had begun to disappear.

The question of the influence of Klein's Erlangen Program has been much debated, with earlier writers and many mathematicians, Garrett Birkhoff among them, believing that it was of great influence. After all, here is an early statement in favour of group theory, and can we not all see that group theory is a staple of the modern curriculum? Besides, many places that teach geometry teach the Kleinian view of geometry, which is exactly the view that various different geometries, all the ones where we represent the transformations with matrices, form into a hierarchy. More recently, a number of historians, Tom Hawkins and Erhard Scholz among them, have argued convincingly against this view. They find that the solid work establishing group theory between 1870 and 1890 (for example, by Camille Jordan, Sophus Lie, and even Poincaré) was done without the Erlangen Program having any impact, and that Klein himself, while he certainly liked to bring groups into his study of geometric problems, did not implement or advance the view of the Program. For an interesting pair of contrasting views of the situation, from which papers giving the original, more strongly contrasting views can be found, one may consult Birkhoff and Bennett [17, pp. 145–176] and Hawkins [110, pp. 34–42].

20.7 Weierstrass and Killing

In 1872 Wilhelm Killing, at that time a student of Weierstrass's in Berlin, learned of Weierstrass's system of coordinates for non-Euclidean geometry. What follows is a modernised summary of what Killing went on to describe in his *Die nicht-Euklidischen Raumformen* (*The non-Euclidean space forms*) [127], concentrating on the two-dimensional case – Killing naturally discussed the n-dimensional situation.

The Weierstrass–Killing approach was to start with a surface with equation $k^2 z^2 + x^2 + y^2 = k^2$, where k^2 is taken as the curvature of the surface. So when $k^2 = 0$ the surface is a plane, when $k^2 = +1$ the surface is a sphere, and when $k^2 = -1$ the surface is a hyperboloid of two sheets. The basic idea was to mimic trigonometry on a sphere.

Unlike Killing, let us take the hyperboloid of two sheets, which has equation $x^2 + y^2 - z^2 = -1$, and consider the group of transformations that map each half to itself, which we shall denote $G(H^+)$. These transformations are given by the matrices of determinant 1 that satisfy $A^T J A = J$, where $J = \begin{pmatrix} 1 & 0 & 0 \\ 0 & 1 & 0 \\ 0 & 0 & -1 \end{pmatrix}$.

Such transformations are composites of ones like these:

$$\begin{pmatrix} \cos\theta & -\sin\theta & 0 \\ \sin\theta & \cos\theta & 0 \\ 0 & 0 & 1 \end{pmatrix}, \begin{pmatrix} \cosh u & 0 & \sinh u \\ 0 & 1 & 0 \\ \sinh u & 0 & \cosh u \end{pmatrix}, \begin{pmatrix} 1 & 0 & 0 \\ 0 & \cosh u & \sinh u \\ 0 & \sinh u & \cosh u \end{pmatrix}.$$

These can be denoted C_θ, A_u and B_u respectively. Because these transformations are linear they map planes through the origin to planes through the origin, a fact which will prove to be very useful. (As an exercise, show that any transformation mapping the hyperboloid to itself is a composite of these.)

The hyperboloid is contained in the cone with vertex the origin and equation $x^2 + y^2 - z^2 = 0$. There is a map from the upper half hyperboloid H^+ (the one for which $z > 0$) to any disc with its centre on the z-axis that lies parallel to the (x, y)-plane, and it is convenient to choose the disc D that touches the upper hyperboloid at its lowest point, where $z = 1$. The map sends the point (x, y, z) on H^+ to the point $(x/z, y/z, 1)$ on D. Notice that planes through the origin that meet H^+ cut the disc D in straight lines.

Today, we would say that Killing introduced a metric on the hyperboloid that was preserved by the group of transformations he was considering. Killing simply used generalised spherical trigonometry. By analogy with the sphere, which has equation $x^2 + y^2 + z^2 = 1$ and is embedded in Euclidean space with metric $ds^2 = dx^2 + dy^2 + dz^2$, Killing considered the "metric" given by $ds^2 = dx^2 + dy^2 - dz^2$. This is not a Euclidean metric on $\mathbb{R}^3$ but it is a true metric when restricted to the hyperboloid, as Killing showed. For, the hyperboloid is parameterised as $(x, y, z) = (\sinh u \cos\theta, \sinh u \sin\theta, \cosh u)$. A brisk calculation shows that in terms of the parameters u and θ, the metric is $ds^2 = du^2 + \sinh^2 u \, d\theta^2$. Killing obtained the same formula from the trigonometry.

We have to show that the transformations preserve this metric. There are three calculations to do, here is a typical one involving A_u. Consider

$$\begin{pmatrix} x' \\ y' \\ z' \end{pmatrix} = \begin{pmatrix} \cosh u & 0 & \sinh u \\ 0 & 1 & 0 \\ \sinh u & 0 & \cosh u \end{pmatrix} \begin{pmatrix} x \\ y \\ z \end{pmatrix}.$$

We have $x' = x \cosh u + z \sinh u$, $y' = y$ and $z' = x \sinh u + z \cosh u$, from which a short calculation shows that $dx'^2 + dy'^2 - dz'^2 = dx^2 + dy^2 - dz^2$, showing that in this case the metric is preserved. The other cases go similarly.

With this metric, we can measure the distance from $C = (0, 0, 1)$ to any point, let us say the one given by this choice of parameter values: $u = u_0$,

$\theta = \theta_0$. This distance is unaltered by the rotation through $-\theta_0$ that sends the given point to the one with parameter values $u = u_0, \theta = 0$. Call this point P_0. Along the arc from C to P_0 that is cut out by the plane that passes through them and the origin, we have $\theta = 0$, so the metric reduces to $ds = du$ and the distance is found (by a simple integration) to be u_0. Along any other path, there will be an extra positive contribution from the now no longer zero $d\theta$ term, so the arc from C to P_0 is indeed the geodesic joining them, and so the distance from C to P_0 is precisely u_0. Or we may say, by using the parameters as polar coordinates, that the distance from the point C to the point with coordinates $(u_0, 0)$ is u_0. We now see the meaning of the parameter u: it measures distance along arcs out of C according to the given metric.

Now we know that planes through the origin and the point C cut out geodesics on the hyperboloid H^+. We also know that the linear transformations we are considering are isometries, so they map geodesics to geodesics. So every plane through the origin cuts out a geodesic on the hyperboloid. Conversely, any geodesic joining two points on the hyperboloid remains a geodesic under a transformation that sends one of these points to the point C. The image of the geodesic under that transformation is now an arc cut out by a plane through the origin, so it always was such a curve. We can conclude that the geodesics on the hyperboloid are exactly the curves cut out on it by planes through the origin.

If we pass from the hyperboloid H^+ to the disc D, the point C is also a point on the disc, and the point P_0 corresponds to the point Q_0 say in the disc. The coordinates of P_0 are $(\sinh u_0, 0, \cosh u_0)$, so Q_0 has coordinates $(\tanh u_0, 0, 1)$. We may record this in polar coordinates in the disc as $(u_0, 0)$. Geodesics out of C in the disc are the images of geodesics on H^+ under the map from H^+ to D: they are straight lines. The distance in the disc according to the induced metric from the centre C to the point Q_0 with coordinates $(v, 0)$ is u where $v = \tanh u$. So we see that the induced metric on the disc gives it non-Euclidean geometry as its induced geometry, and we (after Weierstrass and Killing) have recovered the Beltrami–Klein description of non-Euclidean geometry.

EXERCISES

20.1. Show that the parameterisation of H^+ allows us to introduce intrinsic polar coordinates on H^+ on writing $Z = \begin{pmatrix} 0 \\ 0 \\ 1 \end{pmatrix}$, by showing that any point on H^+ is of the form $C_\theta B_u (Z)$, for a unique θ and u so that we may give the point $C_\theta B_u (Z)$ the coordinates (u, θ).

20.2. Show that if $g \in G(H)$ is such that $g(Z) = Z$, then $g = C_\theta$ for some angle θ, by using the fact that a matrix g such that $g(Z) = Z$ is necessarily of the form $\begin{pmatrix} * & * & 0 \\ * & * & 0 \\ * & * & 1 \end{pmatrix}$.

20.3. Deduce from Exercise 20.2 that if g and g' are such that $g(Z) = P = g'(Z)$, then $g'^{-1}g(Z) = Z$ and so $g = g'C_\theta$ for some angle θ.

20.4. Let P and Q be two points of H, and g be any transformation of H^+ that maps P to Z, so $g(P) = Z$. Deduce that the u-coordinate of $g(Q)$ is determined.

20.5. Hence show that there is a unique transformation in $G(H^+)$ that maps P to Z and Q to a point with coordinates of the form $(x, 0, z)$ and intrinsic polar coordinates $(u, 0)$.

20.6. Deduce that any metric on H^+ for which these transformations are isometries expresses the distance from the point Z to an arbitrary point P as a function of u alone, and explain why the distance from the point Z to the point P can be taken to be u by considering the point U on H^+ with coordinates $(\cosh u, 0, \sinh u) = B_u(Z)$, the point V with coordinates $(\cosh v, 0, \sinh v) = B_v(Z)$, and the effect of the transformation B_v on Z, U and V.

On Writing the History of Geometry – 2

21.1 Assessment questions

Work in the history of mathematics can, and often should, involve understanding some quite hard mathematics, as it does in questions like these.

Question 21.1 (Essay Question 2)

Choose one of the following, and write an essay based on it of about 1,500 words. In the essay you should demonstrate some understanding of the mathematics and demonstrate your ability to situate the people and the ideas in a historical context.

Choose one of the following extracts.

1. Cremona's *Elements of projective geometry*, pp. 148–160 and pp. 239–247 [45]. This is on reciprocal conics and includes a case of Poncelet's porism.

2. Salmon's *A treatise on the higher plane curves*, pp. 22–55, pp. 162–179 and pp. 213–223 [215]. This covers his classification of quartic curves and their singular points.

3. Lobachevskii's *Geometrical researches on the theory of parallels*, pp. 11–45 [153]. This is his presentation of non-Euclidean geometry from 1840.

All these books are available via the Digital Mathematics Library:
`www.mathematik.uni-bielefeld.de/~rehmann/DML/dml_links.html`.

J. Gray, *Worlds Out of Nothing*,
Springer Undergraduate Mathematics Series,
DOI 10.1007/978-0-85729-060-1_21, © Springer-Verlag London Limited 2011

21.2 Advice

The main point of these assessed essays is to put you in the situation of a 19th-century mathematician who has just studied one of the mathematical topics whose history is studied in this book. It is an opportunity to show that you understand the mathematics, and that you can do something "new" with it (by the standards of the time!) by following a piece of mathematics that lies beyond the range of material described here. You can do this in a number of ways: by explaining the proofs in the terms given but in a comprehensible way; by using some mathematics of your own (coordinate methods, for example); by means of well-chosen examples. For some of the extracts, it could be that the author's reasoning is odd, even wrong; if you judge it to be so, and sometimes you should, then prove your claim.

It is better to demonstrate a real understanding of a piece of the mathematics than a superficial understanding of all of the extract. You may use any mathematical methods that seem to you to be valuable (not, necessarily, those of the original authors). Read all the extracts quickly once and find (at least) one you want to proceed with. None of them is altogether easy to read. Let the obscure bits wash over you and wait for something more comprehensible to turn up. You may find that an offending paragraph has an easier second part, or that it is followed by an easier one. Try to form some sense of what the extract is about. You are also invited to stray beyond this book and draw on information accessible in a library or on the Web.

Your secondary, but not negligible task, is to say something historical about the extract: about the author, the topic and the topic's importance. Sources: browse the Web, but keep a critical edge; the *Dictionary of scientific biography* [90]; standard histories of mathematics.

Comments on the individual passages follow.

21.2.1 Cremona

He was a wordy writer, and the first two pages are heavier than they need to be, but not (actually) hard. The conclusion of §322 is much clearer than the text.

An involution in projective geometry is the projective equivalent of a reflection, a map that is its own inverse. It is a theorem of Desargues that given four points A, B, C and D (no three on a line), and a conic through them, if one forms the so-called complete quadrilateral by joining AB and CD and letting them meet at E, with BC and AD meeting at F, and AC and BD meeting at G, then any line that crosses the complete quadrilateral and the conic as shown in Figure 21.1 makes sets of six points in involution. For example, the points P, Q, R, U, V, W are six points in involution. This means that there is a projective transformation which exchanges P and W, Q and V, and R and U. The property of being six points in involution is preserved under a projective transformation.

An involution has a fixed line and a fixed point, but these are not easy to locate in any given figure. They are, however, easy to construct. In the above complete quadrilateral, there is, for example, an involution in the line EG which exchanges A and D and B and C and fixes F. Of course, this is not the above-mentioned involution on the line PW.

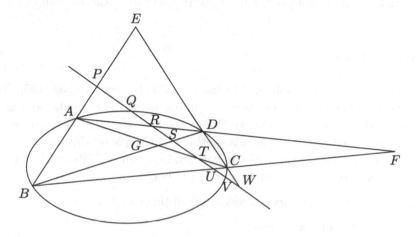

Figure 21.1 Six points in involution

A figure is self-conjugate if the polar line of a vertex is always a line of the figure. If you take as the "auxiliary" conic the conic $x^2 + y^2 = z^2$, can you exhibit a self-conjugate triangle? What if you take the conic $x^2 + y^2 + z^2 = 0$ (making the algebra easier but the picture much worse)?

§326 is central to the passage. First, note that you can talk of the cross-ratio of four lines; it is simply the cross-ratio of the four points you get when you draw a line crossing the four given lines. The paragraph relies on this striking theorem due to Chasles and Steiner independently which is described in the appendix.

Given two points, O and O', and three lines through them, consider a fourth line, l_4, through P. It makes a certain cross-ratio, λ say, with the first three lines through O. Now draw the line l'_4 through O' that makes the same cross-ratio λ with the other lines through O'. The lines l_4 and l'_4 meet, at a point P say. Consider the locus of points P as the line l_4 varies: the theorem is that this locus is a conic through P and P'.

You might be able to prove this result using coordinate methods; make sure you choose the coordinates carefully. If you do read Cremona's account, don't omit the footnote on p. 243!

Page 244 is notable for a case of Poncelet's porism.

Page 245 is notable for a theorem that was, in its day, worthy of a PhD (ever wondered what doctoral-level research might be like?): Hesse's theorem.

You can stop after §336.

Terminology: Figures are said to be correlative if lines in one correspond to vertices in the other, and vice versa. It follows that the cross-ratio of four collinear points in one figure is the cross-ratio of the corresponding lines in the other. Figures are in homology if both are projections of the same figure.

21.2.2 Salmon

Salmon's account closely follows Plücker's in his *Theorie* of 1839 [195]. The challenge here is to meet every type of quartic curve. This would be a good exercise to do if you like curve sketching, want to see what singular points look like, and maybe have access to something like Maple or Mathematica (but doable without – Salmon did it!). If you read on you will discover what his notation means (defined by him earlier in the book), but for reference:

m is the degree of the curve (always four in this case);

δ is the number of double points;

κ is the number of cusps;

n is the class (the number of tangents to the curve from an external point);

τ is the number of bitangents;

ι is the number of inflection points;

D is the deficiency (the modern term is genus).

If you read this over a couple of times it will start to make sense. The extra pages (29, 45–49 and 52–55) will help you with the method he uses and to rumble an oddity about the opening classification. If you can give good illustrations of the curves he draws and satisfy yourself what a complete classification amounts to, you will have done something! Remember, if you draw anything, that a quartic curve can be met by a straight line in at most four points.

Terminology: A crunode is a singular point on a curve where the tangent directions are real and separated; an acnode is an isolated real point where the tangents have complex conjugate directions; and a cusp is a singular point where the tangent directions coincide. All are visible in the various types of cubic curves of the form $y^2 = $ cubic in x.

Salmon went on in his *A treatise on the higher plane curves*, pp. 46–49, to sketch a method that goes back to Newton for finding the form of a curve near a singular point which without loss of generality is taken to be at the origin.[1] It is already clear that if we write the curve in the form $f(x, y) = f_1(x, y) + f_2(x, y) + \cdots$, where f_i is of degree i, the lowest term, j say, that does not vanish identically gives the direction of the tangent lines on setting $f_j(x, y) = 0$. To obtain a more accurate picture, Salmon, following Newton, argued that each branch of the curve through the singular point would have an equation of the form $y = ax^\alpha + bx^\beta + \cdots$, where α is positive and β and all the exponents that follow are greater than α. To find α, the method is to substitute $y = ax^\alpha$ into the equation for the curve, and write down an equation for α that makes two exponents equal and less than or equal to all the other exponents. Once α is found, the coefficient a may be found by equating to zero the terms with equal index. The same method can then be used to find β, b, and so on.

Salmon used the folium of Descartes as an example. The curve has equation $x^3 + y^3 - 3xy = 0$ and, as we know, the coordinate axes are the tangents to the curve at the origin, which is a singular point. We substitute $y = ax^\alpha$ into the equation for the curve, and obtain the equation $x^3 + a^3 x^{3\alpha} - 3ax^{1+\alpha} = 0$. We can write down these equations for α that make two exponents equal: $3 = 3\alpha$, $3 = 1 + \alpha$ and $3\alpha = 1 + \alpha$. The first of these fails to make the equal exponents less than or equal to all the other exponents. The second equation has the solution $\alpha = 2$ and the third equation has the solution $\alpha = \frac{1}{2}$, and these are satisfactory. To find the coefficient a for each value of α, we equate to zero the terms with equal index. In the case when $\alpha = 2$, this gives $1 = 3a$ or $a = \frac{1}{3}$, and in the case when $\alpha = \frac{1}{2}$, this gives $a^3 = 3a$ or $a = \sqrt{3}$, because the solution $a = 0$ gives nonsense. So we find these equations for the branches of the curve through the origin: $y = \frac{1}{3}x^2$ and $y^2 = 3x$, two parabolas which are symmetric and an improvement on the approximation given by the tangents.

21.2.3 Lobachevskii's account in 1840

You are to explain the key points by using the disc model (not available to him, of course). The whole text is too long for present purposes, so I have suggested an extract but you may want to dive in further and select what you want to explain.

---

[1] Salmon referred to Newton, "Methodus fluxionum et serierum infinitorum", which is reprinted as "De methodis serierum et fluxionum", pp. 32–372 in *Mathematical papers* vol. III [177].

There is quite a stretch at the start where all you have to do is draw pictures in the disc, and it is clear that Lobachevskii's essay picks up speed and interest after the opening 15 propositions. But there is a way of reading them which is interesting if the Poincaré disc model is not known to you. Take them as a specification of what the disc model must do as a graphics package: it must be able, where appropriate, to do the itemised things. Spell that out, take it for granted for the moment that it can be done, and then see how the disc model helps. Then see that the disc model does supply all the specified graphical features. It is common practice in mathematics for research mathematicians to satisfy themselves that X will lead to a proof of Y before fully proving X.

Then you might want to skip a bit (about spherical geometry) and then the story becomes three-dimensional. Now you are working inside the unit ball. At the end you get a large number of formulae. Everything now depends on what you know about formulae in non-Euclidean geometry. If that's new to you, look ahead to §25.4 or you could do worse than look at the relevant chapter in *Geometry* (by Brannan, Esplen and Gray) [26]. No need to slog through every derivation of every formula, but see if you can make sense of the enterprise. You might like to compare what you've done with the Beltrami paper as discussed in Chapter 19 of this book.

There are a number of places where you can make Lobachevskii's arguments more plausible without using formulae, and a major challenge here is to see if you can describe the ways in which the work of Beltrami or Poincaré surpasses that of Lobachevskii himself. In what ways is the later work more clear? In what ways is it more rigorous?

Projective Geometry as the Fundamental Geometry

22.1 The rise of projective geometry

By the 1870s, geometry was beginning to be a well-understood subject, with clear research programmes. The study of algebraic curves could be undertaken in several ways, projective geometry was acquiring its status as a fundamental branch of geometry, devoted to the most basic properties of figures (those deeper than the metrical ones), non-Euclidean geometry was acceptable to, and accepted by, most mathematicians (but probably not philosophers). This is indeed the perception of most mathematicians at the time, and most historians of mathematics since, but, as we shall see later, there is another aspect to the story too. But first let us document that projective geometry had become central to many mathematicians' perception of geometry.

It was certainly the view of the English mathematician Arthur Cayley, who in what some regard as his finest paper [29], published in 1859, stated explicitly that descriptive geometry (his term for projective geometry) was all of geometry and that metrical geometry (his term for Euclidean geometry) was but a special case. The remarkable English mathematician James Joseph Sylvester studied briefly at University College London, then at Cambridge where he could not take a degree because he was a Jew, and pursued a difficult career before spending his most successful years at the fledgling Johns Hopkins University

J. Gray, *Worlds Out of Nothing*,
Springer Undergraduate Mathematics Series,
DOI 10.1007/978-0-85729-060-1_22, © Springer-Verlag London Limited 2011

in Baltimore.[1] In 1869 he gave a Presidential address to the British Association for the Advancement of Science Section on Mathematics and Physics. In his "A Plea for the Mathematician" [233] he opposed Thomas Huxley's negative view of mathematics as essentially sterile, a mere handmaiden of the advancing sciences, in these terms:

> I should rejoice to see mathematics taught with that life and animation which the presence and example of her young and buoyant sister (inductive science) could not fail to impart, short roads preferred to long ones, ... projection, correlation, and motion accepted as aids to geometry; the mind of the student quickened and elevated and his faith awakened by early initiation into the ruling ideas of polarity, continuity, infinity, and familiarisation with the doctrine of the imaginary and the inconceivable.

Most of Sylvester's utterances, even when they purport to be proofs, have the hallucinatory aspect to them, but projective geometry is in there somewhere in an honoured role. As the "Plea" makes clear, Sylvester was no friend of Euclidean geometry; he wanted it buried safely out of the schoolboy's reach.

Henry Smith, a quieter figure, but a fine mathematician and Oxford's greatest mathematical light, also spoke in strong terms of projective geometry, calling it a vast field which had been prospected already and which could not fail to repay one's efforts [225, p. 16]. Significantly, the subject was not deductive but proceeded, like science, by careful generalisation and induction. In fact, a number of English writers dealt with the odd features of projective geometry, such as points and lines at infinity, as if they were no more, and no less, mysterious than the latest discoveries of science. Some English mathematicians, though not perhaps the best, adhered to Poncelet's principle of continuity as late as the 1880s, when I think it had more or less died elsewhere.

A drier example is the English mathematician R. S. Ball (admittedly a friend of Klein's, but a good mathematician in his own right) who wrote in 1879 about non-Euclidean geometry (he has just referred to accounts by Cayley, Klein and Lindemann) stating:

> The three works here named are the more conspicuous members of a considerable body of literature in which a remarkable mathematical theory has been developed. The list of writers on the different departments of this theory could be greatly extended, and it would be found to contain the names of many of the most eminent mathematicians, including Gauss, Riemann, Clebsch, and Helmholtz. Although Gauss

---

[1] Biographies of Cayley by Tony Crilly [46], and of Sylvester by Karen Hunger Parshall [183] are published by Johns Hopkins University Press.

in one way and Cayley in another are undoubtedly the founders of the non-Euclidean Geometry, yet it is in Klein's Memoir ... that the systematic development of the theory is to be found. The subject is there treated with singular elegance and completeness. [6]

The point here is that it is the projective approach taken by Klein which Ball found to be the most convincing.

It is true that British mathematics throughout the 19th century has an oddly insular character, but it would not be difficult to match these comments with remarks from French, German or Italian writers. The extract below is from the preface to the English translation of the book on *Elements of projective geometry* by the distinguished Italian geometer, Luigi Cremona, which was put into English at Sylvester's urging and published in 1885 [45]. It gives a fair impression of the high importance he attached to the subject – his book was influential in Italy, too – and also of the often complicated history of projective geometry before 1800. You may assume that the names mentioned are all such as to lend lustre to the cause.

22.2 Cremona

Luigi Cremona is one of those mathematicians whose life involved politics as much as science, and whose influence was as much political as scientific. He participated actively in the unification of Italy, an event which did much to awaken intellectual life throughout the many states that came together to form the new country in 1860. He was born in Pavia in December 1830, and in the revolutionary year of 1848, when many countries saw militant revolt against the current order, he supported Milan and Venice against Austria, at that time the country to which they belonged. This struggle went well for a time but ended with the defeat of Venice in August 1849. Cremona was by now a sergeant with the troops defending Venice, and such was the bravery with which the Venetians fought that they were allowed to leave defeated but with honour. Cremona returned to his native Pavia, and studied mathematics under Brioschi and others. As he later wrote (quoted in G. Loria "Luigi Cremona et son oeuvre mathématique", *Bibliotheca Mathematica* (1904) [157, pp. 125–195]: "The years that I passed with Brioschi as pupil and later as colleague are a grand part of my life; in the first portion of these years I learned to love science and in the other how to transfer it to a large circle of auditors." 'It is worth noting that Cremona's father had died when he was only 11, and he had only been able to attend the higher school forms and go to university at Pavia because of his stepfather's support. On his return from the war he found that

Figure 22.1 Cremona

his mother had died, and he again needed financial support from his family to finish his degree.

Cremona took a doctorate in civil engineering from Pavia in 1853, but the Austrian government made it difficult for him to work because of his war record, and instead he became a private tutor of mathematics. This in turn led to him publishing mathematical papers, and in 1856 he became a mathematics teacher at the Ginnasio (scientific high school) in Cremona. Politics intervened again in 1859 when, after a complicated series of battles, the Austrians lost control of Lombardy (Cremona is a town in the Lombardy region, south-east of Milan) which passed to Piedmont, then supported by, but not part of, France. The new Lombardy Government promoted Cremona to a major lycée in Milan. In 1860 he was appointed a professor at the University of Bologna by Victor Emmanuel II, who was officially proclaimed the king of the newly united Italy on 17 March 1861, by a parliament sitting in Turin.

In 1866 Cremona moved to teach higher geometry and graphical statics[2]

---

[2] Graphical statics is an attempt to apply the methods of projective geometry to the design of structures held together by rods, such as bridges. Cremona's work in this area developed that of Maxwell, the famous theorist of magnetism and electricity. Maxwell had created a theory of reciprocal figures; Cremona showed

at the Polytechnic of Milan, and also that year he shared the Steiner prize of the Berlin Academy with J. C. F. Sturm for his long memoir on cubic surfaces, *Mémoire... sur les surfaces du troisième ordre* [44]. The memoir is one of his high points as a mathematician, along with his theory of what are still called Cremona transformations and are used to resolve the singularities of plane curves. There are analogous transformations of space, but their theory is still a matter of deep research.

In 1873 – the year Cremona published his *Elementi di geometria projettiva* – he moved to Rome as professor of higher mathematics, with the task of reorganising the college of engineering. He now found that his new responsibilities put an end to his chances of writing any more original mathematics. In 1879, the year he was elected a fellow of the Royal Society of London, he became a senator, and eventually Minister of Public Education and Vice-President of the Senate. He used his influence to promote a reform of school education in Italy that advocated the teaching of Euclid's *Elements* in the original Greek. Not surprisingly, mathematics proved an unpopular subject for most people, but a very powerful and original group of mathematicians also emerged in Italy who regarded Cremona as a major figure.

22.2.1 Cremona's projective geometry

In the introduction to his book on projective geometry [45], Cremona wrote:

> It is, I think, desirable that theoretical instruction in geometry should have the help afforded it by the practical constructing and drawing of figures. I have accordingly laid more stress on *descriptive* properties than on *metrical* ones; and have followed rather the methods of the *Geometrie der Lage* of Staudt than those of the *Géométrie supérieure* of Chasles. It has not however been my wish entirely to exclude metrical properties, for to do this would have been detrimental to other practical objects of teachings. I have therefore introduced into the book the important notion of the *anharmonic ratio* [Chasles' term for cross-ratio], which has enabled me, with the help of the few abovementioned propositions of the ordinary geometry, to establish easily the most useful metrical properties, which are either consequences of the projective properties, or are closely related to them.
>
> I have made use of *central projection* in order to establish the idea of *infinitely distant elements*; and, following the example of Steiner and of Staudt, I have placed the law of *duality* quite at the beginning of

how to interpret this as a duality in three-dimensional space.

the book, as being a logical fact which arises immediately and naturally from the possibility of constructing space by taking either the point or the plane as element. The enunciations and proofs which correspond to one another by virtue of this law have often been placed in parallel columns; occasionally however this arrangement has been departed from, in order to give to students the opportunity of practising themselves in deducing from a theorem its correlative. Professor Reye, remarks, with justice, in the preface to his book, that Geometry affords nothing so stirring to a beginner, nothing so likely to stimulate him to original work, as the principle of duality; and for this reason it is very important to make him acquainted with it as soon as possible, and to accustom him to employ it with confidence.

Cremona was indeed so enamoured of duality that he wrote out material in two columns when appropriate, as Figure 22.2, taken from later in his book [45, p. 76], shows. What theorem is here described? (Note the typographical error.)

If a hexagon (six-point) $AB'C\,A'BC'$ (Fig. 60) has its vertices of odd order (1st, 3rd, and 5th)	If a hexagon (six-side) $ab'ca'bc'$ (Fig. 61) be such that its sides of odd order (1st, 3rd, and 5th)

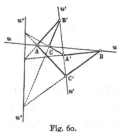 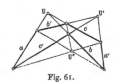

Fig. 60.   Fig. 61.

on one straight line u, and its vertices of even order (2nd, 4th, and 6th) on another straight line u', then the three pairs of opposite sides (AB' and AB' , $B'C$ and BC', CA' and $C'A$) meet in three points lying on one and the same straight line u''*.	meet in one point U, and its sides of even order (2nd, 4th, and 6th) meet in another point U', then the three straight lines which join the pairs of opposite vertices (ab' and $a'b$, $b'c$ and bc', ca' and $c'a$) pass through one and the same point U''.

Figure 22.2 Cremona presents a theorem and its dual

Cremona then defended the name projective geometry on the grounds that:

the subjects are to a great extent of venerable antiquity, matured in the minds of the greatest thinkers, and now reduced to that form of extreme simplicity which Gergonne considered as the mark of perfection in a scientific theory. In my analysis I shall follow the order in which the various subjects are arranged in the book. The conception of *elements lying at an infinite distance* is due to the celebrated mathematician Desargues; who more than two centuries ago explicitly considered parallel straight lines as meeting in an infinitely distant point, and parallel planes as passing through the same straight line at an infinite distance.

The same idea was thrown into full light and made generally known by Poncelet, who, starting from the postulates of the Euclidean Geometry, arrived at the conclusion that the points in space which lie at an infinite distance must be regarded as all lying in the same plane.

[...]

The law of duality, as an independent principle, was enunciated by Gergonne; as a consequence of the theory of reciprocal polars (under the name *principe de réciprocité polaire*) it is due to Poncelet. The geometric forms (range of points, flat pencil) are found, the names excepted, in Desargues and the later geometers. Steiner has defined them in a more explicit manner than any previous writer.

[...]

Harmonic section was known to geometers of the most remote antiquity; the fundamental properties of it are to be found for example in Apollonius.

[...]

The complete theory of the anharmonic ratios is due to Möbius, but before him Euclid, Pappus, Desargues, and Brianchon had demonstrated the fundamental proposition of Art. 63 [that cross-ratio is projectively invariant].

[...]

The theory of pole and polar was already contained, under various names, in the works already quoted of Desargues and de la Hire; it was perfected by Monge, Brianchon, and Poncelet. The last-mentioned geometer derived from it the theory of polar reciprocation, which is essentially the same thing as the law of duality, called by him the *principe de réciprocité polaire*.

The principal properties of *conjugate diameters* were expounded by Apollonius in books ii and vii of his work on the Conics.

[...]

Those who desire to acquire a more extended and detailed knowledge of the progress of Geometry from its beginnings until the year 1830 (which is sufficient for what is contained in this book) have only to read that classical work, the *Aperçu historique* of Chasles.

One great merit of the work of Cremona was to make clear and public
that projective geometry had finally rid itself of the embarrassment of its Euclidean origins, and could indeed claim to be the fundamental geometry. Its
claim to such an exalted status rests, as has been said, on the fact that every
projective property is also a Euclidean one, but the reverse is not true. But
the claim is tarnished if, on a closer reading, it would seem that the definitions of the most important figures and proofs of the most important theorems
require the use of Euclidean arguments. For example, if, as in Chapter 3, a
projective transformation is used to reduce a figure to a special case and a
theorem is then derived, however elegantly, by Euclidean means. Cremona's
book made it clear that all of the key definitions, such as points in involution and more especially that of a conic, could be given in purely projective
terms, and the proofs of all the theorems conducted likewise without resort
to Euclidean geometry. Much of the groundwork had been done by Chasles,
Steiner and von Staudt, but Cremona's work pulled it all together very ef-
fectively. Projective geometry was presented independently of Euclidean geometry, which now stood out as a body of ideas concerned with extra structure.

22.3 Salmon

Finally, if deeds may stand for thoughts, the mathematical life work of George
Salmon in Dublin is testimony to the high opinion he had of projective geometry. Salmon was a friend of Cayley's, and the author of some of the best
textbooks written on geometry in English in the 19th century. They were so
good that they were translated into German.[3]

In Gow's paper you may read that in Klein's opinion, Salmon's books
are "like refreshing and instructive walks through wood and field and cultivated gardens, where the guide draws attention now to this beauty, now to
that strange appearance, without forcing everything into a rigid system of
faultless perfection ... In this flower garden we have all grown up, here we
have gathered the foundation of knowledge on which we have to build." [92,
p. 39]

This is very fair. Salmon instructs through examples, developing techniques
as he goes along, not at all the heavy theorist in whose pages examples are but
dimly seen.

---

[3] The best single source of information on Salmon is R. Gow [92].

22.4 Anxiety – Pasch

Signs of anxiety about the nature of geometry run like fissures through late 19th-century mathematics. A scrupulous reading of Euclid's *Elements* showed that it is far from being a gapless chain of reasoning. It is necessary to stipulate such omitted "obviousnesses" as that circles cannot pass through one another without meeting, and that a line which enters a triangle through a side exits through one of the other sides or the opposite vertex ([184, §2, Grundsatz IV] – Pasch's axiom, to be discussed below). The first of these had been added to the *Elements* by a 10th-century Arab commentator, the second was a late 19th-century observation, and they must stand here for a number of similar comments.

Moritz Pasch's *Vorlesungen über neuere Geometrie* of 1882 [184] is the first book in which a thorough resolution of these difficulties is proposed, and a putatively gapless chain of reasoning in synthetic geometry put forward. Indeed, Pasch himself wrote of the disturbing state of geometry that was one of his reasons for writing his celebrated book of 1882. In the preface he compared the new, advanced, synthetic geometry with the old, elementary geometry. "Elementary geometry", he wrote [184, p. 2], "cannot only be reproached for its clumsiness, but also for incompletenesses and obscurities, which the ideas and proofs still retain in extended measure. The repair of this defect is an incessant struggle, in manifold ways, and if one examines the results one can come to the opinion that the struggle is hopeless." He then went on to show how, in his opinion, the task was not hopeless after all, and indeed that his book finally put the matter to rest. The point here is that the criticism of elementary, intuitive geometry from the standpoint of late 19th-century criteria of rigour was always finding faults, so much so that that whole branch of introductory mathematics could seem riddled with flaws and perhaps beyond repair. This is a most uncomfortable position for a mathematician to be in, and it was to that anxiety that Pasch responded with his book.

Pasch rejected the idea of reducing geometry to analytic geometry, finding that the methods of the latter subject were in opposition to the basic concepts. By this he meant that the methods of proof in analytic geometry were heavily arithmetical, but geometry was only occasionally so. As he said in his introduction, he presented his analysis largely in terms of projective geometry, presumably on the grounds that it is more basic than Euclidean, because it does not involve the concept of distance. His approach was to formulate rigorously every fact about plane geometry that he felt was necessary. He started with the undefined or primitive concept of the straight line segment between two points. Any result or property about segments which he felt necessary to assume without proof he called a Grundsatz [184, p. 17]. All Grundsätze were, he said, immediately grounded in observation, and he cited Helmholtz's paper "On the origin and significance of the axioms of geometry" of 1870 at this point. Results he could deduce from the Grundsätze he called Lehrsätze. There were eight Grundsätze needed to base the theory of line segments, of which the first is "there is always a unique segment joining any two points". In general, Grundsätze should be laid down until the mathematician could henceforth reason logically and without further appeal to sense perceptions. The rest of the book is devoted to showing that that can be done. A further four Grundsätze enabled Pasch to conduct plane geometry (the famous axiom now named after him is the fourth of these [184, p. 21]); three-dimensional geometry followed without further ado. To obtain what are loosely called the points at infinity in projective geometry, or the ideal points (Klein's term), Pasch showed that one could make rigorous the idea that the two kinds of pencils of lines in a plane are equivalent. The first is the pencil of lines through a point, the second a pencil of parallel lines.

The first kind may be taken to define a (finite) point, the second kind a point at infinity.

The same division into Grundsätze and Lehrsätze characterised Pasch's treatment of projective transformations and of congruence. The assumptions now all explicit, he then made sure that all of elementary projective geometry was now established: duality; the special case of geometry with respect to a fixed conic, including Klein's version of non-Euclidean geometry; and cross-ratio. As Pasch pointed out, the axioms of congruence permit one to coordinatise projective space using a projective net, such as Möbius and then von Staudt had proposed (let me just note that later mathematicians have preferred weaker axioms for which this is not always the case).

The concepts of collinearity and concurrence present no difficulty. The essential projective invariant called cross-ratio is subtler; following the lead of von Staudt, Klein had shown how this can be defined other than as a ratio of a ratio of distances, which would seem to close a vicious circle. Pasch's treatment was careful and complicated, and will not be discussed here.

22.5 Helmholtz

Hermann von Helmholtz was drawn to geometry through his work in physiological optics. He wanted to characterise those n-dimensional spaces or manifolds that admit the free motion of bodies but, finding even the case $n = 3$ was difficult, he obtained from Riemann's successor, Schering, at Göttingen copies of Schering's notes about Riemann's work which he used to give his own account, first in his "On the facts underlying geometry" [111], and then in "On the origin and significance of the axioms of geometry" [112]. The requirement that the free mobility of bodies must be allowed was imposed by many who wrote on the subject; it cuts down the infinitude of Riemannian spaces to a much smaller number. It was supported by tradition: Euclid required the superposition of figures, experience and (Helmholtz's reason) the belief that rigid body motions are essential to congruence and hence to measurement (an argument Riemann had refuted). Riemann had observed that free mobility implies constant curvature, and Helmholtz regarded free mobility as true in any case.

Helmholtz's first paper was marred by his ignorance of non-Euclidean geometry, as Beltrami pointed out to him (Helmholtz then published a note correcting the omission). The second paper not only allows non-Euclidean geometry, it gives good reasons for suggesting that we could have intuitive knowledge of it (a key position in any argument against Kant). Helmholtz gave the analogy with the world seen in a spherical mirror. One might look at the image in the mirror and say that the distances between points are those in the outside world, or that they are the distances in the mirror. Either way, imagine that you might live in the mirror, so to speak, and confront the choice. Now infinite (Euclidean) space is to be analysed or experienced in terms of a decidedly non-Euclidean metric.

Helmholtz's two papers are resolutely empiricist: unlike Riemann, who based his work on hypotheses, he believed that there are facts that underlie geometry, and our acquaintance with them leads us to accept some axioms for geometry. The mirror analogy suggests that the choice between Euclidean and non-Euclidean geometry cannot be resolved "by pure geometry", as Helmholtz put it. (As we shall see, this makes Helmholtz sound like Poincaré.) But Helmholtz observed that the laws of mechanics would differ in the two spaces. "If to the geometrical axioms we add propositions relating to the mechanical properties of natural bodies, ... such a system has a real import which can be confirmed or refuted by experience." [112, p. 25] Somewhat later, in 1877, he held the straightforward position that empirical measurements of triangles will decide the question.

The insistence that facts underlie geometry, rather than axioms (almost all of which should be Euclid's) places Helmholtz nearer to Riemann than Euclid. A fact, after all, is a well-verified hypothesis. What this philosophy admits, but the contemporary attitude to axioms did not, was that the very foundations could shift; whereas an axiom was a self-evident truth, so basic that it could not be proved.

22.5.1 Free mobility

A warning note should be struck here about a topic and subtlety that are often omitted from accounts like these at the price of introducing error.[4] The topic goes by the name of space forms. It can be introduced by a simple example. Consider the usual Euclidean plane, in which we shall denote points by complex numbers. Act on it by the group $G := \mathbb{Z} \times \mathbb{Z}$ of Gaussian integers according to the rule that the element (m, n) of G moves points as follows: $z \mapsto z + m + in$. The corresponding quotient space (the set of equivalence classes of points under this action) is obtained from the unit square $S := \{x + iy, 0 \le x \le 1, 0 \le y \le 1\}$ by identifying top and bottom (equivalent under $z \mapsto z + i$) and the two sides (equivalent under $z \mapsto z + 1$). So the corresponding quotient space is the torus, and in small regions the geometry is obviously still Euclidean, so the induced metric is the Euclidean metric. This space is sometimes called the flat torus, to distinguish it from the more familiar inner tube, which induces its metric from the ambient $\mathbb{R}^3$ and is very obviously of variable curvature.

Now a rotation of a body in $\mathbb{R}^2$ naturally extends to a rotation of the whole of $\mathbb{R}^2$, and this gives a very natural way to speak mathematically about the rotations of a body. But a rotation of the corresponding body in the flat torus does not extend to a rotation of the whole of the torus. Free mobility of figures, if this is taken to mean the existence of a suitably large supply of maps of the space containing the figure to itself, is not a property of all spaces of constant curvature.

This problem turns up a lot. Every time there is a space of constant curvature on which a group acts isometrically and discretely (roughly, equivalent points are always a finite distance apart and do not accumulate) there is a quotient space of the same dimension and with the same constant curvature. But free mobility of figures in the two settings means very different things.

It seems that Clifford may have been the first to appreciate this point, and the discovery of the flat torus is due to him, although not by the method described above. Klein caught on quickly, and then engaged in lengthy discussions with Killing about it in the course of which exactly the subtlety just described was clarified. For more detail, the reader is referred to Epple's paper [69].

---

[4] I am indebted to Moritz Epple for gently reminding me of this. See his paper of 2002 [69].

Hilbert and his Grundlagen der Geometrie

Figure 23.1 Hilbert

23.1 Hilbert

David Hilbert, who came to dominate German mathematics between 1890 and the 1920s, and who many would say was the leading mathematician in the world of his generation, was born in Königsberg on 23 January 1862.[1] Königsberg was a small town in East Prussia, best known for being the home town of the philosopher Immanuel Kant, and Hilbert's schooling there was unremarkable. He later said that "I did not particularly concern myself with mathematics at school because I knew that I would turn to it later". He then went to the

---

[1] On the life and work of Hilbert, see Constance Reid's *Hilbert* [207], with the invaluable reprint of Hermann Weyl's obituary of Hilbert, and Gray [102].

J. Gray, *Worlds Out of Nothing*,
Springer Undergraduate Mathematics Series,
DOI 10.1007/978-0-85729-060-1_23, © Springer-Verlag London Limited 2011

university in Königsberg in 1880. Although small, it had a strong tradition in mathematics and physics, beginning with Carl Jacobi and the physicist Franz Neumann, who pioneered the teaching of experimental physics. A steady stream of mathematicians and scientists came to or graduated from Königsberg: the physicist Gustav Kirchhoff, the geometer Otto Hesse and Heinrich Weber, who held the chair in mathematics at Königsberg from 1875 to 1883 and was succeeded by Ferdinand Lindemann, who had just become famous for his proof that π is transcendental.

Heinrich Weber had, with Richard Dedekind, published the posthumous edition of Riemann's *Werke* in 1876, and through Weber, Hilbert came into contact for the first time with the strong current in German mathematical life that led back to Gauss. But apart from a term in Heidelberg, Hilbert did not travel. He did not go to Berlin, for example, as was customary in those days. However, at Königsberg Hilbert met his two lifelong influences: Hermann Minkowski, a fellow student who was two years younger, but already a term ahead (Minkowski, unlike Hilbert, was a prodigy) and Adolf Hurwitz. Hurwitz was only three years older than Hilbert when he was appointed to an extraordinary professorship at Königsberg. Hilbert later recalled to Blumenthal (a student of his, and author of a memoir of Hilbert's life) that "Minkowski and I were totally overwhelmed by his knowledge and we never thought we would ever come that far" [18, p. 390]. It was through Hurwitz that Hilbert gained entry to the circle Felix Klein was building around him.

Hilbert finished his doctorate in 1885, and now he did travel. To Leipzig to meet Klein, and then to Paris, where he met Henri Poincaré, who was some eight years older than him, but without finding a kindred spirit. Back in Königsberg, Hilbert took his Habilitation and began to lecture, but Königsberg was a small university, and although he became a very good lecturer, he often had an audience of only two or three. On one occasion, the winter semester 1891/1892, Hilbert had an audience of exactly one for a course on complex function theory, and his first lectures on projective geometry at Königsberg attracted two students plus the director of the royal art school. At Easter 1888, Hilbert was introduced by Klein to Paul Gordan at Erlangen and came alive as a research mathematician. He moved rapidly to write a series of papers on what is called invariant theory, a subject renowned then for its technical complexity and which Hilbert rewrote in a language of great conceptual generality, abstractness and clarity. This had been Gordan's speciality, and for a time he is said to have regarded Hilbert's new version of the subject as "not mathematics, but theology", but he eventually came round and conceded that even theology has its uses.

In 1890 a number of elderly German professors began to resign, and new positions opened up. Klein had moved to Göttingen in 1886 and was keen to get Hilbert there, but he only succeeded at the second attempt, in 1895. Some of Klein's colleagues criticised his choice, presuming that he wanted to appoint an easy going, amenable younger man. To this Klein replied [18, p. 399]: "I want the most difficult of all." Hilbert's arrival marks the start of the rise of Göttingen to become the greatest university for mathematics in the world. The contrast between Hilbert and Klein was marked, but they made a productive partnership. Both were very hard workers. But Hilbert's colleagues did not think of him as an organiser, nor did they regard him as particularly well read – in those respects he was outstripped by Felix Klein. They thought of Hilbert as "a man of problems", someone with a tremendous will, who truly believed every problem can be solved.

The confidence Hilbert had in his own abilities was widely shared. The German Mathematical Society (the Deutsche Mathematiker-Vereinigung or DMV) had been founded in 1890 with Georg Cantor as its first president. In 1893 they asked Hilbert and Minkowski to report on the theory of numbers. The choice of Minkowski was obvious. He had established himself as a world authority while still a student, and had worked energetically on the subject ever since. But Hilbert had little to show for himself except his work on invariant theory and his expressed interest in the new subject. Minkowski, however, withdrew from the project pleading pressure of other work, so the final *Zahlbericht* [117] or *Report on the theory of numbers*, for all its 371 pages, is only a torso. Nonetheless, it became a classic, and shaped research in number theory for a decade. Hermann Weyl, whose own masterful control of the German language was probably unmatched by any other mathematician, regarded the preface of Hilbert's *Report* as one of the great works of prose in the mathematical literature, and described how as a young man he had heard "the sweet flute of the Pied Piper that Hilbert was, seducing so many rats to follow him into the deep river of mathematics." [247, p. 132]

23.2 Hilbert and geometry

Almost at once, in what became a trademark move, Hilbert abandoned number theory, just as he had earlier abandoned invariant theory. His new love was the foundations of elementary geometry, which shocked his Göttingen colleagues. They did not know that Hilbert had already been drawn to the subject in his years at Königsberg. He had not at first been attracted to the teaching of projective geometry, a staple subject he felt he did not have much to say about. But he soon enough found that there were things to say.

Hilbert discovered that the experts disagreed about a number of things in synthetic projective geometry. In particular, they disagreed about which theorems followed from which others. In September 1891 he heard a lecture by Hermann Wiener in which he raised the possibility of developing projective geometry axiomatically from Pascal's and Desargues' theorems. Blumenthal reported that this was the occasion for Hilbert to say [18, p. 403] "One should always be able to say, instead of 'points, lines, and planes', 'tables, chairs, and beer mugs'." In other words, one should be able to purge one's arguments of words that might smuggle in meanings and imply a validity that the logical structure of the argument cannot actually carry. The point was not original with Hilbert, but he grasped its significance more profoundly.

In early January 1898 Klein passed a letter on to Hilbert that he had received from Friedrich Schur. In this letter Schur raised the possibility of deriving Pappus's theorem without using the Archimedean axiom (which says that given any two lengths $a < b$ there is always an integer such that $na > b$). This opened the way for a non-Archimedean geometry and, thus inspired, Hilbert lectured on geometry at Göttingen in the winter semester, much to the surprise of Blumenthal, who wrote [18, p. 402] "This caused astonishment among the students, for even we older people, participants in the stroll through the number fields, had never known Hilbert to concern himself with geometrical questions. Astonishment and wonder were caused when the lecture began and a completely novel content emerged."

23.2.1 The *Grundlagen der Geometrie*

Hilbert's ideas were first published in 1899, as part of a book marking the unveiling of the statue to Gauss and Wilhelm Weber in Göttingen. It seems that it was Klein who suggested to Hilbert that the subject matter of his lectures at the time would make an appropriate contribution. Hallett and Majer have reproduced those lectures [107], and they illustrate very well how Hilbert worked, because the treatment evolved as the lectures went by. The reader is referred to their book for the details, which covers such things as a shift in attitude to the parallel axiom and some rethinking of the order and mutual independence of the axioms Hilbert proposed. In the lectures the various uses of the parallel axiom were investigated, whereas in the published *Grundlagen der Geometrie* [118] it was put to more use. Also, some results that were axioms in the lecture courses became theorems in the book.[2]

---

[2] In his lectures Hilbert gave some references to previous literature that did not appear in the book: Pasch's [184], which he called a very clear but also very extensive (or far-reaching, weitläufiges) book, and Killing's two-volume *Grundlagen*

The *Grundlagen der Geometrie* (*Foundations of geometry*) [118] was to run to ten editions, seven in Hilbert's lifetime, and to be translated into several languages. It came to epitomise for many what Hilbert's ideas about mathematics were, and was perhaps more easily understood because it presented them in an elementary arena. Nonetheless, as we shall see, it carried a number of interesting flaws to match its insights.

Hilbert had not forgotten the beer mugs. The book begins: "We think of three different systems of things: the things of the first system we call points and denote A, B, C, ... ; the things of the second system we call lines and we denote a, b, c, ... ; the things of the third system we call planes and denote α, β, γ, ... ". There then followed five kinds of axiom which determine what one can say about these things. There are axioms of incidence (which enable you to say "this point lies on this line"), order ("this point lies between these two"), congruence, parallels and continuity (which is where Hilbert located the Archimedean axiom). Curiously, Hilbert did not insist in the axioms of incidence that any two lines meet in a point. So it is not possible to say that he axiomatised projective geometry and then added more axioms to obtain other geometries; his simplest plane geometry lacks any duality between points and lines, presumably because of his greater interest in metrical and arithmetic concerns.

23.2.2 Desargues' theorem

At each successive stage Hilbert was concerned to show what theorems can now be deduced, thus making it clear which axioms are actually required, and also to show that the axioms are mutually independent and consistent. He was not entirely successful at establishing independence, but that is a minor flaw. He did show, for example, that Pappus's theorem is independent of Desargues' theorem, and the admission or exclusion of the Archimedean axiom has a decisive effect on the validity of Pappus's theorem. In fact, as Hilbert nearly showed, Pappus's theorem implies Desargues' theorem.[3] A lot of the work in the book is done by what he called segment arithmetic. Hilbert based it on Pappus's theorem (which Hilbert confusingly called Pascal's theorem). This gave him ways analogous to addition and multiplication of combining segments to get others. It followed that if the geometry was coordinatised, the coordinates would have to obey certain rules. In fact, he showed that when Pappus's theorem is true the coordinates must lie in a system where multiplication is commutative, but Desargues' theorem can be true in a system where multiplication is not com-

---

der Geometrie, which was "less clear but very rich".
[3] This is Hessenberg's theorem, see [116].

mutative. Hilbert also gave a convoluted example of a plane geometry in which Desargues' theorem was false.[4] The lectures show how Hilbert worked his way to showing that Desargues' theorem in the plane is a necessary and sufficient condition for the plane to be embeddable in a three-dimensional projective space.

Hilbert's example was improved by an American astronomer in 1902, and Hilbert included Moulton's example in later editions of his book. Moulton divided the familiar plane into two by a line called the Axis (see Figure 23.2), and then defined lines to be either a conventional line parallel to the Axis, or a conventional line crossing the Axis but sloping to the left, or a new kind of line obtained from a line crossing the Axis and sloping to the right. If such a line makes an acute angle, β, with the Axis, it is replaced by a broken line composed of the half of the original line below the Axis and a line segment, h, above the Axis making an angle α at P with the Axis, where $\tan \alpha = \frac{1}{2} \tan \beta$.

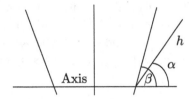

Figure 23.2   Moulton's lines

The usual fundamental axioms of projective geometry are satisfied, for example two lines meet in at most one point, but in this geometry, Desargues' theorem is false. The real force of this result is that it shows that Moulton's plane cannot be embedded in any three-dimensional space (if it could, Desargues' theorem could be proved). See Figure 23.3.

Max Dehn's contributions   Max Dehn was a student of Hilbert's at Göttingen who wrote his PhD thesis there in 1900. In the same year he solved the third of the Hilbert Problems that Hilbert had just proposed in Paris (see [102]). He shared Hilbert's interests in the foundations of geometry, and noted that there are two theorems about the angle sum of a triangle that could be named after Legendre and make no assumption about parallels: first, one might claim that the angle sum of a triangle cannot exceed two right angles; second, one might claim that if the angle sum of one triangle equals two right angles then the angle sum of every triangle is two right angles. Legendre had proved both of these

---

[4] In fact, in his lectures in 1898–1899 Hilbert gave a detailed account of which axioms do or do not imply Desargues' theorem, as Hallett and Majer discuss in [107, p. 214].

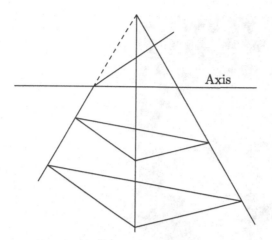

Figure 23.3 The failure of Desargues' theorem

results, but his proof made essential use of the Archimedean axiom. Hilbert asked Dehn to investigate this question in the light of the various possible axiom systems for geometry (as we know from a letter from Hilbert to Hurwitz of 5 November 1899, which records that Dehn had by then succeeded.[5] Dehn showed that the second of the above theorems follows from Axiom Groups I, II, IV alone, but that the first theorem is false without the Archimedean axiom.

Hilbert's ideas went beyond geometry, and suggested that axiomatising all of mathematics might be productive. One community of mathematicians that picked up this idea was the young American mathematicians around E. H. Moore at Chicago. Several had studied in Göttingen; Moore himself had attended the ceremony for the unveiling of the Gauss–Weber memorial in 1899, and in 1901 he led a seminar on the foundations of geometry and lectured on Hilbert's work. Hilbert's axioms had already been criticised by Friedrich Schur for their lack of independence, and Moore discovered that although Schur was right, the redundancies were incorrectly identified. His presentation of this discovery in class attracted the gifted young mathematician Oswald Veblen to the study of geometry. In 18 months Veblen wrote his PhD thesis, in which he based Euclidean geometry on 12 axioms based on the notions of point and order. He then went on to spend many years on various aspects of the foundations of geometry, and eventually became one of the leading American geometers and a major influence on the Institute for Advanced Study at Princeton.

---

[5] The letter is quoted in Toepell [236, p. 257]. For Dehn's work see [50]. This account follows that of Hallett and Majer [107, p. 214].

Figure 23.4 Veblen

23.3 Impact

The success of Hilbert's *Grundlagen der Geometrie* enabled Hilbert to add to
it. Some later editions, including the standard English translation (*Foundations of geometry* [119]) carry an axiomatisation of non-Euclidean geometry
(see [119, Appendix III]) that Hilbert first published in volume 57 of *Mathematische Annalen* in 1903. It is worth noting that the original intention of the
Grundlagen der Geometrie was not to describe Euclidean geometry and non-
Euclidean geometry as a contrasting pair of axiom systems. However, as the
qualities of an axiomatic approach to mathematics were more and more vividly
perceived, it did become customary to see the discovery of non-Euclidean geometry as the discovery of an axiom system different from Euclid's. It certainly
is such a thing, but the original discovery was done in a different way and
for different reasons. After all, a consistent axiom system different from Euclid's had been available all along: spherical geometry. Even to see the history
of non-Euclidean geometry as the search for an axiom system differing from
Euclidean geometry only in respect of the parallel axiom is to overstate the
importance of axioms and to understate other questions about the nature of
geometry: the role of transformations, the use of formulae, its applicability to
space.

23.4 References

This is the place to mention Hartshorne's lovely and stimulating book *Geometry: Euclid and beyond* [109] for its lucid accounts of Euclid's *Elements* and Hilbert's axioms for geometry – and indeed for its account of non-Euclidean geometry and much else.

24

The Foundations of Projective Geometry in Italy

When the distinguished mathematician and historian of mathematics Hans Freudenthal analysed Hilbert's *Grundlagen* he argued that the link between reality and geometry appears to be severed for the first time in Hilbert's work. However, he discovered that Hilbert had been preceded by the Italian mathematician Gino Fano in 1892. Recent historians of mathematics have shown that, in Italy at least, Fano's point of view on the nature of geometrical entities had been a generally accepted theory for at least a decade, but it was not in fact axiomatic in Hilbert's manner; other Italian mathematicians were, however, ahead of Hilbert in this regard. How did this come about, what did they do, and why did they lose out?

We shall see that one answer to the last question is that the Italian school considered that giving foundations for geometry was all that had to be done, whereas Hilbert saw it as exemplifying what could be done with profit across mathematics. Another reason, which Klein emphasised, is that the training of mathematicians in Italy was too closely tied to the teaching of first-year students, thus emphasising the basics at the expense of the frontier. And of course, Hilbert was Hilbert, a major mathematician, working in Göttingen (the world centre of mathematics) and a lucid writer unlike, for example, Peano, as we shall see.

J. Gray, *Worlds Out of Nothing*,
Springer Undergraduate Mathematics Series,
DOI 10.1007/978-0-85729-060-1_24, © Springer-Verlag London Limited 2011

There were in fact many Italian mathematicians, and – to over simplify, but not by much – they range from Giuseppe Peano and his followers to Corrado Segre and his followers along a scale with rigour and formalism at one end and imaginative, intuitive research at the other.

24.1 Peano and Segre

In the mid-1880s, Giuseppe Peano and Corrado Segre were in their early twenties and starting their highly successful careers at the University of Turin. Their work was very different in spirit, however, and led to a very public clash at the end of the decade.

Peano started work in analysis, and came quickly to the conclusion that the subject was riddled with illicit appeals to intuition. This pushed him towards formalising what he did – he is a prominent figure in the story of the origins of a vector space – and enabled him to come to the profound, highly counter-intuitive discovery of a space-filling curve. This is a curve that goes through every point of the plane. His own account was later simplified by the American mathematician E. H. Moore and by Hilbert. He also gravitated to the Esperanto movement, which had been launched in the 1890s, and came to advocate a different would-be international language of his own, *latine sine flexione* (Latin without inflections) which is in many ways easier to read. He also advocated using symbols in place of words, and some of the logical and set-theoretic symbols we use today are due to him. But it must be said that the combination is almost impossible to read, and that as he got older he got keener and keener on this, even lecturing this way and promoting a journal devoted to it. Nonetheless, he was a sharp logical critic of mathematical arguments.

Segre was in many ways the opposite. A gifted mathematician whose career got off to an early start, he saw it as one of his tasks to move Italian geometers into spaces of higher dimensions. This path had been opened up by another mathematician, Veronese, who when in Leipzig with Klein in 1881, had shown that just as curves in the plane with complicated singular points could be seen as projections of non-singular curves in space (proving that claim is another story!) so too could surfaces in space with complicated singular points sometimes be seen as projections of non-singular surfaces in some higher-dimensional projective space. This passage to spaces of arbitrary dimension was an intuitive step for Segre, defended, if at all, on the grounds that the algebra generalised, but he retained a firmly geometric sense of these new spaces.

To quote from his dissertation:

> The geometry of n-dimensional spaces has now found its place amongst the branches of mathematics. And ... it is impossible not to acknowledge the fact that it is a science, in which all propositions are rigorous, since they are obtained with essentially mathematical reasoning. The lack of a representation for our senses does not matter greatly to a pure mathematician. Born, as it were, out of Riemann's famous work of 1854 ... n-dimensional geometry develops along two separate lines: the first one deals with the curvature of spaces and is therefore connected with non-Euclidean geometry, the second one studies the projective geometry of linear spaces ... and in my work I am to focus on the latter. This path opens for keen mathematicians an unbound richness of extremely interesting research. [222, p. 3]

Segre's ideas were not always expressed precisely. He wrote of homogeneous coordinates in any number of dimensions, but forgot to say that $[0, 0, \ldots, 0]$ is not allowed. This was the kind of sloppiness (harmless as it is in this case) that upset Peano. Peano's account of a vector space is much more rigorous. But for Peano such definitions tended to appear as summaries or conclusions, whereas for Segre they were springboards to future work, much of it of high quality, especially in algebraic geometry.

In 1889, following Pasch's example, Peano produced an important systematisation of the foundations of geometry. His point of view was extremely abstract:

> We have therefore a category of entities called points. These entities are not defined. Furthermore, given three points, we consider a relation between them, which we indicate with the expression $c \in ab$. This relation is not defined either. The reader can understand by the sign 1 any category of elements, and by $c \in ab$ any relation between three elements of that category. All the following definitions will always be true ... If a certain group of axioms is true, then all the propositions deduced from it will also be true. [187, p. 77]

But also self-limited:

> The treatment of the problem of definitions ... is coherent with the idea that the axiomatic method can fulfill its ordering function only on theories that have been thoroughly tried and tested, such as those of classic mathematics. [187, p. 80]

So when Segre called for mathematicians to go boldly and take up the challenge of climbing peaks through "untraced paths and dangerous slopes" without waiting for unlikely and remote "secure ways that lead to [the peaks] without danger" [5, p. 385] Peano objected. He was attached to an empiricist conception of mathematics as a "perfected logic" and therefore:

> Each author is allowed to accept the experimental laws that more appeal to him, and he can also propose any hypothesis he likes ... [but] If an author starts from hypotheses that are contrary to experience, or from hypotheses that cannot be verified by experience, nor their consequences, he will be able, for sure, to infer some wonderful theory that will lead others to cry: what gain if the author had applied his reasoning to practical hypotheses! (Quoted in Avallone et al. [5, p. 385] from Peano [185, p. 67].)

Peano therefore rejected higher-dimensional geometry, not just because he considered it lacking in rigour (he could have tried to remedy that defect) but because he thought it "useless". This limited his impact on the future foundations of geometry.

One of Segre's students, Gino Fano, stands as it were midway between the two. His early work was on n-dimensional geometry, and was clear without being rigorous. He did nonetheless try to codify his use of geometry in a system of postulates and, taking his lead from Segre, was interested in demonstrating the independence of his postulates (that is, the n-th postulate is shown to be independent of the preceding $n-1$). Thus, the first three postulates imply that a plane is identified by any three of its non-collinear points, i.e. by any two of its lines that intersect each other, etc. He defined the higher-dimensional spaces so that each one was the projection of the previous from a point outside it and demonstrating that an S^r (a hyperplane of dimension r) is identified by any $r+1$ of its points, provided that they are independent. Next he proved various propositions about the intersections of spaces, etc.; none of which required a new postulate [72, p. 111].

Fano's 4th postulate asserts that each line contains more than two points. Its necessity is demonstrated by proposing a model where each line contains exactly two points and each S^r exactly r points. This model satisfies the first three postulates but not the 4th, so the 4th postulate is independent. These postulates allowed Fano to construct the fourth harmonic point D, of three collinear points A, B, C. But is the fourth harmonic point D distinct from C or not? Fano showed by a model consisting of seven points and seven lines (with three points on each line) that D is not always distinct. His counter-example is nowadays presented as a projective plane over the field of two elements and called a Fano plane. In fact, Fano presented, for each prime n, a geometry coherent with the postulates introduced up to now, and such that the sequence fourth harmonic point D, of A, B, C, fourth harmonic point E, of B, C, D, etc., eventually returns to the point A. Fano verified that these geometries satisfied all the postulates he had introduced and wrote down r-dimensional spaces with $\frac{p^{r+1}-1}{p-1}$ points and as many hyperplanes.

Accordingly, another postulate, the 5th postulate, is needed to ensure that there are harmonic series that do not fold back on themselves

Notice that Fano is not interested in any psychological interpretation of his geometries. They are not, however, defined using coordinates, the whole approach is, ultimately, intuitive. Perhaps surprisingly, the issue of the independence of postulates arises with him (and Segre) and not with Peano. In particular, there was some attention paid to the idea of creating systems with specified properties and as few axioms as possible. Nonetheless, his new geometries were merely counter-examples, destined to show the independence of the axioms. They were not presented as starting points for the development of new geometrical research. (That was to be the view of the American geometers in the early 1900s.)

24.2 Enriques

Segre's finest student, the Italian mathematician who did most to advance algebraic geometry, even eclipsing his former master, was Federigo Enriques. He was only 22 when, in January 1894, he began his lectures on projective and descriptive geometry at the University of Bologna. This made him rethink the foundations of projective geometry to his own satisfaction, and in those years he corresponded a great deal with his friend, fellow geometer and future brother-in-law Guido Castelnuovo.[1] He had an open-minded attitude to the fundamental terms in geometry:

> As to those intuitive concepts, we do not intend to introduce anything other than their logical relations, so that a geometry thus founded can still be given an infinite number of interpretations, where an arbitrary meaning is ascribed to the elements called "points". [However,] we think that the experimental origin of geometry should not be forgotten while researching those very hypotheses on which it is founded. [24, p. 142, n. 5]

---

[1] The correspondence has been published as Bottazzini [24].

Figure 24.1 Enriques

Or, to quote from the appendix to his *Lezioni di geometria proiettiva* published four years later:

> We have tried to show how projective geometry refers to intuitive concepts, psychologically well defined ... On the other hand, however, we have warned that all deductions are based only on those propositions immediately inferred from intuition, which are stated as postulates. From this point of view, the geometry that we have presented, looks like a logical organism, where the elementary concepts of point, line and plane (and those defined through these) are simply elements of some primitive logical relations (postulates) and of other logical relations that are inferred. The intuitive content of these concepts is totally irrelevant. This observation originates from a very important principle that affects all modern geometry: the principle of replaceability of geometrical elements. ... projective geometry can be considered as an abstract science, and it can therefore be given interpretations different from the intuitive one, by stating that its elements (points, lines, planes) are concepts determined in whichever way, which verify logic relations expressed by the postulates. A first corollary of this general principle is the law of duality of space. [66, pp. 347–348]

By the 1890s, the axiomatic conception of mathematics was advancing in much of mathematics (think of the definition of continuity, for example). This was especially true in Germany, and Enriques's position was close to Klein. This helps explain why a German translation of Enriques's book came out in early 1903, with a foreword by Klein himself. Klein praised the book in these terms:

> Over the last two decades Italy has been the true centre of advanced research in the field of projective geometry. Among the specialists, this is well known ... But the Italian researchers have gone far beyond also on a practical level: they have not disdained to draw some didactic conclusions from their own studies ... And this is all the more desirable in Germany since our didactic literature has lost all touch with recent research achievements ... The presentation is always intuitive, but completely rigorous, as it could only be after the clever researches on the foundations of projective geometry presented in earlier essays by the same author. I would like to draw particular attention to the presentation of the metric: the clear and explicit treatment of its foundation by means of the absolute (in the plane by means of a circle with a known centre). [66, p. iii]

Typically, Klein drew Enriques into one of his major enterprises, his (ultimately 23-volume, incomplete) *Encyklopädie der mathematischen Wissenschaften*. He asked Enriques to write the chapter on the foundations of geometry which Klein then edited in 1907 [68]. Enriques took the opportunity to develop his thought in an organic way, and also to insert vast historical digressions, which were very important to his ideas about how research in mathematics can be done. He contrasted Hilbert and Peano in these terms: Hilbert studied any system of postulates, without regard to physics or psychology, provided they had mathematical interest; Peano emphasised formal, logical considerations. It was Hilbert and his followers who were, therefore, more and more abstract.

Enriques discussed the work of Hilbert and his followers carefully and at length, observing the various counter-examples, the construction of non-Pappian (i.e. Hilbert's non-Pascalian) geometries and their link to the question of the non-commutativity and the construction of non-Archimedean bodies. But, interestingly enough, neither Enriques, nor any other of the more brilliant Italian mathematicians turned this generic interest into a proper research topic.

24.3 Pieri

Mario Pieri[2] graduated from Pisa in 1884, and then went to Turin to study with Segre and Peano. He learned his rigour from Peano, while from Segre he took the notion that an axiomatic structure completely independent of experience can still be valuable.

The significant novelty in Mario Pieri's work, which distinguishes it from Pasch's, is the complete abandonment of any intention to formalise what is given in experience. Instead, as he wrote in 1895 [190], he treated projective geometry "in a purely deductive and abstract manner ..., independent of any physical interpretation of the premises". Primitive terms, such as line segments, "can be given any significance whatever, provided they are in harmony with the postulates which will be successively introduced". In the paper [191], Pieri presented foundations for projective geometry which were entirely independent of intuition both in the premises (there were 19 axioms) and in the method. By using the notion of class and membership as a logical tool, and so writing in the arid fashion of Peano and by exploiting the predicate calculus, Pieri proved theorems while excluding any resort to external intuition or implicit perceptual, linguistic or cultural experience. As he put it:

> A good ideographic algorithm is generally acknowledged as a useful tool to discipline thought, to eliminate ambiguities, mental limitations, unexpressed assumptions and other faults which are an integral part of both spoken and written language, and that are so harmful to speculative investigation. Therefore it is vitally important to make use of the method of algebraic logic. Neglecting it, especially in this type of study, seems to me a deliberate rejection of the most valid tool for the analysis of ideas that we have at our disposal today. [191, p. 4]

Pieri went well beyond Peano in allowing the mathematician to create a geometry, indeed, one can wonder what is geometric in Pieri's work at all (although usual projective geometry is a possible interpretation). But he was well aware that, for example, lines in one geometry may be taken as points in another, so he wanted to be free of any given interpretation of the basic terms.

Pieri's work was very influential. Russell and Couturat regarded him as the founder of mathematics as a hypothetico-deductive science. One might cite earlier figures, but few have his attention to rigour, which was unrivalled by any other Italian geometers, except perhaps Peano. But, by writing in the

---

[2] On Pieri, see Elena Marchisotto's paper "The projective geometry of Mario Pieri – a legacy of Georg Karl Christian von Staudt" and the references cited there in *Historia Mathematica* 2006, 33, 277–314.

unfriendly style of his mentor, and because he died young and before he had attracted any students, he had no influence other than through his work.

The language question is far from trivial. For example, Klein wrote to him (and he merely expressed what many Italian mathematicians felt):

> Is it not possible to express your thoughts in a simple language without Peano's symbols? My general experience shows that in Germany works written in this symbolic language have got a very limited readership; indeed they are rejected a priori. I do not intend to question the principle of this symbolic language, on the contrary I do believe that it can be very useful in purely deductive research such as yours, to avoid errors that could be all too easy to make. Although this can be of assistance to the researcher when he summarises his results, he should however be able to express in ordinary language not only the results, but also the reasoning behind such results. [5, p. 418]

24.4 Conclusions

So indeed the Italians preceded Hilbert with many of the approaches to mathematics that his *Grundlagen der Geometrie* is taken to exemplify. They failed however to take the next step, and display the general value of the axiomatic approach. Instead, for them, axiomatising geometry was a matter of tidying it up. Even Enriques saw no value in exotic geometries, and took his research elsewhere (to great effect, be it said). This was exacerbated by a crisis in the Peano school and its gradual isolation in the Italian cultural milieu, and by a tradition that developed of seeing the foundations of geometry as a topic divorced from research and confined to pedagogy.

24.5 Veronese's theory of projection and section in higher dimensions

Back at the start of this book, we studied the central projection of a figure in one plane onto another. In his paper, Veronese generalised this idea to higher dimensions [243]. There is nothing difficult in this, once one accepts that the centre of projection needs no longer be a point, and that the domain and image spaces may also have different dimensions. This example, one of his most successful, may help to persuade you of the merits of his approach. He called his approach the method of projection and section because the lines from the centre of projection are cut (whence "section") by the image space.

Veronese did not introduce the idea of higher-dimensional spaces just for its own sake, but to make the projective geometry of the plane and of projective three-space easier to understand. For example, consider the cusp with equation $x^2 = y^3$. It can be parameterised this way: $(x, y) = (t^3, t^2)$. Now consider the curve in space parameterised by $(x, y, z) = (t^3, t^2, t)$. It is clearly a non-singular curve with a well-defined tangent everywhere, and its projection onto the (x, y)-plane is the cusp. So one can hope to study in this way a singular curve by passing to a non-singular version of it. To give another example, where we shall also find the parameterisation, the singular cubic with equation $y^2 = x^2(x + 1)$ has a double point at the origin. The line through the origin with equation $y = mx$ meets the curve again where $m^2x^2 = x^2(x + 1)$ and $x \neq 0$. This is the point $(x, y) = (m^2 - 1, m(m^2 - 1))$. So the curve in space parameterised by $(x, y, z) = ((m^2 - 1), m(m^2 - 1), m)$ is a non-singular curve which projects down to a singular curve in the plane.

In 1882, Veronese, who was visiting Klein at the University of Leipzig at the time, submitted a paper to *Mathematische Annalen* [243] in which he described how the method of projection and section familiar from work in three dimensions could be extended to any number of dimensions. In the course of this paper he showed that some surfaces with singularities could be studied in the same way as singular curves, by seeing them as projections of non-singular surfaces in a higher-dimensional space. The implication was that perhaps any surface could be treated in this manner, and that is true, but it was to be very hard to prove this and a convincing proof was not to be found until the late 1930s. Nonetheless, Veronese's example was suggestive, and we give it here.

It concerns Steiner's Roman surface, so-called because Steiner discovered it while in Rome in 1844.[3] Steiner had asked if there could be a surface, other than a quadric surface (one of degree two) all of whose plane sections are conics. The sections could be pairs of conics, but never, for example, a cubic curve and a line. As these remarks indicate, he was looking among surfaces of degree four, and he found one. One equation for it is

$$\left(x^2 + y^2 + z^2 + w^2 - 2xy - 2xz - 2xw - 2yz - 2yw - 2zw\right)^2 = 64xyzw.$$

Another version derives from this parameterisation of this affine part of the surface

$$\left[\frac{rs}{r^2 + s^2 + t^2}, \frac{st}{r^2 + s^2 + t^2}, \frac{tr}{r^2 + s^2 + t^2}, 1\right].$$

This defines a map of the sphere with equation $r^2 + s^2 + t^2 = 1$ to the surface. This map sends antipodal points on the sphere to the same map on the Roman

---

[3] Web browsing will prove very interesting at this point.

surface, so it defines a map of the projective plane onto the Roman surface. The surface contains the lines $[t, 0, 1, 0]$, $[1, t, 0, 0]$, $[0, 1, t, 0]$ and in fact intersects itself along these lines, so the singularities of this surface are in fact three straight lines.

Veronese came to the Roman surface by thinking about conics in the plane. A conic has an equation of the form $ax^2 + by^2 + cz^2 + 2fyz + 2gzx + 2hxy = 0$. This suggested to him that there was a map from the projective plane to projective five-dimensional space, P^5, defined by $(x, y, z) \mapsto (x^2, y^2, z^2, yz, zx, xy)$. This maps onto the Roman surface by the map

$$(u_0, u_1, u_2, u_3, u_4, u_5) \mapsto (\frac{u_3}{u_2}, \frac{u_4}{u_2}, \frac{u_5}{u_2}),$$

and the map is

$$(x, y, z) \mapsto (x^2, y^2, z^2, yz, zx, xy)$$
$$\mapsto (x^2, x^2 + y^2, x^2 + y^2 + z^2, yz, zx, xy)$$
$$\mapsto (\frac{yz}{x^2+y^2+z^2}, \frac{zx}{x^2+y^2+z^2}, \frac{xy}{x^2+y^2+z^2}).$$

Now, the map $(u_0, u_1, u_2, u_3, u_4, u_5) \mapsto (\frac{u_3}{u_2}, \frac{u_4}{u_2}, \frac{u_5}{u_2})$ is a projection! Here's how. Take the line of points $L := (a, b, 0, 0, 0, 0)$ in P^5 and define a map from P^5 to $P^3 := \{(0, 0, p, q, r, s)\}$ as follows. The point Q and the line L define a plane in P^5 that meets the three-dimensional projective space P^3 in a point; take this point to be the image of Q. In projective coordinates, let Q have coordinates (u, v, w, x, y, z) then the plane defined by Q and L consists of these points: $\lambda(u, v, w, x, y, z) + (1 - \lambda)(a, b, 0, 0, 0, 0)$, and it meets the P^3 at $(0, 0, p, q, r, s)$ if and only if $\lambda(u, v, w, x, y, z) + (1 - \lambda)(a, b, 0, 0, 0, 0) = (0, 0, p, q, r, s)$. Evidently $p = \lambda w$, $q = \lambda x$, $r = \lambda y$, $s = \lambda z$, so the map is

$$(u, v, w, x, y, z) \mapsto (w, x, y, z).$$

The space of points of the form $(x^2, x^2 + y^2, x^2 + y^2 + z^2, yz, zx, xy)$ is mapped onto the space of points of the form

$$(x^2 + y^2 + z^2, yz, zx, xy),$$

which without loss of generality is the surface in three-dimensional ordinary space whose coordinates are

$$(\frac{yz}{x^2 + y^2 + z^2}, \frac{zx}{x^2 + y^2 + z^2}, \frac{xy}{x^2 + y^2 + z^2})$$

– the Roman surface!

In recognition of Veronese's pioneering achievement, the image of the projective plane under the map $(x, y, z) \mapsto (x^2, y^2, z^2, yz, zx, xy)$ is called the Veronese surface in P^5.

Henri Poincaré and the Disc Model of non-Euclidean Geometry

25.1 Poincaré

Jules Henri Poincaré was born in Nancy, a town in Lorraine in the East of France, on 29 April 1854. His father was professor of medicine at the university there; his mother, a very active and intelligent woman, consistently encouraged him intellectually, and his childhood seems to have been very happy, at least until the war intervened. The town was surrendered to the Germans as part of the settlement of the Franco-Prussian war 1870–1871, and Poincaré remembered seeing German troops occupying it. This may have been one reason for his choosing the military school, the École Polytechnique, over the increasingly popular civilian École Normale, when the time came. As a child he did not at first show an exceptional aptitude for mathematics, but towards the end

J. Gray, *Worlds Out of Nothing*,
Springer Undergraduate Mathematics Series,
DOI 10.1007/978-0-85729-060-1_25, © Springer-Verlag London Limited 2011

of his school career his brilliance became apparent, and he entered the École Polytechnique at the top of his class. Even then he displayed what were to be lifelong characteristics: a capacity to immerse himself completely in abstract thought, seldom bothering to resort to pen and paper, a great clarity of ideas, a dislike for taking notes so that he gave the impression of taking ideas in directly, and a perfect memory for details of all kinds. When asked to solve a problem he could reply, it was said, with the swiftness of an arrow. He had a slight stoop, he could not draw at all, which was a problem more for his examiners than for him, and he was totally incompetent in physical exercises.

He graduated only second from the École Polytechnique because his inability to draw cost him marks on descriptive geometry, and proceeded to the École des Mines in 1875.[1] In 1878 he presented his doctoral thesis to the Faculty of Sciences in Paris on the subject of partial differential equations. Darboux said of it that it contained enough ideas for several good theses, although the methods in the thesis often fell short of rigorous proof, and he urged Poincaré to tighten it up. In the event, the thesis was not published (until it appeared in the first volume of his *Oeuvres*, in 1928 [199]). This fecundity and inaccuracy is typical of Poincaré; ideas spilled forth so fast that, like Gauss, he seems not to have had the time to go back over his discoveries and polish them. The thesis permitted Poincaré to give a course in analysis at the Faculté des Sciences at Caen, and he was officially released from his duties as a mine inspector on 1 December 1879. Poincaré thus set off on a path leading from the École Polytechnique and the École des Mines to a university teaching career that had been worn by some of the professors Poincaré most admired, including Camille Jordan and Alfred Cornu.

25.1.1 A prize competition

1880 was a busy time for him. On 22 March, he deposited his essay "Mémoire sur les courbes définies par une équation differentielle" with the Académie des Sciences as his entry in their current prize competition. That essay considered first-order non-linear differential equations of the form $\frac{dx}{X(x,y)} = \frac{dy}{Y(x,y)}$, where X and Y are real polynomial functions of real variables x and y, and investigated the global properties of their solutions. He later withdrew the essay, on 14 June 1880, without the examiners reporting on it, perhaps wanting to concentrate on the theory of complex differential equations. He had by then submitted an essay on that theme, on 29 May 1880. Like the doctoral thesis,

---

[1] See Paul Appell's biography (*Henri Poincaré*) [3, p. 28], who also suggested that visionary reasoning may have been a cause, and Bellivier [12], who noted in his biography that Poincaré lost crucial points in topography, drawing and architecture.

this essay was also only to be published in the first volume of his *Oeuvres* [199, pp. 578–613].

The prize competition asked for any essay improving some point in the theory of differential equations. It had been posed by Charles Hermite, who was also chairman of the panel of judges. Hermite was the most influential French mathematician of his generation, alongside Bertrand. Bertrand occupied more prestigious positions, but Hermite's research carried greater weight, and between them they could more or less decide who was to get the call to Paris and who was to languish in the provinces. Prize competitions run by the Académie des Sciences in Paris were one way in which Hermite exerted his influence. It was the custom throughout the 19th century for the Académie to announce various prizes in mathematics. Typically, a title would be announced, with a panel of judges, and a cut-off date some two years hence. The system was difficult to work. There were occasions when no entry was thought worthy. Then the essay might be re-announced, or perhaps the prize would go to someone for their work, whether or not it fitted the title. Sometimes the questions would be devised with a likely winner in mind, as was the case when Kovalevskaya won the Prix Bordin (see Cooke [42]).[2]

Despite adding three lengthy supplements to his essay, each marking significant progress, Poincaré did not win the competition, and was only awarded second prize, behind Halphen. (The supplements themselves were lost for a long time and only recently republished, in Poincaré [201]). Of more significance for us, however, is the way in which Poincaré came to appreciate and harness non-Euclidean geometry in order to understand the problem he had chosen.

25.1.2 Poincaré's discovery of non-Euclidean geometry

Poincaré himself has left us one of the most justly celebrated accounts of the process of mathematical discovery, which concerns exactly this point. Poincaré gave this account in a lecture he gave to the Société de Psychologie in Paris in 1908, and it was later published in 1909 as the third essay in his volume *Science et méthode* [198].

Poincaré began by doubting that a certain type of function (which he later called Fuchsian functions) could exist, but shortly came to the opposite view. He tells us in the lecture that:

> For two weeks I tried to prove that no function could exist analogous to those I have since called the Fuchsian functions: I was then totally

---

[2] On mathematical prizes in general, see Gray [104] on the Clay prize.

ignorant. Every day I sat down at my desk and spent an hour or two there: I tried a great number of combinations and never arrived at any result. One evening I took a cup of coffee, contrary to my habit; I could not get to sleep, the ideas surged up in a crowd, I felt them bump against one another, until two of them hooked onto one another, as one might say, to form a stable combination. In the morning I had established the existence of a class of Fuchsian functions those which are derived from the hypergeometric series. I had only to write up the results, which just took me a few hours. [198, p. 50]

Somewhat later, he saw in a moment of intense insight how non-Euclidean geometry could illuminate the study of these functions. His problem had involved him in studying how triangles with circular-arc sides might fit together. To simplify matters, he had found a way (to be discussed below) of straightening out their sides, but he was still confined to a few simple cases.

At that moment I left Caen where I then lived, to take part in a geological expedition organised by the École des Mines. The circumstances of the journey made me forget my mathematical work; arrived at Coutances we boarded an omnibus for I don't know what journey. At the moment when I put my foot on the step the idea came to me, without anything in my previous thoughts having prepared me for it; that the transformations I had made use of to define the Fuchsian functions were identical with those of non-Euclidean geometry. I did not verify this, I did not have the time for it, since scarcely had I sat down in the bus than I resumed the conversation already begun, but I was entirely certain at once. On returning to Caen I verified the result at leisure to salve my conscience. [198, pp. 51, 52]

The first supplement sheds considerable light on Poincaré's grasp of Fuchsian functions and non-Euclidean geometry at this time (it was received by the Académie on 28 June 1880). Poincaré observed that these transformations form a group [201, p. 9], and remarked:

There are close connections with the above considerations and the non-Euclidean geometry of Lobachevskii. In fact, what is a geometry? It is the study of a group of operations formed by the displacements one can apply to a figure without deforming it. In Euclidean geometry the group reduces to rotations and translations. In the pseudogeometry of Lobachevskii it is more complicated.

To study the group [involved in this problem] is therefore to do the geometry of Lobachevskii. Pseudogeometry will consequently provide us with a convenient language for expressing what we will have to say about this group.

Poincaré's realisation on boarding the bus at Coutances can be described very simply. He realised that the straightened versions of the figures described at the end of his prize essay were identical with the figures in Beltrami's description of non-Euclidean geometry; that therefore the original figures were conformally accurate representations of non-Euclidean figures; and finally that this meant the transformations he was using were non-Euclidean isometries. Beltrami's detailed discussion of the non-Euclidean differential geometry of the disc enabled Poincaré to give a new meaning to his previously analytical transformations.

Poincaré's insight was still confined to triangles. Liberation came from an unexpected source, arithmetic.

> I then undertook to study some arithmetical questions without any great result appearing and without expecting that this could have the least connection with my previous researches. Disgusted with my lack of success, I went to spend some days at the sea-side and thought of quite different things. One day, walking along the cliff, the idea came to me, always with the same characteristics of brevity, suddenness, and immediate certainty, that the arithmetical transformations of ternary indefinite quadratic forms were identical with those of non-Euclidean geometry.
>
> Once back at Caen I reflected on this result and drew consequences from it; the example of quadratic forms showed me that there were Fuchsian groups other than those which correspond to the hypergeometric series. [198, pp. 52, 53]

This shows that the model of non-Euclidean geometry on one hyperboloid of a two-sheeted hyperboloid (defined by the ternary indefinite quadratic form $x^2 + y^2 - z^2 = -1$) allowed Poincaré to deal with polygons of any number of sides.

25.1.3 The Poincaré and Beltrami discs

It's an open question how Poincaré first heard of non-Euclidean geometry.[3] Hoüel's translations of Lobachevskii's *Geometrische Untersuchungen* and Beltrami's *Saggio* and *Teoria* had come out in 1866 and 1869, and Poincaré could have picked up the idea from staff or students at the École Polytechnique when he was a student there from 1873 to 1875. Paul Tannery, a friend of Hoüel, published articles about it in the *Revue philosophique* in 1876 and 1877,

---

[3] I thank Scott Walter for this information.

and a French translation of Helmholtz's "Über den Ursprung und die Bedeutung der geometrischen Axiome" was also published in the *Revue scientifique*
in 1877. But if we must remain uncertain how Poincaré ever came to learn non-
Euclidean geometry in the first place – and a connection to Klein's work seems
unlikely in view of the letters the two men were later to exchange – the mathematical connection between the Beltrami model of non-Euclidean geometry in
the *Saggio* and Poincaré's new model in which triangles have circular-arc sides
is easy to describe as Poincaré saw it.[4] Place the Beltrami model on the equatorial disc of a sphere, and project the disc vertically down onto the southern
hemisphere (see Figure 25.2).

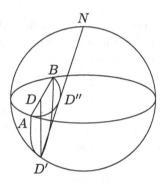

Figure 25.2 Poincaré's projections

A line in the disc is mapped onto a semicircular arc hanging vertically
down from the equator, which it meets at right angles. Now project the southern hemisphere stereographically (from the north pole) back onto the equatorial
plane. The semicircular arc is projected onto the arc of a circle, or a straight line,
(because stereographic projection sends circles to circles or straight lines) meeting the equator at right angles (because stereographic projection is conformal).
So the image of a non-Euclidean geodesic is a straight line or an arc of a circle
in the disc, meeting the boundary circle at right angles.

Now Poincaré already knew that the transformations that moved triangles around in the disc were conformal (they came from a problem in complex
function theory), so he now knew that they were conformal transformations in
non-Euclidean geometry. But non-Euclidean triangles cannot be similar without being congruent, and so Poincaré deduced that that means these conformal
maps are isometries. So he was able to use geometric arguments from non-
Euclidean geometry in order to prove results about problems in the theory of

---

[4] Strictly speaking, this is Poincaré's later account, found in his *Oeuvres* [199, vol. 5,
p. 8], and in translation in Stillwell [232, p. 121]. His first account, given in 1881,
like Beltrami's account in his *Teoria* [14], was rooted more strongly in formulae.

differential equations. He may not have won the prize of the Académie des Sciences, but in numerous short papers and five very long papers in the opening issues of a new mathematical journal, *Acta Mathematica*, Poincaré put together a remarkable new blend of group theory, non-Euclidean geometry and complex function theory. The result not only established him as the leading mathematician of his day in France, it was the first success for non-Euclidean geometry, and made mathematicians look more favourably upon it as a topic that was not only exciting in its own right but capable of being applied in other parts of mathematics.

25.2 Poincaré and Klein

Almost as soon as Poincaré began to publish, he was drawn into a correspondence with Felix Klein, who had begun to come to some of the same ideas but was not sure how to proceed with them. The correspondence was later published, and makes fascinating reading.[5] Poincaré seems to have read much less mathematics than Klein, who was something of a scholar about mathematical literature, but to have been more adept at making up new mathematics. In one letter, Klein observed that all the figures they drew were made up of circular arcs perpendicular to a fixed circle, and that there are many figures made up of circular arcs that are not perpendicular to a fixed circle. Poincaré replied very quickly that such a figure can be thought of as if it was cut out on the plane by a set of hemispheres, so it is a picture in three-dimensional non-Euclidean geometry! (Can you see how?)

Klein later said that the struggle to keep up with Poincaré cost him his health. He suffered a nervous collapse in autumn 1882 from which he never fully recovered, and switched over by the 1890s to managing other people's research projects. Once he arrived in Göttingen he set about building up the mathematics department there, throwing himself into fund raising and channelling the money into mathematics education, applied mathematics, engineering and technology, while making sure that pure mathematics was in the hands of the best mathematicians he could hire. .

To end with a remark and a figure by way of a footnote: the conformal model of non-Euclidean geometry that Poincaré discovered is mathematically exactly what Riemann described. But there is no reason to believe Poincaré knew this, and even if he had read Riemann's *Habilitationsschrift*, he would

---

[5] In, for example, Poincaré *Oeuvres* [199, vol. 11, pp. 13–25] and Klein *Abhandlungen* [137, vol. 3, pp. 587–621]. There is a discussion in Gray [101].

have had to unpack all the mathematics, because Riemann, as you will recall, gave only the formula for the metric but no description of the picture it
implies.

Figure 25.3 shows one of the most famous in non-Euclidean geometry: the
tessellation of the non-Euclidean disc by triangles, all of whose angles are
$\pi/2, \pi/3$ or $\pi/7$ – note that this means the triangles are all (non-Euclidean)
congruent. It was used by Klein in his work before he realised that it was a
picture in non-Euclidean geometry.

Figure 25.3 The tessellation $(2, 3, 7)$

25.3 Circumcircles

At this point it is interesting to take up a result that Wolfgang Bolyai proved
in the *Tentamen*, and that Lobachevskii also discussed in his *Geometrische
Untersuchungen* of 1840. This is the striking result that, in the presence of the
other axioms of Euclid's *Elements* the parallel postulate is equivalent to the
result that through any three points not on a line there passes a circle.

How could this be false? In the Poincaré disc model, any three points lie on
either a straight line or a (Euclidean) circle. But if this straight line does not
pass through the centre of the disc, then it does not form what is called a *d*-line,

and yet the three points on it cannot lie on a circle. More generally, if the three points lie on an arc of a circle that touches the boundary circle (a horocycle) or that crosses the circle, then they cannot lie on a non-Euclidean circle.

Now the Euclidean argument that shows that three non-collinear points have a circumcircle, and locates its centre by taking the perpendicular bisector of two of the sides of the triangle formed by the points and showing that they meet. What Lobachevskii showed in [153, §29 and §30] is how such perpendicular bisectors can fail to meet: they may be parallel, or divergent. It is a useful exercise to draw all these different cases in the Poincaré disc.

25.4 Inversion and the Poincaré disc

The Poincaré disc model becomes much more useful and convincing mathematically when we understand how to introduce transformations of it that are angle and distance preserving. To do this, we introduce the concept of inversion in a circle.

We have seen that the Poincaré disc model of non-Euclidean geometry consists of a space – the points inside the unit disc – upon which a metric is defined. With respect to this metric, geodesics are either diameters of the unit disc or else arcs of (Euclidean) circles perpendicular to the unit circle. Angles are represented conformally in this disc model. In this section, the mathematical details are presented via a set of exercises which are intended to show that Poincaré's claims are entirely justified, but they do not recapitulate his own mathematical analysis. For the details of that, the reader is referred to Poincaré [200] for an English translation of almost all of Poincaré's original papers with a modern commentary, to Poincaré [201] for his first accounts and to Gray [101] for a further discussion. A diligent reader will discover that there are no significant differences in the mathematical accounts.

The first eight exercises present the definition and the major properties of the geometric transformation called inversion, and show how an inversion can be described algebraically. In particular, they establish that inversions are angle preserving (up to sign, like reflections). Exercises 25.9–25.14 direct attention to the inversions that map the unit disc to itself, and show that they permit a definition of distance between two points in the disc. This distance is not altered by an inversion. Exercises 25.15–25.18 present an argument leading to a formula for the distance between any point and the centre of the disc. Exercise 25.19 defends the idea that with respect to this metric the geodesics are indeed the diameters of the unit circle and the arcs of circles perpendicular to the unit circle. Exercise 25.20 solves the problem of finding all the curves equidistant from a geodesic in the non-Euclidean disc, and Exercise 25.21 shows how to find the circumference of a non-Euclidean circle of given radius. It is instructive to compare these results with their counterparts in spherical geometry.

25.4.1 Inversion

Inversion in the circle C centre O and radius r is a map of the plane, with the point O deleted, to itself, which is defined by the following rule: $P \mapsto P'$, where O, P and P' lie on the same half-line out of O and $OP.OP' = r^2$.

Exercise 25.1

Show that:

(a) points on the circle are mapped to themselves;

(b) points inside the circle are mapped to points outside the circle and points outside to points inside;

(c) if P maps to P' then P' maps to P (so inversion is its own inverse).

Exercise 25.2

(a) Show that if the point O is taken to be the zero of the complex plane, then inversion is the map $z \mapsto \frac{r^2}{\bar{z}} = \frac{r^2}{z\bar{z}} z$.

(b) More generally, show that if the point O is taken to be the point α of the complex plane, then inversion is the map $z \mapsto \frac{r^2}{\bar{z} - \bar{\alpha}} + \alpha$. (Hint: the map is a sequence: translate α to the origin, perform the inversion in a circle centre the origin and radius r, translate the origin back to α.) It is helpful to rewrite this as $z \mapsto \frac{\alpha \bar{z} - c}{\bar{z} - \bar{\alpha}}$.

Exercise 25.3

(a) Use the fact that inversion fixes this circle pointwise to deduce that the equation of the circle centre α and radius r is $z = \frac{r^2 - \alpha\bar{\alpha} + \alpha\bar{z}}{\bar{z} - \bar{\alpha}}$, which simplifies to $z\bar{z} - \bar{\alpha}z - \alpha\bar{z} + c = 0$, where $c = \alpha\bar{\alpha} - r^2$.

(b) Show that this is the same equation as $x^2 + y^2 - 2ax - 2by + c = 0$, where $\alpha = a + ib$.

It is sometimes convenient to speak of the circle (α, c), which has radius r given by $r^2 = \alpha\bar{\alpha} - c$.

We shall be interested in the effect of inversion on lines and circles and on angles between lines and circles. Without loss of generality, we may assume we are inverting in the unit circle, $z\bar{z} = 1$, so inversion is given by $z \mapsto 1/\bar{z}$.

Exercise 25.4

Consider the locus defined by $kz\bar{z} - \bar{\alpha}z - \alpha\bar{z} + c = 0$, where $k = 0$ or 1, which defines a circle when $k = 1$ and a straight line when $k = 0$. Show that under inversion, this transforms to the locus $k\frac{1}{\bar{z}}\frac{1}{z} - \bar{\alpha}\frac{1}{\bar{z}} - \alpha\frac{1}{z} + c = 0$, which simplifies to $k - \bar{\alpha}z - \alpha\bar{z} + cz\bar{z} = 0$. This yields four cases:

(1) $k = 1, c \neq 0$: a circle not through the origin goes to a circle not through the origin;

(2) $k = 1, c = 0$: a circle through the origin goes to a straight line not through the origin;

(3) $k = 0, c \neq 0$: a straight line not through the origin maps to a circle through the origin;

(4) $k = 0, c = 0$: a straight line through the origin maps to itself.

In each case, these statements need to be modified to take note of the fact that the origin has been deleted – we shall henceforth assume that this has been done.

Notice that in case (1), which is the case of most interest, we may write the transformed equation as $\frac{1}{c} - \frac{\bar{\alpha}}{c}z - \frac{\alpha}{c}\bar{z} + z\bar{z} = 0$. This makes it clear that the image circle has centre $\frac{\alpha}{c}$ and radius $\frac{r}{c}$.

Exercise 25.5

Show that, however, the image of the centre of the original circle does not go to the centre of the transformed circle.

Exercise 25.6

Find the angle between two circles by applying the cosine rule to the triangle formed by their centres and the relevant point of intersection. If the circles are (α, c) and (α', c') with radii r and r' respectively, show that the distance between their centres is $|\alpha - \alpha'|^2$ and that the cosine of the angle between their radii is $\frac{\alpha\bar{\alpha}' + \bar{\alpha}\alpha' - (c + c')}{2rr'} = \frac{2\Re(\alpha\bar{\alpha}') - (c + c')}{2rr'}$.

Exercise 25.7

(a) Deduce that the circles are perpendicular if and only if $\alpha\bar{\alpha}' + \bar{\alpha}\alpha' = c + c'$.

(b) Deduce that a circle (α', c') is perpendicular to the unit circle if and only if $c' = 1$.

Exercise 25.8

Consider the circles (α, c) and (α', c'), which we assume intersect, with radii r and r' respectively. Show that inversion is angle preserving (up to sign, so strictly speaking one should say angle reversing) by showing that inversion in the unit circle sends them to the circles $(\alpha/c, 1/c)$ and $(\alpha'/c', 1/c')$, and that their radii are likewise transformed to r/c and r'/c' respectively. Deduce that the angle between the transformed circles is the same as the angle between the original circles, and so inversion is angle preserving.

Exercise 25.9

Let C be a circle meeting the unit circle at right angles.

(a) Show that inversion in the unit circle maps C to itself by checking explicitly that a point z on C is mapped to a point w that satisfies the equation for C.

(b) We may establish the same result another way, as follows. The circle C' meets the unit circle at two points, P and Q say, that are mapped to themselves. The circle C' is mapped to a circle, and this circle also meets the unit circle at right angles (because C' does, and inversion is angle preserving). But there is only one circle that meets the unit circle at P and Q at right angles, and that is the circle C' itself, so C' is mapped to itself.

Exercise 25.10

Deduce that any map of the form $z \mapsto \frac{\alpha \bar{z}-1}{\bar{z}-\bar{\alpha}}$, for which $|\alpha| > 1$, maps the unit circle to itself, and the interior to the interior. What is the explanation of the condition on α?

Exercise 25.11

(a) Show that the composite of the map $z \mapsto \frac{\alpha\bar{z}-1}{\bar{z}-\bar{\alpha}}$ followed by the map $z \mapsto \frac{\beta\bar{z}-1}{\bar{z}-\bar{\beta}}$ is $z \mapsto \frac{(\bar{\alpha}\beta-1)z+\alpha-\beta}{(\bar{\alpha}-\bar{\beta})z+\alpha\bar{\beta}-1}$. This can be written as $z \mapsto \frac{\gamma z+\delta}{\bar{\delta}z-\bar{\gamma}}$ where the determinant $\gamma\bar{\gamma} - \delta\bar{\delta}$ is non-zero.

(b) Show that when this is done $\left|\frac{\delta}{\gamma}\right| < 1$.

Exercise 25.12

(a) Show, conversely, that any map of that form is a composite of two inversions in circles, provided $\left|\frac{\delta}{\gamma}\right| < 1$ and $\gamma \neq 0$.

(b) Show also that maps of this form with $\gamma = 0$ map the unit circle to itself.

We now know that maps of the form $z \mapsto \frac{\gamma z+\delta}{\bar{\delta}z-\bar{\gamma}}$ with $\left|\frac{\delta}{\gamma}\right| < 1$ map the unit disc to itself. These maps of the unit disc to itself are easier to study than single inversions (because they do not involve $\bar{z}$) and attention will henceforth generally be confined to them.

Exercise 25.13

(a) Show that the map $z \mapsto \frac{\gamma z + \delta}{\bar{\delta} z - \bar{\gamma}}$, with non-zero determinant, maps the point $\frac{-\delta}{\gamma}$ to zero.

(b) Show that the map $z \mapsto \frac{\gamma z + \delta}{\bar{\delta} z - \bar{\gamma}}$, with non-zero determinant, fixes zero if $\delta = 0$ and is then of the form $z \mapsto (\gamma/\bar{\gamma}) z = e^{2i\theta} z$, where $\gamma = e^{i\theta}$. Such a map is a rotation of the unit disc.

Exercise 25.14

Let f and f' be two maps sending z_0 to zero, and consider $f' f^{-1}$.

(a) Show that it maps 0 to 0, and deduce that $f' f^{-1}(z) = e^{2i\theta} z$ for some θ.

(b) Deduce that for each $z_1 \neq z_0$ there is a unique $t \in \mathbb{R}$ with $t > 0$ and an f such that $f(z_0) = 0$ and $f(z_1) = t$, and moreover that any other f' with the properties that $f'(z_0) = 0$ and $f'(z_1) \in \mathbb{R}$ differs from f by at most a rotation, so $f'(z_1) = f(z_1)$.

It follows from Exercise 25.14 that t is determined by z_0 and z_1 alone and so it is independent of any choice of transformation of the unit disc to itself. It may therefore be taken as a measure of the separation between z_0 and z_1. The most useful measure of separation, one worthy of the name of distance, would be additive on segments, that is, given a segment of the real line from 0 to t in the disc and a segment from t to s in the disc, with $t < s < 1$, the length of the segment from 0 to t plus the length of the segment from t to s would be the length of the segment from 0 to s.

Exercise 25.15

Let the length of the segment from 0 to t be given by a function $\phi(t)$. Use the transformation $z \mapsto \frac{z - t}{tz - 1}$ that maps t to 0 and s to $\frac{s-t}{ts-1}$, to show that the additivity condition on the function ϕ is $\phi(t) + \phi(\frac{s-t}{ts-1}) = \phi(s)$.

To solve this equation, we first write $\frac{t-s}{1-st} = u$, so $t = \frac{u+s}{1+us}$, and the equation becomes $\phi(s) + \phi(u) = \phi(\frac{u+s}{1+us})$. It turns out that it is easier to solve if we introduce the inverse function $f = \phi^{-1}$. This function exists if, as we may surely assume, ϕ is a monotonic increasing function of t. If $s = f(x)$ and $u = f(w)$, the equation we have to solve becomes $f(x + w) = \frac{f(x) + f(w)}{1 + f(x)f(w)}$.

Exercise 25.16

Verify that $f(x) = k\,\tanh(x)$ satisfies the equation $f(x+w) = \frac{f(x)+f(w)}{1+f(x)f(w)}$.

Exercise 25.17

To solve the equation in Exercise 25.16, assume that the function f is differentiable.

(a) Show that $f(x + w) - f(x) = f(w)\frac{1-f(x)^2}{1+f(x)f(w)}$.

(b) Divide both sides by w and let w tend to 0, and hence show that this expression yields $\frac{df}{dx} = \frac{df}{dx}(0)\cdot\left(1 - f(x)^2\right)$ (on the reasonable assumption that $f(0) = 0$).

Exercise 25.18

Set $\frac{df}{dx}(0) = k$, write $y = f(x)$ and the equation becomes the differential equation $\frac{dy}{dx} = k(1-y^2)$. Show that this equation has the unique solution with $y(0) = 0$, that is $y = \tanh(kx)$. In terms of our original problem, this says that the additive function of the separation between 0 and t is $\frac{1}{k}\tanh^{-1}(t)$. Without loss of generality, we may take $k = 1$.

We now have a good measure of distance in the disc, and we may now say that a map of the unit disc to itself of the form $z \mapsto \frac{\gamma z+\delta}{\delta z-\bar{\gamma}}$ is an isometry of the disc. It is also clear, since rotations are isometries, that for each non-zero z_0 in the disc the set of points of the form $z_0 e^{2i\theta}$ is a circle with respect to this sense of distance. Notice that it looks exactly like a Euclidean circle. Our next task is to find the geodesics with respect to this definition of distance.

Exercise 25.19

Show that the geodesics through the point 0 are precisely the diameters of the unit disc. For this to be so, it is enough to show that the positive real axis is part of a geodesic (because the other lines through 0 differ from this one only by a rotation, and a rotation is an isometry). But this is trivial, for if it were not, let the geodesic from 0 to t say pass through a point P not on the real axis. The disc centre 0 and passing through P and the disc centre t passing through P both contain a common segment on the real axis, so the length from 0 to P plus the length from P to t is greater than the length along the real axis from 0 to t.

It remains to give the general formula for the distance. Notice first that the usual formula for the metric on a surface is a quadratic form:

$$ds^2 = Edx^2 + 2Fdx\,dy + Gdy^2 = (dx, dy)^T Q(dx, dy),$$

where Q is the 2 by 2 matrix $\begin{pmatrix} E & F \\ F & G \end{pmatrix}$. It follows that a rotation, say

$$A = \begin{pmatrix} \cos\theta & -\sin\theta \\ \sin\theta & \cos\theta \end{pmatrix},$$

around the origin maps (dx, dy) to $A(dx, dy)$ and so the quadratic form becomes

$$\begin{pmatrix} \cos\theta & \sin\theta \\ -\sin\theta & \cos\theta \end{pmatrix} \begin{pmatrix} E & F \\ F & G \end{pmatrix} \begin{pmatrix} \cos\theta & -\sin\theta \\ \sin\theta & \cos\theta \end{pmatrix}.$$

For the rotation to be an isometry, we have

$$\begin{pmatrix} \cos\theta & \sin\theta \\ -\sin\theta & \cos\theta \end{pmatrix} \begin{pmatrix} E & F \\ F & G \end{pmatrix} \begin{pmatrix} \cos\theta & -\sin\theta \\ \sin\theta & \cos\theta \end{pmatrix} = \begin{pmatrix} E & F \\ F & G \end{pmatrix}.$$

After a little work, this equation is seen to be true if and only if

$$\begin{pmatrix} \cos\theta & \sin\theta \\ -\sin\theta & \cos\theta \end{pmatrix} \begin{pmatrix} E & F \\ F & G \end{pmatrix} = \begin{pmatrix} E & F \\ F & G \end{pmatrix} \begin{pmatrix} \cos\theta & \sin\theta \\ -\sin\theta & \cos\theta \end{pmatrix},$$

from which it follows that $E = G$ and $F = 0$. This supplies the value of the functions E, F and G at the origin.

To find explicit expressions for these functions at ζ, we consider the transformation sending ζ to the origin: $z' = \frac{z-\zeta}{1-\bar\zeta z}$. This implies that

$$dz' = \frac{(1 - \bar\zeta z)dz + (z - \zeta)\bar\zeta dz}{(1 - \bar\zeta z)^2} = \frac{(1 - \zeta\bar\zeta)dz}{(1 - \bar\zeta z)^2}.$$

So at $z = \zeta$, we have $dz' = \frac{dz}{1-\zeta\bar\zeta}$. This formula implies that $dx' = \frac{dx}{1-\zeta\bar\zeta}$ and $dy' = \frac{dy}{1-\zeta\bar\zeta}$, because $1 - \zeta\bar\zeta$ is real, and so $E(\zeta) = \frac{E(0)}{1-\zeta\bar\zeta}$, $F(\zeta) = \frac{F(0)}{1-\zeta\bar\zeta} = 0$ and $G(\zeta) = \frac{G(0)}{1-\zeta\bar\zeta}$. The result is that the metric in the disc for which the transformations we have been studying are isometries is given by

$$ds^2 = k^2 \frac{dx^2 + dy^2}{(1 - (x^2 + y^2))^2}.$$

This expression for a metric is easy to use to find the corresponding curvature. The formula for curvature, K, is to be found in any good differential geometry book near you, for example, Dubrovin et al. [57, vol. 1, p. 114]. Given a metric $ds^2 = g(u,v)(du^2 + dv^2)$, it helps to write this as $ds^2 = g(z, \bar z)\, dz\, d\bar z$. The formula for the curvature is then $K = -\frac{2}{g} \cdot \frac{\partial^2}{\partial z \partial \bar z}(\ln g)$. The metric we have just found can be written with $g(z, \bar z) = k^2 \frac{dz d\bar z}{(1 - z\bar z)^2}$. Routine differentiation now shows that in this case $K = -\frac{2}{g} \cdot \frac{\partial^2}{\partial z \partial \bar z} \ln g = -\frac{4}{k^2}$. So if we choose $R = 1$ and, for the constant, $k = 2$, the metric we have introduced into the disc has constant negative curvature -1.

We can also find the equation of a curve everywhere equidistant from a geodesic. We take the geodesic with equation $y = 0$, which is mapped to itself by all the transformations of the form $z \mapsto \frac{z+t}{1+tz}$ for $|t| < 1$. We consider the images of the point $i\lambda$, which are $z = \frac{t+i\lambda}{1+i\lambda t}$.

Exercise 25.20

(a) Show that $z\bar{z} = \frac{t^2 + \lambda^2}{1 + \lambda^2 t^2}$.

(b) Show that $z - \bar{z} = \frac{2i\lambda(1-t^2)}{1+\lambda^2 t^2}$, and deduce that $z\bar{z} + ib(z - \bar{z}) = 1$ if and only if $b = \frac{\lambda^2 - 1}{2\lambda}$.

(c) Deduce that the curve everywhere equidistant from the geodesic $y = 0$ and passing through the point $i\lambda$ is the part of the circle with equation $z\bar{z} + i\frac{\lambda^2 - 1}{\lambda}(z - \bar{z}) - 1 = 0$ lying inside the unit disc.

So we see that arcs of circles (and also straight line segments) not perpendicular to the boundary circle also have a meaning in the Poincaré disc model: they are curves everywhere equidistant from the geodesic with the same endpoints on the boundary. The same result can be proved by noting that circles through the points 1 and -1 are perpendicular to the geodesics that are perpendicular to the line $y = 0$. The isometries of the disc that fix the points 1 and -1 are inversions in a geodesic perpendicular to the line $y = 0$, so they map each circular arc through 1 and -1 to itself.

We can also find the expression for the circumference of a non-Euclidean circle in the disc model. For convenience we choose the circle centre O and with Euclidean radius r.

Exercise 25.21

(a) Show that the formula for the metric (you may assume $k = 4$) can be written in polar coordinates as

$$ds^2 = \frac{k(dr^2 + r^2 d\theta^2)}{(1 - r^2)^2}.$$

(b) Deduce that the circumference of this circle is given by

$$\frac{kr}{1 - r^2} \int_0^{2\pi} d\theta = \frac{2\pi kr}{1 - r^2}.$$

(c) Use the fact that $r = \tanh \frac{\rho}{k}$, where ρ is the non-Euclidean radius of the circle, to deduce that the circumference is $k\pi \sinh 2\frac{\rho}{k}$, and check that this agrees closely with the Euclidean formula when ρ is very small.

25.5 References

As a way of showing how lively this branch of mathematics continues to be, I can do no better than to recommend one of the truly lovely books in mathematics, which will repay indefinite amounts of study: *Indra's pearls: The vision of Felix Klein* by David Mumford, Caroline Series and David Wright [173]. Wonderful pictures are combined with real explanations of otherwise difficult mathematics to mutual advantage.

Is the Geometry of Space Euclidean or Non-Euclidean?

26.1 How to decide?

From 1870 to 1914, public interest grew in the idea that space might not be Euclidean. This interest was inextricably linked with the idea that space might be four-dimensional, which was also mixed up with the idea that time could be considered as a dimension. All these ideas are separate, as matters of mathematical fact, but they attracted a tremendous amount of interest, not least because of the prospect that one could travel in the fourth dimension, or in time, in ways denied to us hitherto.

The most mundane of these ideas is that space might be non-Euclidean, but it was also seen as the one with the most chance of being true. We have seen already that professional astronomers such as Gauss and Bessel had no trouble accepting the idea, although they could see straight away that space could be only very slightly non-Euclidean. Once non-Euclidean geometry became accepted as a possibility among mathematicians it became more generally recognised that space might be described better by non-Euclidean geometry than by Euclidean geometry, for a value of the constant that crops up in the formulae of non-Euclidean geometry that could be determined by experiment and observation.

I am not aware of a serious attempt (involving equipment, experimental protocol, etc.) to conduct such an experiment (a research trawl through the journals of the day needs to be done) but it would not surprise me if efforts in this direction were "armchair" because the actual practical consequences are so slight. Nonetheless, the view from the armchair needs to be set out.

J. Gray, *Worlds Out of Nothing*,
Springer Undergraduate Mathematics Series,
DOI 10.1007/978-0-85729-060-1_26, © Springer-Verlag London Limited 2011

It seems hopeless to choose a plane, a line and a point in that plane, and to look at all the lines through that point to see if they meet the given line. So one focuses on corollaries on the assumption that geometry might be non-Euclidean. Of these, the angle sum of a triangle suggests itself. If it can be established that the angle sum of a triangle is always less than π, then the geometry of space is non-Euclidean geometry, and this conclusion follows from sufficiently accurate measurements of just one triangle. In an era before space flight, all triangles would necessarily be terrestrial, and therefore small, so the measurement of the angles would have to be very precise.

A second method would be the failure to draw squares: one draws a segment of a fixed length, turns through a right angle, draws the segment again, turns through a right angle, draws the segment again, turns through a right angle, draws the segment again, and compares the starting and finishing points. They agree in plane Euclidean geometry, but not in non-Euclidean geometry (nor on the sphere), and the amount by which they disagree is proportional to the curvature.

A third method, astronomical in nature, would be the one suggested by Lobachevskii. It requires a truly infinite universe, which we can see all of, and a sufficient degree of certainty that one has looked at big enough triangles.

All such attempts were called into question in a series of papers by Poincaré around 1900, and gave rise to a philosophy of geometry, and more generally of science, called conventionalism. This philosophy continues to be of interest to the present day.

26.2 Poincaré's conventionalism

By 1890, Poincaré was the leading mathematician of his day. He divided his future career between mathematics and physics, being the leading expert of complex function theory, differential equations, non-Euclidean geometry, dynamical systems and topology (all subjects he did much to reformulate and even create), as well as becoming the French expert on electromagnetic theory and anticipating Einstein with his theory of the moving electron (another story, again!). He was also a lucid writer, and wrote a large number of popular essays on the implications of contemporary science. He was the natural person to turn to elucidate the question: is space Euclidean or non-Euclidean?

As the extracts below make clear, Poincaré's surprising answer was that while non-Euclidean geometry made sense, there was no way of telling if space was Euclidean or non-Euclidean. Any experiment would involve an interpretation. One could always say that light rays (or whatever played the role of straight lines in the experiment) were indeed straight, and so space was non-Euclidean (if that is what the measurements seemed to indicate). Or, one could say that space was Euclidean, and that rays of light were curved. There was no possibility, however, of deciding between these alternatives on logical grounds. So the only way forward was an arbitrary choice based on human convenience. There would be a collective agreement or convention, but nothing was forced.

In his essay "Space and geometry" (reprinted in *Science and hypothesis*) [196, pp. 51–71] Poincaré invited his readers to consider a uniform sphere of radius R in which the temperature at all points r from the centre is proportional to $R^2 - r^2$. He showed that sentient beings in the sphere (made of the same uniform material) would find the natural geometry of their world was one that looked to us exactly like non-Euclidean geometry. Geodesics would be arcs of circles at right angles to the boundary sphere, and so on. But who is to say we are right and they are wrong, that they "really" live in Euclidean space but have a heating problem? What would we say to ourselves if we really did find it looked as if the geometry of physical lines in space was non-Euclidean?

Naturally, Poincaré expected the conventional choice to be that space was Euclidean – but not because it was a vector space but because, in his opinion, Euclidean geometry was more or less hard wired into our brains!

26.2.1 Enriques disputes

Poincaré's philosophical claims were not always accepted. They were disputed head-on by the Italian mathematician Federigo Enriques. By 1907 Enriques was the Italian mathematician of stature who was involved in issues of mathematics education and philosophy. His major work in this connection is the *Problemi della scienza* of 1906 which, after some delays, was translated into English as *Problems of science* and published in 1914 [67].

Enriques objected to Poincaré's account on the grounds that it denies us any way of saying that the changes in length of rulers in the plane are due to changes in temperature. Our experience of heat is that it is a localised phenomenon, and moreover one to which different bodies respond differently. In the Poincaré model, we cannot say these things that characterise what we say about heat, so, said Enriques, heat is not playing the role of a physical concept but a geometrical one. The same is true of light rays, which demonstrably depart from straightness in inhomogeneous mediums. Therefore, Enriques concluded, "in this other world, geometry would be really and not merely apparently different from ours" [67, p. 178]. To think otherwise is to make the contrast between appearance and reality into something transcendental.

If in some strange way large regions of space were to appear to us to be more and more non-Euclidean, as measured by the behaviour of light rays, then

Enriques would have argued as follows. Either we discover some way in which the light rays are being pulled out of their predicted path, by some process we can at least quantify and perhaps even control, or we do not, after exhaustive searching. On the first alternative, a new physical process has been discovered, and our idea of geometry is left unchanged. On the second alternative, we would have to say that geometry was to be altered.

Enriques cheerfully and deliberately walked into a number of deep philosophical issues in his book *Problems of science*. Against what he called the nominalism maintained by Poincaré, which says that there is nothing involved in talking about space other than talk about bodies in space, Enriques argued that geometry is a part of physics [67, p. 174]. Spatial relations, he said, have actual physical significance and therefore, he concluded, give us knowledge of reality. Poincaré, on the other hand, as Enriques reminded his readers with several quotations from *La science et l'hypothèse* [196], had stopped at the experience of the mutual relation of bodies, thus being a nominalist in Enriques' sense of the term, and Enriques regarded this as the antithesis of realism. Enriques believed that claims about space were claims about all the real (and presumably possible) relations between bodies, and, it is clear, among these claims Enriques would have put the homogeneity of space.

For Enriques, claims to knowledge about space – many such claims, anyway – reduce to claims about measurements, for example the isosceles theorem [67, p. 182]. Following Klein's lectures Enriques argued that the real meaning of the isosceles theorem is that if the side lengths differ by less than a given amount, ε, then the base angles differ by less than an amount depending on that ε in a specified way. In this way the theorems of geometry can be turned into statements about bodies. Another example was how claims about the angle sums of non-Euclidean triangles could be tested astronomically and shown to depend, within stated limits of accuracy, upon infeasibly large regions of space [67, p. 192].

Geometry is also a theory, an organised body of knowledge, and mathematicians had at their disposal at least two geometrical theories of space (Euclidean and non-Euclidean geometry). Enriques' opinion of axioms in geometry and the hypothetico-deductive side of the subject was possibly unexpected. He disagreed strongly with Sartorius's remark (*Gauss zum Gedächtnis* [217, p. 81]) that "Gauss regarded geometry as a logical structure, only in case the theory of parallels was conceded as an axiom".[1] On the contrary, said Enriques, none of the postulates of geometry had the character of a logical axiom, and all the definitions of the fundamental entities of geometry were logically defective but

---

[1] Enriques may have been the victim of a mistranslation. The German word rendered here as "logical" is "consequent" which means something more like "coherent" than "logical" in the modern sense of the terms.

instead made assumptions about reality. In Enriques' opinion, geometry was not a matter of writing down some axioms (plausible or not, but in any case mutually consistent) and then reasoning purely logically. Geometrical deductions were concealed talk about physically possible systems. Different systems of postulates, forming various geometries, he said, "express different physical hypotheses" [67, p. 197], although he allowed that mathematicians could construct geometric concepts with no close link with sense data. To explain how strikingly abstract concepts can have a certain objectivity, Enriques asserted that these abstract concepts give "a possible representation of reality" [67, p. 191].

Unlike a number of his Italian contemporaries, or even Hilbert, Enriques was no formalist. According to Enriques, because of the way knowledge is acquired, all knowledge is subject to a continual process of revision, so that what may be "known" at one time may be found to be false later on. Unlike Popper later, he did not regard this fallibility as the characteristic feature of scientific knowledge, and indeed the example he had in mind was the exactitude of geometry, but he did write: "There is no reason however why it may not be disproved if untrue" [67, p. 184]. Empirical verification may be definitive if it comes up with a negative result (a counter-example), but is merely probable when it comes up with a positive (confirmatory) result [67, p. 155]. He distinguished between explicit and implicit hypotheses in a theory (the explicit hypotheses are those particular to the theory in question, the implicit hypotheses are those needed to connect the theory to the object of study, and may well be the explicit hypotheses of logically prior theories) and wrote:

> The progress of science is a process of successive approximations, in which new and more precise, more probable and more extended inductions result from partially verified deductions, and from those contradictions that correct the implicit hypotheses.
>
> In this process certain primary and general concepts, such as those of geometry and mechanics, give us some guiding principles that are but slightly variable if not absolutely fixed. Therefore we should turn our attention to these concepts in order to explain their actual value and their psychological origin. [67, p. 166]

Concepts, he observed [67, p. 117], come in two kinds: those appropriate to a certain physical reality, and those which are not tied down in that way. The second kind, Enriques regarded as psychological. Logic he regarded as operating in a meaning-independent way on psychologised concepts. In keeping with his fallibilism, Enriques practised what has been called "meaning finitism": the idea that the meaning of a term or concept is established upon only finitely many instances and is therefore necessarily vague.

26.3 Poincaré on the subjective experience of a non-Euclidean space

From "Space and geometry", in *Science and hypothesis*

Suppose, for example, a world enclosed in a large sphere and subject to the following laws:– The temperature is not uniform; it is greatest at the centre, and gradually decreases as we move towards the circumference of the sphere, where it is absolute zero. The law of this temperature is as follows:– If R be the radius of the sphere, and r the distance of the point considered from the centre, the absolute temperature will be proportional to $R^2 - r^2$. Further, I shall suppose that in this world all bodies have the same coefficient of dilatation, so that the linear dilatation of any body is proportional to its absolute temperature. Finally, I shall assume that a body transported from one point to another of different temperature is instantaneously in thermal equilibrium with its new environment. There is nothing in these hypotheses either contradictory or unimaginable. A moving object will become smaller and smaller as it approaches the circumference of the sphere. Let us observe, in the first place, that although from the point of view of our ordinary geometry this world is finite, to its inhabitants it will appear infinite. As they approach the surface of the sphere they become colder, and at the same time smaller and smaller. The steps they take are therefore also smaller and smaller, so that they can never reach the boundary of the sphere. If to us geometry is only the study of the laws according to which invariable solids move, to these imaginary beings it will be the study of the laws of motion of solids deformed by the differences of temperature alluded to.

No doubt, in our world, natural solids also experience variations of form and volume due to differences of temperature. But in laying the foundations of geometry we neglect these variations; for besides being but small they are irregular, and consequently appear to us to be accidental. In our hypothetical world this will no longer be the case, the variations will obey very simple and regular laws. On the other hand, the different solid parts of which the bodies of these inhabitants are composed will undergo the same variations of form and volume. [196, pp. 65–66]

Poincaré then described how light would behave in such a space, and how sentient beings in the space would build up their sense of geometry exactly as we do. He then concluded as follows.

If they construct a geometry, it will not be like ours, which is the study of the movements of our invariable solids; it will be the study of the changes of position which they will have thus distinguished, and will be "non-Euclidean displacements", and this will be non-Euclidean geometry. So that beings like ourselves, educated in such a world, will not have the same geometry as ours. [196, p. 68]

26.4 Poincaré's arguments

The following extracts are on experimental tests of space and geometry.

From "Experiment and geometry", in *Science and hypothesis*

1. I have on several occasions in the preceding pages tried to show how the principles of geometry are not experimental facts, and that in particular Euclid's postulate cannot be proved by experiment. However convincing the reasons already given may appear to me, I feel I must dwell upon them, because there is a profoundly false conception deeply rooted in many minds.

2. Think of a material circle, measure its radius and circumference, and see if the ratio of the two lengths is equal to 2π. What have we done? We have made an experiment on the properties of the matter with which this roundness has been realised, and of which the measure we used is made.

3. Geometry and Astronomy:– The same question may also be asked in another way. If Lobachevskii's geometry is true, the parallax of a very distant star will be finite. If Riemann's [the old, confusing name for spherical geometry] is true, it will be negative. These are the results which seem within the reach of experiment, and it is hoped that astronomical observations may enable us to decide between the two geometries. But what we call a straight line in astronomy is simply the path of a ray of light. If, therefore, we were to discover negative parallaxes, or to prove that all parallaxes are higher than a certain limit, we should have a choice between two conclusions: we could give up Euclidean geometry, or modify the laws of optics, and suppose that light is not rigorously propagated in a straight line. It is needless to add that every one would look upon this solution as the more advan-

tageous. Euclidean geometry, therefore, has nothing to fear from fresh experiments.

4. Can we maintain that certain phenomena which are possible in Euclidean space would be impossible in non-Euclidean space, so that experiment in establishing these phenomena would directly contradict the non-Euclidean hypothesis? I think that such a question cannot be seriously asked. To me it is exactly equivalent to the following, the absurdity of which is obvious:– There are lengths which can be expressed in metres and centimetres, but cannot be measured in toises [a measure of length in France, obsolete even in Poincaré's day], feet, and inches; so that experiment, by ascertaining the existence of these lengths, would directly contradict this hypothesis, that there are toises divided into six feet. Let us look at the question a little more closely. I assume that the straight line in Euclidean space possesses any two properties, which I shall call A and B; that in non-Euclidean space it still possesses the property A, but no longer possesses the property B; and, finally, I assume that in both Euclidean and non-Euclidean space the straight line is the only line that possesses the property A. If this were so, experiment would be able to decide between the hypotheses of Euclid and Lobachevskii. It would be found that some concrete object, upon which we can experiment – for example, a pencil of rays of light – possesses the property A. We should conclude that it is rectilinear, and we should then endeavour to find out if it does, or does not, possess the property B. But it is not so. There exists no property which can, like this property A, be an absolute criterion enabling us to recognise the straight line, and to distinguish it from every other line. Shall we say, for instance, "This property will be the following: the straight line is a line such that a figure of which this line is a part can move without the mutual distances of its points varying, and in such a way that all the points in this straight line remain fixed?" Now, this is a property which in either Euclidean or non-Euclidean space belongs to the straight line, and belongs to it alone. But how can we ascertain by experiment if it belongs to any particular concrete object? Distances must be measured, and how shall we know that any concrete magnitude which I have measured with my material instrument really represents the abstract distance? We have only removed the difficulty a little farther off. In reality, the property that I have just enunciated is not a property of the straight line alone; it is a property of the straight line and of distance. For it to serve as an absolute criterion, we must be able to show, not only that it does not also belong to any other line than the straight line and to distance, but also that it does not belong to

any other line than the straight line, and to any other magnitude than distance. Now, that is not true, and if we are not convinced by these considerations, I challenge any one to give me a concrete experiment which can be interpreted in the Euclidean system, and which cannot be interpreted in the system of Lobachevskii. As I am well aware that this challenge will never be accepted, I may conclude that no experiment will ever be in contradiction with Euclid's postulate; but, on the other hand, no experiment will ever be in contradiction with Lobachevskii's postulate. [196, pp. 72–75]

Summary: Geometry to 1900

Such have been the transformations in geometry between the 1840s, when we last took stock, and 1900, that it seems advisable to pass it briefly under review before moving to some final considerations. In Chapter 13, we looked at Möbius's algebraic version of projective geometry and his introduction of barycentric coordinates. All of projective geometry was thereby turned into algebra, including the line at infinity in the projective plane, and projective transformations.

Plücker resolved the duality controversy when he showed that either the original curve or its dual is singular (or both may be) and that singular points bring down the degree of the dual: by two for a double point and three for a cusp. Bitangents and inflectional tangents on a curve give rise to double points and cusps on the dual. The Plücker formulae enabled Plücker to make a profound study of cubic and quartic curves. The price to pay eventually – and after some delay – is that points on algebraic curves must be allowed to be complex, and complex curves are almost impossible to draw.

Riemann was interested in studying space in order to understand the transmission of forces at a distance, and he generalised this to the study of geometries in any number of dimensions. In his view, geometry was possible on any space where the methods of the calculus allowed one to define lengths, and he argued that the fundamental geometric properties of a figure are those that are intrinsic to it, rather than to some space in which the figure is found. He therefore distinguished intrinsic properties of a space (such as curvature, which he generalised from Gauss's two-dimensional setting to any number of dimensions) from those it acquired by being embedded in some other (ambient, or surrounding)

J. Gray, *Worlds Out of Nothing*,
Springer Undergraduate Mathematics Series,
DOI 10.1007/978-0-85729-060-1_27, © Springer-Verlag London Limited 2011

space, and argued that a space could have genuine geometric (i.e. intrinsic) properties even if it was not embedded in a Euclidean space. This destroyed the paramount status of Euclidean geometry and Euclidean space as the source of geometry.

Riemann only skirted the topic of non-Euclidean geometry, and seems not to have known much, if anything, of the work of Bolyai and Lobachevskii. But inspired by his ideas, the Italian geometer Beltrami showed how to change the formulae for geometry on a sphere to obtain formulae for a surface of constant negative curvature, which corresponded to drawing the surface on a disc of arbitrary radius in such a way that the non-Euclidean geometry geodesics appeared straight (compare the geodesic projection of a hemisphere onto a tangent plane). This provided the first conclusive proof that non-Euclidean geometry was indeed logically possible and free of contradiction. In 1880 Poincaré came up with his conformal model of non-Euclidean geometry in a disc, which makes its metrical character more evident.

Klein went to the University of Bonn when not yet 17 to study physics under Plücker. But Plücker had switched back to projective geometry, and so Klein began to study mathematics as well. With Plücker's unexpected death, Klein was pushed into writing up the older man's posthumous work, and in a career of amazing rapidity he soon became a professor at the small, sleepy University of Erlangen in 1872. There he published an address outlining the unification of all the new geometries: all were to be special cases of projective geometry. In particular, non-Euclidean geometry appeared as a special case of projective geometry confined to the interior of a conic. The Beltrami metric of non-Euclidean geometry is a geometric invariant because it is defined in terms of cross-ratio.

By the 1880s, projective geometry was generally regarded as the fundamental geometry, as numerous English writers and the successful book by Cremona testify [45]. It was the hegemonic geometry in Klein's vision of geometry, and was singled out by Pasch. Pasch's book [184] set out to show how, by carefully abstracting from experience various families of assumptions (about line segments, about whole lines, about plane regions, etc.), the whole of projective geometry can then be deduced logically from these axioms.

David Hilbert, the dominant mathematician from the 1890s to the 1920s (with Poincaré) embarked on a study of projective geometry in the 1890s which led him to the study of axiom systems per se. He showed, for example, that there was a plane projective geometry in which Desargues' theorem is false, although in any three-dimensional geometry it is an easy consequence of the axioms (later he replaced his argument with a much easier one, due to Moulton, see p. 264). His *Grundlagen der Geometrie* [118, 1st edn., 1899] not only axiomatised several geometries (including Euclidean geometry, which in later editions he referred to as Cartesian geometry; later editions axiomatised non-Euclidean geometry) and showed what theorems followed from what axioms, it suggested the value of thinking axiomatically in other domains of mathematics, including mathematical physics.

The comparison with the Italians (Peano, Pieri, Fano, Segre, Enriques) is interesting, because it shows that although these mathematicians investigated finite geometries, axiomatic structures and independence results, the more axiomatically minded were held back by the negative role of teaching, which confined their study of geometry to the needs of first-year students and led their research elsewhere.

27.1 References

There are very few source books with extensive material in English on geometry in the 19th and 20th centuries but happily there is Peter Pesic's *Beyond geometry: classic papers from Riemann to Einstein*, [188]. It carries papers by, among others, Riemann, Klein, Poincaré and Einstein.

What is Geometry? The Formal Side

We have seen a number of types of argument used in geometry: algebraic, differential geometric, projective and, especially, axiomatic arguments. An interesting paper published by the American historian and philosopher of mathematics Ernest Nagel, in 1939, made the provocative suggestion that the principle of duality in plane projective geometry was an important stimulus to thinking of geometry in a purely logical way, independent of any appeal to intuition [174].

28.1 Nagel's thesis

Nagel argued that if points and lines are to be treated on a par, then intuition is ultimately denied a role in geometry. We can accept intuitively that a line is made up of points, but not that the fundamental unit of the plane is the line. We can only accept that intellectually. But if the principle of duality is fundamental in projective geometry, which it is, then the intellect must be allowed a more fundamental role than intuition, and so, according to Nagel, logical reasoning about objects defined only abstractly took over from intuitive reasoning in geometry.

One might go further. Projective geometry allows one to take objects from one space and treat them as points in another. For example, a plane conic has an equation of the form

$$ax^2 + bxy + cy^2 + dx + ey + f = 0,$$

where not all of a, b, c, d, e, f may be zero. Moreover, the equations

$$kax^2 + kbxy + kcy^2 + kdx + key + kf = 0, \ k \neq 0,$$

define the same conic. So each conic may be represented by the six-tuple (a, b, c, d, e, f), that is to say, as a point in five-dimensional projective space. Morally (as mathematicians sometimes say when they are about to say something slightly false, which means in this case forgetting about the degenerate conics) a conic is a parabola if and only if $b^2 - 4ac = 0$. So the space of all parabolas in the plane (a possibly alarming thing to contemplate) becomes a harmless subspace of five-dimensional projective space.

Or, to take another example, the space of all lines in three-space becomes a quadric in five-dimensional projective space (but I won't describe how). The point is that the term "point of a projective space" has become something that one can reason about abstractly, and then interpret in any one of a number of ways. The correctness of the reasoning must be guaranteed in advance, by abstract methods, and then the interpretation is valid.

Nagel did not discuss non-Euclidean geometry, but the subject only reinforces the point. Since the two geometries are competing for the honour of being the true geometry of space, and because they conflict, it would seem that geometry must be developable as a coherent subject, independent of its physical meaning.

The example of Hilbert's *Grundlagen der Geometrie*, and of Italian work, shows that by 1900 there was a growing body of reasoning that was syntactic (driven by the grammar or rules of reasoning) rather than semantic (controlled by the meanings of the terms it employs). Such reasoning is (could be? should be?) formal or mechanical. It is, strictly speaking, meaningless. It does require that the axioms be self-consistent (a question that early writers ducked, and which is not as simple as perhaps they thought).

The example taken from Hilbert's *Grundlagen der Geometrie* shows what happens. It's rather dry, but in principle derived impeccably from the initial assumptions (in practice, Hilbert wasn't too precise, as the footnotes show).

However, this does suggest a way of formulating other questions, not necessarily in geometry, about objects which are not intuitive and which might be best defined implicitly by axioms. Such rules don't say what the objects are, they simply govern what you can say about them. Significant examples of this way of doing mathematics are groups (coming into mathematics in a big way by about 1900) and commutative algebra (promoted most successfully by Emmy Noether in the 1920s).

28.2 From Hilbert's *Grundlagen der Geometrie*

The following extract is from the 1st edition in an authorised translation by
E. J. Townsend [118, pp. 2–5].

THE FIVE GROUPS OF AXIOMS.
§1. THE ELEMENTS OF GEOMETRY AND THE FIVE GROUPS
OF AXIOMS.

Let us consider three distinct systems of things. The things composing
the first system, we will call *points* and designate them by the letters A,
$B, C, \ldots$; those of the second, we will call *straight lines* and designate
them by the letters $a, b, c, \ldots$; and those of the third system, we will call
planes and designate them by the Greek letters $\alpha, \beta, \gamma, \ldots$ The points
are called the *elements of linear geometry*; the points and straight lines,
the *elements of plane geometry*; and the points, lines, and planes, the
elements of the geometry of space or the *elements of space*.

We think of these points, straight lines, and planes as having certain mutual relations, which we indicate by means of such words as
"are situated," "between," "parallel," "congruent," "continuous," etc.
The complete and exact description of these relations follows as a consequence of the *axioms of geometry*. These axioms may be arranged
in five groups. Each of these groups expresses, by itself, certain related fundamental facts of our intuition. We will name these groups as
follows:

I,	1–7.	Axioms of *connection*.
II,	1–5.	Axioms of *order*.
III.		Axiom of *parallels* (Euclid's axiom).
IV,	1–6.	Axioms of *congruence*.
V.		Axiom of *continuity* (Archimedes's axiom).

§2. GROUP I. AXIOMS OF CONNECTION.

The axioms of this group establish a connection between the concepts
indicated above; namely, points, straight lines, and planes. These axioms are as follows:

I, 1. *Two distinct points A and B always completely determine a
straight line a. We write $AB = a$ or $BA = a$.*

Instead of "determine," we may also employ other forms of expression; for example, we may say A "lies upon" a, A "is a point of" a,
a "goes through" A "and through" B, a "joins" A "and" or "with"
B, etc. If A lies upon a and at the same time upon another straight

line b, we make use also of the expression: "The straight lines" a "and" b "have the point A in common," etc.

I, 2. *Any two distinct points of a straight line completely determine that line; that is, if $AB = a$ and $AC = a$, where $B \neq C$, then is also $BC = a$.*

I, 3. *Three points A, B, C not situated in the same straight line always completely determine a plane α. We write $ABC = \alpha$.*

We employ also the expressions: A, B, C, "lie in" α; A, B, C "are points of" α, etc.

I, 4. *Any three points A, B, C of a plane α, which do not lie in the same straight line, completely determine that plane.*

I, 5. *If two points A, B of a straight line a lie in a plane α, then every point of a lies in α.*

In this case we say: "The straight line a lies in the plane α," etc.

I, 6. *If two planes α, β have a point A in common, then they have at least a second point B in common.*

I, 7. *Upon every straight line there exist at least two points, in every plane at least three points not lying in the same straight line, and in space there exist at least four points not lying in a plane.*

Axioms I, 1–2 contain statements concerning points and straight lines only; that is, concerning the elements of plane geometry. We will call them, therefore, the *plane axioms of group I*, in order to distinguish them from the axioms I, 3–7, which we will designate briefly as the *space axioms* of this group.

Of the theorems which follow from axioms I, 3–7, we shall mention only the following.

> THEOREM 1. Two straight lines of a plane have either one point or no point in common; two planes have no point in common or a straight line in common; a plane and a straight line not lying in it have no point or one point in common.

> THEOREM 2. Through a straight line and a point not lying in it, or through two distinct straight lines having a common point, one and only one plane may be made to pass.

§3. GROUP II. AXIOMS OF ORDER.[1]

The axioms of this group define the idea expressed by the word "between," and make possible, upon the basis of this idea, an *order of sequence* of the points upon a straight line, in a plane, and in space.

---

[1] Hilbert added a footnote: These axioms were first studied in detail by M. Pasch in [184]. Axiom II, 5 is in particular due to him.

The points of a straight line have a certain relation to one another which the word "between" serves to describe. The axioms of this group are as follows:

II, 1. *If A, B, C are points of a straight line and B lies between A and C, then B lies also between C and A.*

II, 2. *If A and C are two points of a straight line, then there exists at least one point B lying between A and C and at least one point D so situated that C lies between A and D.*

II, 3. *Of any three points situated on a straight line, there is always one and only one which lies between the other two.*

Later editions carried this proof of Axiom II, 3, which is due to A. Wald.

Let A not lie between B and C and let also C not lie between A and B. Join a point D that does not lie on the line AC with B and choose by Axiom II, 2 a point G on the connecting line such that D lies between B and G. By an application of Axiom II, 4 [below] to the triangle BCG and to the line AD it follows that the lines AD and CG intersect at a point E that lies between C and G. In the same way, it follows that the lines CD and AG meet at a point F that lies between A and G. If Axiom II, 4 is applied now to the triangle AEG and to the line CF it becomes evident that D lies between A and E, and by an application of the same axiom to the triangle AEC and to the line BG one realises that B lies between A and C.

We return to the original edition.

II, 4. *Any four points A, B, C, D of a straight line can always be so arranged that B shall lie between A and C and also between A and D, and, furthermore, that C shall lie between A and D and also between B and D.*

This axiom was recognised by E. H. Moore [170], to be a consequence of the plane axioms of incidence and order formulated above. Compare also the works subsequent to this by Veblen [241] and Schweitzer [221]. A thorough investigation of independent sets of axioms of order that postulate ordering on straight lines is found in E. V. Huntington [124].

Here is the proof. It is taken from the second English translation of Hilbert's *Foundations of geometry* [119] and uses a result called Theorem 4 in later editions of Hilbert's book, which says that of any three points A, B, C on a line, there is always one that lies between the other two. Let A, B, C, D be four points on a line g. The following will now be shown:

1. If B lies on the segment AC and C lies on the segment BD then the points B and C also lie on the segment AD. By Axioms I, 3 and II, 2 choose a point E that does not lie on g, and a point F such that E lies between C and F. By repeated applications of Axioms II, 3 and II, 4 it follows that the segments AE and BF meet at a point G, and moreover, that the line CF meets the segment GD at a point H. Since H thus lies on the segment GD and since, however, by Axiom II, 3, E does not lie on the segment AG, the line EH, by Axiom II, 4, meets the segment AD, i.e. C lies on the segment AD. In exactly the same way one shows analogously that B also lies on this segment.

2. If B lies on the segment AC and C lies on the segment AD then C also lies on the segment BD and B also lies on the segment AD. Choose one point G that does not lie on g, and another point F such that G lies on the segment BF. By Axioms I, 2 and II, 3 the line CF meets neither the segment AB nor the segment BG and hence, by Axiom II, 4 again, does not meet the segment AG. But since C lies on the segment AD, the straight line CF meets then the segment GD at a point H. Now by Axiom II, 3 and II, 4, again the line FH meets the segment BD. Hence C lies on the segment BD. The rest of assertion 2 thus follows from 1.

Now let any four points on a line be given. Take three of the points and label Q the one which by Theorem 4 and Axiom I, 3 lies between the other two and label the other two P and R. Finally label S the last of the four points. By Axiom I, 3 and Theorem 4 again it follows then that the following five distinct possibilities for the position of S exist:

1. R lies between P and S;

2. P lies between R and S;

3. S lies between P and R simultaneously when Q lies between P and S;

4. S lies between P and Q;

5. P lies between Q and S.

The first four possibilities satisfy the hypotheses of assertion 2 and the last one satisfies those of assertion 1. Wald's result is thus proved. We now return to Hilbert's original account.

DEFINITION. We will call the system of two points A and B, lying upon a straight line, a *segment* and denote it by AB or BA. The points lying between A and B are called the *points of the segment* AB or the *points lying within the segment* AB. All other points of the straight line

are referred to as the *points lying outside the segment AB*. The points
A and B are called the *extremities* of the segment AB.

II, 5. *Let A, B, C be three points not lying in the same straight line and
let a be a straight line lying in the plane ABC and not passing through
any of the points A, B, C. Then, if the straight line a passes through a
point of the segment AB, it will also pass through either a point of the
segment BC or a point of the segment AC.*

Axioms II, 1–4 contain statements concerning the points of a straight
line only, and, hence, we will call them the *linear axioms of group II*.
Axiom II, 5 relates to the elements of plane geometry and, consequently,
shall be called the *plane axiom of group II*.

§4. CONSEQUENCES OF THE AXIOMS OF CONNECTION AND ORDER.

By the aid of the four linear Axioms II, 1–4, we can easily deduce the
following theorem:

THEOREM 3. Between any two points of a straight line, there always
exists an unlimited number of points.

29
What is Geometry? The Physical Side

Although we shall not discuss the point in any detail, you might ask if non-Euclidean geometry passes or fails this test: can there be a mechanics in such a space? Can one set up such things as an inverse square law of gravity, or conservation of energy and momentum? Could one recover, doubtless in an altered form, everything that physicists had discovered about Euclidean space if it turned out that space obeyed non-Euclidean geometry? The answer is, as you might suspect, yes, and this was shown by Lipschitz, a follower of Riemann, in the 1870s.

29.1 Geometry and physics

What is the connection between geometry and physics? The standard answer, from the time of Descartes and Newton, was that geometry was neutral. It was simply the arena in which events take place. Space itself neither helps nor hinders this process, and it doesn't enter the explanation in any way. Descartes had imagined that space was full of swirling vortices of a novel kind of matter that pushed the planets round in their orbits (thus explaining why they all go round in the same direction). Newton had discredited that idea, and replaced it with an unexplained idea of gravity, a force somehow acting at a distance (a very considerable distance). Space (God's sensorium, Newton called it) remained a neutral place, although it now housed a novel, and highly mysterious entity, the force of gravity. With the remarkable string of predictive successes that the

J. Gray, *Worlds Out of Nothing*,
Springer Undergraduate Mathematics Series,
DOI 10.1007/978-0-85729-060-1_29, © Springer-Verlag London Limited 2011

theory of Newtonian gravity accumulated throughout the 18th century, people grew less suspicious of it and became reconciled to forces acting at a distance. Even when the forces of magnetism and electricity were discovered, the mode of action was not always thought to involve the nature of space, simply to happen in space.

The question becomes obscure with the transmission of electromagnetic forces. Once light was taken to be a wave phenomenon, after the work of Fresnel and Young at the start of the 19th century, this was often supposed to happen through an aether (waves, presumably, are waves in something). One could take the view that space was neutral, but filled with this aether, or that space was the aether. The failure of aether theories is well known, and might as well allow us not to enter this debate, but there is another reason not to go this far. One can ask: does 19th-century electromagnetic theory assume that lengths and angles are affected by the aether (or, if you prefer, that the measurement of lengths and angles is affected by the aether)? For as long as the answer is no, we may say that these physical theories are not geometric.

29.2 Einstein

Everything changes, of course, with the theories of special and general relativity, where the measurement of, and one might say the nature of, space and time are directly at stake. Debates about the possibly non-Euclidean nature of space may have helped open the way to relativity theory. One of the authors Einstein read carefully when he was at the University of Zürich was Poincaré, and his popular essays were part of Einstein's mental background. Moreover, because the popular impact of the very idea of non-Euclidean geometry was so great around 1900, the public was in some sense prepared for Einstein's very different ideas when they came along after the First World War.

29.2.1 The special theory of relativity

Let us look first at Einstein's special theory of relativity, introduced in 1905. This is a physical theory to do with measurement. It proposes nothing less than a radical revision of our ideas about time and space. Einstein's purpose was to explain away certain problems that had come up with objects, such as electrons, that move at very nearly the speed of light. To explain how clocks at distinct points A and B may tell the same time, he said that one must first synchronise the clocks. To do this, one sends a light signal from A to B

and back. One must assume that the time the light takes to travel from A to B is the same as the time it takes to go from B to A. On this assumption, if t_A and t'_A are times measured at A, and t_B a time measured at B, then for synchronised clocks one has the equation $t_B - t_A = t'_A - t_B$. The speed of light, c, is a constant in this theory (a fact for which there was almost too much experimental evidence) and indeed $c = \frac{2AB}{t'_A - t_A}$. In this way, one establishes a network of synchronised clocks, all at rest with respect to one another.

Einstein now imagined a moving rod passing, with velocity v, through a network of stationary clocks. The rod is travelling along the direction it points in, and it has clocks, also called A and B, one at each end, and which are synchronised in the opinion of observers on the rod and at rest with respect to it. For these clocks

$$t_B - t_A = \frac{r_{AB}}{c - v} \quad \text{and} \quad t'_A - t_B = \frac{r_{AB}}{c + v},$$

where r_{AB} is the length of the moving rod as measured in the stationary frame. It is clear that stationary observers must say that the clocks on the moving rod are not synchronised. What then can observers with such clocks say about what they measure? Since there is no way in physics of distinguishing between two observers in constant relative motion, the two systems of measurement must be equivalent – this is the basis of Einstein's equivalence principle (and the reason that the theory is called a theory of relativity).

Einstein let the observer on the rod use coordinates ξ, η, ζ and τ. He asserted that the transformation between the coordinate systems must be linear because space and time are homogeneous; it would have been simpler to say that observers in uniform relative motion agree on when an object is moving along a straight line. Then, by a rather clumsy argument, he deduced that, if x and ξ are coordinates measured along the rod and its direction of motion,

$$\tau = \beta\phi(v)(t - vx/c^2) \quad \text{and} \quad \xi = \beta\phi(v)(x - vt),$$

where $\beta = \frac{1}{\sqrt{(1 - v^2/c^2)}}$ and ϕ was some as yet undetermined function of v.

He then showed that the principle of relativity and the constancy of the speed of light were compatible, which they are if $\phi(v) = 1$ for all values of v. He thus obtained these formulae, since very familiar:

$$\tau = \beta(t - vx/c^2) \quad \text{and} \quad \xi = \beta(x - vt),$$

where $\beta = \frac{1}{\sqrt{(1 - v^2/c^2)}}$.

The physical meaning of these formulae is that observers in a state of constant relative motion disagree about their measurements of lengths and times. These formulae are best understood by first re-deriving them in a simpler way.

Suppose a rod AB passes through, travelling in the x-direction as far as we are concerned. The end A moves along the line ℓ_A, the end B along the line ℓ_B.

These lines are straight because the rod is in uniform motion with respect to us. Light emitted from A as it passes through the origin of our coordinate system travels along the straight line c_+ in the opinion of ourselves and the observers on the rod. It meets B at P, where it is reflected and travels back to A. The moving observer says that the light travelled a distance c arriving therefore at B, when it is at the point Q and has coordinates $(c, 1)$. It then returns to A, arriving there when it has coordinates $(0, 2)$. What do we (the static observers) say?

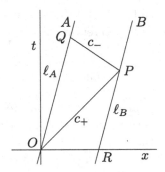

Figure 29.1   Changing coordinates

We work out the equation of the outgoing light, marked c_+ in Figure 29.1, find the coordinates of P in our frame, work out the equation of the returning light, marked c_- in the diagram, and find the coordinates of Q in our frame. Leaving these as exercises, which involve nothing more than the fact that lines ℓ_A and ℓ_B have equations $x = vt$ and $x = vt + \lambda$ for some as yet unknown λ, and that the equation of a ray of light is always of the form $x = \pm ct + m$ for a constant m that depends on where the light passes through, here are the conclusions: we say

P has coordinates $\left(\frac{c\lambda}{c-v}, \frac{\lambda}{c-v}\right)$;

Q has coordinates $\left(\frac{2c\lambda v}{c^2-v^2}, \frac{2c\lambda}{c^2-v^2}\right)$.

Now consider the points which, abusing the notation only slightly, may be called $\frac{1}{2}Q$ and $\frac{1}{c}(P - \frac{1}{2}Q)$. The moving observer says they have coordinates $(0, 1)$ and $(1, 0)$ respectively – which is why those points were chosen. By simple linear algebra, we say those points have coordinates $\left(\frac{cv\lambda}{c^2-v^2}, \frac{c\lambda}{c^2-v^2}\right)$ and $\left(\frac{c\lambda}{c^2-v^2}, \frac{\lambda v/c}{c^2-v^2}\right)$. So we can write down the matrix converting all the moving observer's coordinates to coordinates in our frame, and it is $\begin{pmatrix} \frac{c\lambda}{c^2-v^2} & \frac{cv\lambda}{c^2-v^2} \\ \frac{\lambda v/c}{c^2-v^2} & \frac{c\lambda}{c^2-v^2} \end{pmatrix}$. This matrix has determinant $\frac{(c^2-v^2)\lambda^2}{(c^2-v^2)^2}$.

Consider now what the moving observer thinks of us. This observer thinks he or she is at rest and we are moving, but all the above calculations are calculations this observer would do, except that the velocity is reversed. The corresponding matrix is $\begin{pmatrix} \frac{c\lambda}{c^2-v^2} & \frac{-cv\lambda}{c^2-v^2} \\ \frac{-v\lambda/c}{c^2-v^2} & \frac{c\lambda}{c^2-v^2} \end{pmatrix}$. This matrix must be the inverse of the one found by us, and for this to be the case the determinant of the original matrix, which is $\frac{(c^2-v^2)\lambda^2}{(c^2-v^2)^2}$, must be equal to 1. We deduce that $\lambda^2 = c^2 - v^2$, so $\lambda = \sqrt{c^2 - v^2}$, and the matrix takes the familiar form $\begin{pmatrix} \frac{c}{\sqrt{c^2-v^2}} & \frac{cv}{\sqrt{c^2-v^2}} \\ \frac{v/c}{\sqrt{c^2-v^2}} & \frac{c}{\sqrt{c^2-v^2}} \end{pmatrix}$.

Notice that as $c \to \infty$ the matrix tends to $\begin{pmatrix} 1 & v \\ 0 & 1 \end{pmatrix}$, which is as it should (you might like to check this). This limiting process is sometimes called passing to the Newtonian limit; it expresses what physics would be like if the speed of light really were infinite.

Now we can see that we think the length of the moving rod, which is OR in Figure 29.1, is $\lambda = \sqrt{c^2 - v^2}$, which compares with the length of the rod as determined by the moving observer, which is c. The length has been shrunk in the ratio $c : \sqrt{c^2 - v^2}$. It has been contracted. For the moving observer and ourselves to agree that light has the same speed c for each of us, this implies that the observer's time intervals also appear shorter to us, their time is said to be dilated.

29.2.2 The paradoxes of special relativity

Length contraction is a paradoxical phenomenon. A super-train rushes towards a small gap in the rails at almost the speed of light. The driver is not bothered, because the gap in the rails appears massively contracted to him. But those standing by the gap see a massively contracted train, much smaller than the gap, rushing very fast into the gap, and predict disaster. What does happen?

Time contraction is worse. Jill leaves Jo behind on a high-speed flight to a nearby star, and returns in the same fashion. Each calculates that the other has hardly aged – what happens when they meet? In fact, time contraction is confirmed every day in the behaviour of cosmic rays, which would otherwise not have enough time to reach the ground in the numbers that they do, and even by high-speed satellites (note that that calculation involves accelerating reference frames – it's much harder).

29.3 Minkowski

Einstein's arguments were entirely about physical quantities, and indeed at that stage in his career he had a typical physicist's disdain for mathematics (it's what you do when you have to, but the truth is elsewhere). In 1908 Hermann Minkowski, the distinguished mathematician and close friend of David Hilbert, put Einstein's ideas into geometric dress, essentially by going over Einstein's ideas in a geometric fashion. His address to the meeting of German scientists in Cologne in September 1908 begins with these famous words: "The views of space and time which I wish to lay before you have sprung from the soil of experimental physics, and therein lies their strength. They are radical. Henceforth space by itself, and time by itself, are doomed to fade away into mere shadows, and only a kind of union of the two will preserve an independent reality."[1] Einstein was not at first impressed. It seemed to him to be extra mathematics to no extra purpose in physics. He was shortly to change his mind, as he tried to incorporate gravity into his relativistic scheme.

29.4 Einstein, gravity and the rotating disc

After 1905, when Einstein had published his theory of special relativity, he turned his mind to the nature of gravitation. In 1909 he wrote to Sommerfeld, a former assistant of Klein and now a professor of physics in Munich, to say that he had looked at the case of the geometry on a rotating disc. Special relativity theory predicts that if an object is moving with respect to another, it will appear contracted along its direction of motion. So to an observer at the centre of the disc, a metre stick lying out towards the boundary will appear contracted. This contraction will be greatest along the direction at right angles to the radius joining the observer to an end of the stick. The situation is very similar to the cooled sphere that Poincaré had described (and Einstein had studied Poincaré's popular essays with great care). If the disc is such that its outer edge is rotating at the speed of light, Einstein showed (not in the extract below, alas) that the observer at the centre will indeed think that the geometry on it is a non-Euclidean geometry (but not, note, the non-Euclidean geometry of Bolyai and Lobachevskii, simply a geometry different from Euclid's).

Einstein first wrote about the rotating disc in two papers published in 1912 [60, 61]. These papers are discussed in an interesting paper by John Stachel, who remarks that they strike a tentative note, suggesting that even

---

[1] Minkowski [163, pp. 75–88], English translation in [59].

Einstein found the subject matter difficult.[2] The problem that bothered Einstein on and off for many years was the relationship between coordinates on the one hand and measurements with rods and clocks on the other. Einstein then returned to the matter much more confidently in his book of 1916 [62], which is quoted from below, by which time his theory of general relativity was much more fully developed – and published. Stachel [226] argues that in 1912 Einstein was able to deal with a gravitational field that is constant everywhere and for all time by allowing the speed of light to vary. He then moved on to stationary gravitational fields: ones that vary from point to point but not over time. The simplest example is the rotating disc. It was enough to alert him to the idea that coordinates have no simple meaning in such contexts, and to apply his famous equivalence principle only infinitesimally (as we would say, on the tangent space). This probably suggested to him that he should think of coordinates as Gauss had suggested. We know that Einstein had learned this material from Geiser when a student at the Eidgenössische Technische Hochschule in Zürich in 1905, that he regarded the course as a "masterpiece of the pedagogical art", and that he found the material useful when creating the general theory of relativity (a long, complicated process that took him three years). We also know that in 1912 Einstein knew only Gauss's work, but not that of Riemann (and still less that of Ricci or Levi-Civita). But he had come to appreciate Minkowski's four-dimensional geometric account of special relativity, and saw, in 1912, that he needed a four-dimensional generalisation of Gauss's ideas. This he obtained with the help of his friend Marcel Grossmann. As Stachel notes, if this account of Einstein's route to his famous discovery is broadly correct, then so too is Einstein's own theoretical account in his book *Relativity* of 1916 broadly fair to the historical development.

What Einstein described in the second extract below is the idea of gravity as a non-uniform force, varying from point to point, but affecting all bodies equally, that changes lengths and so is inevitably geometric. Thus Einstein's theory of general relativity is a geometric theory of physics.

In Chapter 26 we considered Poincaré's ideas about space, and his conventionalist position that one cannot say where geometry ends and physics begins. This idea merits further consideration. His idea was that the length of an infinitesimal ruler would seem to us (outside the sphere, or, if you prefer, disc, of radius R) to be given by $ds = \frac{dr}{R^2 - r^2}$. This integrates (use the substitution $r = R \tanh \rho$) to give $s = \frac{\rho}{R} = \frac{1}{R} \tanh^{-1} \frac{r}{R}$, confirming that this is indeed non-Euclidean geometry. But what are we to make of the idea that the universe cools everything as it moves outwards? This is to be a truly universal force, applicable to every body alike, acting instantaneously, over which we have no control, which we cannot mimic, which we do not find has "hot spots" and "cold

---

[2] See Stachel [226], where he discusses two papers by Einstein [60, 61].

spots". It is unlike the other forces of nature, because it cannot be manipulated by us in the way that we can create electromagnetic or gravitational fields. For these reasons, many people have banished it from physics, even as a hypothesis. The alternatives are Euclidean geometry for space with this "force", or non-Euclidean geometry for space and no "force", which resolve themselves into a simple choice: non-Euclidean geometry and no "force".

Einstein's rotating disc is another thought experiment. He never seriously maintained that space might be rotating. But when he published the equations that describe how space is curved by matter and therefore gravitation exerts its effect, it was noted that for two-dimensional space there is a solution which resembles a paraboloid. It is called Flamm's paraboloid after its discoverer. The paraboloid is everywhere negatively curved, but it gets less curved as one moves outwards. In the same way, the gravitational effect of a mass diminishes with distance. Flamm's paraboloid shows that our part of space may indeed be taken to be non-Euclidean, but not exactly as Bolyai might have thought. Instead it is markedly non-Euclidean near the sun, and less so as one moves outwards. But it is, in any case, not Euclidean.

29.5 From Einstein's *Relativity: The special and general theory* [62]

We start off again from quite special cases, which we have frequently used before. Let us consider a space–time domain in which no gravitational field exists relative to a reference-body K whose state of motion has been suitably chosen. K is then a Galilean reference-body as regards the domain considered, and the results of the special theory of relativity hold relative to K. Let us suppose the same domain referred to a second body of reference K', which is rotating uniformly with respect to K. In order to fix our ideas, we shall imagine K' to be in the form of a plane circular disc, which rotates uniformly in its own plane about its centre. An observer who is sitting eccentrically on the disc K' is sensible of a force which acts outwards in a radial direction, and which would be interpreted as an effect of inertia (centrifugal force) by an observer who was at rest with respect to the original reference-body K. But the observer on the disc may regard his disc as a reference-body which is "at rest"; on the basis of the general principle of relativity he is justified in doing this. The force acting on himself, and in fact on all other bodies which are at rest relative to the disc, he regards as the effect of a gravitational field. Nevertheless, the space-distribution of this gravitational field is of a kind that would not be possible in Newton's theory of gravitation. (The field disappears at the centre of the disc and increases proportionally to the distance from the centre as we proceed outwards.) But since the observer believes in the general theory of relativity, this does not disturb him, he is quite in the right when he believes that a general law of gravitation can be formulated by a law which not only explains the motion of the stars correctly, but also the field of force experienced by himself.

The observer performs experiments on his circular disc with clocks and measuring-rods. In doing so, it is his intention to arrive at exact definitions for the signification of time- and space-data with reference to the circular disc K', these definitions being based on his observations. What will be his experience in this enterprise?

To start with, he places one of two identically constructed clocks at the centre of the circular disc, and the other on the edge of the disc, so that they are at rest relative to it. We now ask ourselves whether both clocks go at the same rate from the standpoint of the non-rotating Galilean reference-body K. As judged from this body, the clock at the centre of the disc has no velocity, whereas the clock at the edge of the disc is in motion relative to K in consequence of the rotation. According to a result obtained in Section XII, it follows that the latter clock goes at a rate permanently slower than that of the clock at the centre of the circular disc, i.e. as observed from K. It is obvious that the same effect would be noted by an observer whom we will imagine sitting alongside his clock at the centre of the circular disc. Thus on our circular disc, or, to make the case more general, in every gravitational field, a clock will go more quickly or less quickly, according to the position in which the clock is situated (at rest). For this reason it is not possible to obtain a reasonable definition of time with the aid of clocks which are arranged at rest with respect to the body of reference. A similar difficulty presents itself when we attempt to apply our earlier definition of simultaneity in such a case, but I do not wish to go any farther into this question.

Moreover, at this stage the definition of the space co-ordinates also presents insurmountable difficulties. If the observer applies his standard measuring-rod (a rod which is short as compared with the radius of the disc) tangentially to the edge of the disc, then, as judged from the Galilean system, the length of this rod will be less than 1, since, according to Section XII, moving bodies suffer a shortening in the direction of the motion. On the other hand, the measuring-rod will not experience a shortening in length, as judged from K, if it is applied to the disc in the direction of the radius. If, then, the observer first measures the circumference of the disc with his measuring-rod and then the diameter of the disc, on dividing the one by the other, he will not obtain as quotient the familiar number $\pi = 3.14\ldots$, but a larger number (throughout this consideration we have to use the Galilean (non-rotating) system K as reference-body, since we may only assume the validity of the results of the special theory of relativity relative to K (relative to K' a gravitational field prevailed)) whereas of course, for a disc which is at rest with respect to K, this operation would yield exactly. This proves that the propositions of Euclidean geometry cannot hold exactly on the rotating disc, nor in general in a gravitational field, at least if we attribute the length 1 to the rod in all positions and in every orientation. Hence the idea of a straight line also loses its meaning. We are therefore not in a position to define exactly the co-ordinates x, y, z relative to the disc by means of the method used in discussing the special theory, and as long as the co-ordinates and times of events have not been defined, we cannot assign an exact meaning to the natural laws in which these occur. [62, ch. XXIII]

In Chapter XXIV Einstein explained how to lay out a Cartesian grid of little squares on a surface (imagined as a marble table), and went on:

If everything has really gone smoothly, then I say that the points of the marble slab constitute a Euclidean continuum with respect to the little rod, which has been used as a "distance" (line-interval). By choosing one corner of a square as "origin", I can characterise every other corner of a square with reference to this origin by means of two numbers. I only need state how many rods I must pass over when, starting from the origin, I proceed towards the "right" and then "upwards", in order to arrive at the corner of the square under consideration. These two numbers are then the "Cartesian co-ordinates" of this corner with reference to the "Cartesian co-ordinate system" which is determined by the arrangement of little rods.

By making use of the following modification of this abstract experiment, we recognise that there must also be cases in which the experiment would be unsuccessful. We shall suppose that the rods "expand" by an amount proportional to the increase of temperature. We heat the central part of the marble slab, but not the periphery, in which case two of our little rods can still be brought into coincidence at every position on the table. But our construction of squares must necessarily come into disorder during the heating, because the little rods on the central region of the table expand, whereas those on the outer part do not.

With reference to our little rods – defined as unit lengths – the marble slab is no longer a Euclidean continuum, and we are also no longer in the position of defining Cartesian co-ordinates directly with their aid, since the above construction can no longer be carried out. But since there are other things which are not influenced in a similar manner to the little rods (or perhaps not at all) by the temperature of the table, it is possible quite naturally to maintain the point of view that the marble slab is a "Euclidean continuum". This can be done in a satisfactory manner by making a more subtle stipulation about the measurement or the comparison of lengths.

But if rods of every kind (i.e. of every material) were to behave *in the same way* as regards the influence of temperature when they are on the variably heated marble slab, and if we had no other means of detecting the effect of temperature than the geometrical behaviour of our rods in experiments analogous to the one described above, then our best plan would be to assign the distance *one* to two points on the slab, provided that the ends of one of our rods could be made to coincide with these two points; for how else should we define the distance without our proceeding being in the highest measure grossly arbitrary? The method of Cartesian co-ordinates must then be discarded, and replaced by another which does not assume the validity of Euclidean geometry for rigid bodies. The reader will notice that the situation depicted here corresponds to the one brought about by the general postulate of relativity (Section XXIII). [62, ch. XXIV]

Einstein added a footnote at the end of Section XXIV, in which he discussed the geometry on the surface of an ellipsoid, which is, of course, not Euclidean. Then he commented: "Gauss indicated the principles according to which we can treat the geometrical relationships in the surface, and thus pointed out the way to the method of Riemann of treating multi-dimensional, non-Euclidean *continua*. Thus it is that the mathematicians long ago solved the formal problems to which we are led by the general postulate of relativity."

30

What is Geometry? Is it True? Why is it Important?

30.1 Truth

It is oddly hard to think of true statements once you allow doubt to play a role. What could another person say to you that you would absolutely have to believe? Statements people make about themselves are surely not acceptable (any follower of detective fiction will know that that is hopeless). The weather? The date? Are you sure you haven't been hoodwinked, kidnapped, drugged? If we decide to cut these speculations short of paranoia and allow a reasonable degree of interaction with the world in good faith, we enter an old and vexed distinction between certain and merely probable knowledge. Merely probable knowledge was once known as scientia, the word from which science derives, and it had an aspect of unreliability about it. Today, statements of science are among the most highly regarded (and may well figure in any answer to the opening question).

Without venturing too far into the philosophy of science, it is worth noting that there are problems here. The truths of science are typically reached after a number of inferences, they are generalisations (hydrogen and oxygen make water, objects attract under an inverse square law), they are not observation statements. You can't just go into a laboratory, let alone a kitchen, and check them. More worryingly, they are not always true, and indeed a number of scientific advances depend on reversing some beliefs that were previously held to be true (anyone for unbreakable atoms?).

J. Gray, *Worlds Out of Nothing*,
Springer Undergraduate Mathematics Series,
DOI 10.1007/978-0-85729-060-1_30, © Springer-Verlag London Limited 2011

30.1.1 Mathematical truths

The question for us is where to put the theorems of mathematics. Traditionally, and that means since Greek times, the truths of mathematics were absolute and certain, everything that scientia wasn't. They were deduced impeccably from indubitable premises. There were always difficulties in maintaining this view (proofs in algebra are not couched in the mode of Euclidean geometry, the arguments in Euclid's *Elements* are not in Aristotelian syllogistic form, let us not even mention the foundations of the calculus from its inception) but there was also a consensus. If it was imperilled at all, it was by the slightly doubtful status of the parallel postulate whence the attention the parallel postulate drew down the ages. In this context, the thought of non-Euclidean geometry arrives with destructive force.

Gauss knew this very well. If, as he said, geometry is not to be ranked with arithmetic but with mechanics, this means that unlike arithmetic, which is certain, geometry requires some empirical element. It is as true as the rest of science, perhaps it is the truest part, but the total certainty that pure logic conveys is to be denied to geometry. Bolyai and Lobachevskii were both more and less explicit: non-Euclidean geometry would have to be confirmed or refuted by actual measurements of actual space, but they made scientific statements of that kind, rather than philosophical ones.

Thereafter, mathematicians tended to confine themselves to such remarks, but the philosophical position is patent. Once there are two geometries with equal logical status but incompatible theorems, it is an empirical matter to decide which is true, and geometry can no longer be certain knowledge. Geometry, and with it a large amount of mathematics, ceases to be true. It can no longer be an absolute standard, a piece of perfection, although it may be better on some scale of things. Arithmetic would presumably remain as the paradigm.

Suppose that there was precisely a simple dichotomy: Euclidean or non-Euclidean geometry. Suppose that one turned out to be true (to correspond exactly to the way the world is, whatever that might mean). What then would be the status of the other one? It could only be a logically well-ordered system of statements that didn't apply to the real world. It lacks objects to which the words refer, because the terms straight line and the like only have meaning in the true geometry. If it is well organised, it must be because of the quality of the arguments, not because of our mental grasp of the meanings of the terms. Unless and until some other interpretation can be found, it is a skeleton, interesting or irrelevant according to taste.

There was, of course, another geometry of that kind: projective geometry. I do not know of anyone who claimed that parallel lines meet at infinity in anything like the sense that it could be claimed that such and such a speck of

light is actually from a distant galaxy. In principle, one can go to the galaxy, but
not to infinity. Projective geometry was an intellectual abstraction, valid in its
own way, more fundamental on some interpretations than Euclidean geometry,
but an abstraction nonetheless. So there are geometries which are impeccable
logically, but without empirical content. (To be sure, real existing geometries
were imperfect, as Pasch and others commented, but the task of putting them
right was not hopeless, even if it might be daunting.)

The Euclidean–non-Euclidean dichotomy is worth thinking about a little
more. There is only one point of disagreement, and it comes early on: the parallel postulate. Thereafter the subjects diverge. Before the break come all the
absolute theorems, after it all the theorems true in one geometry and false in
the other. So these geometries share a common commitment to deduction, but
have one different assumption. It begins to look as if they are both okay as exercises in logic, although some other intellectual exercise altogether discriminates
between them when deciding if one is true.

30.2 Proof

All these considerations promoted the view that mathematics is a colossal exercise in thinking straight. Now, this was a widely held view among educators
in the 19th century. What raised it beyond the level of a commonplace was the
idea that mathematics succeeded even though, or even precisely because, it had
no subject matter, no objects of which we had knowledge. This had not been
the view around 1800. In 1800, most people held a spread of views along these
lines: mathematics is an exercise in logical reasoning, about objects which we
understand perfectly (number and magnitude, point, line, curve, plane, etc.).
At its most sophisticated, this view was expounded by Kant, and cannot be
discussed even briefly here (he envisaged a category of truths which were synthetic, meaning they were about the world, and yet a priori, meaning they
were certainly true; these included the categories of thought that enabled people to do mathematics, for example). By 1900, confidence in our access to these
mental objects was waning, and only the logical aspect was left.

The consequence is that large parts of mathematics may no longer be true,
but if logical, this mathematics may yet be applicable, perhaps in novel ways.
What it is, or must be, is proved. The aim of the enterprise is not true statements, but proved ones. And since a deduction is only as good as its premisses,
one must have initial assumptions, rules of deduction, and a check that these
rules are scrupulously followed. However, now that mathematics is no longer
able to be, or constrained to be, true, these assumptions may be arbitrary,

provided only that they are self-consistent. The example of arithmetic suggested, however, that some assumptions might be better than others, those,
that is, that could be said to follow from pure logic alone (should there be any
– doubt could set in anywhere!).

I do not claim that geometry and only geometry promoted the philosophy of mathematics that sees it as a huge game (a comparison with chess is
often made in such accounts). The other main source was the nature of the
real numbers, and the growing role of set theory. But geometry was undoubtedly a major factor. Nor indeed did the best mathematicians think of it as a
game, a purely arbitrary exercise. Hilbert, for example, who is often thought
of as a prime mover in this enterprise, was insistent that there are aesthetic
criteria (applications, use, elegance, profundity) which are essential. This actually rather sophisticated view is probably the orthodoxy these days, and most
mathematicians don't think about it, but it caused problems when it came in.

The real problems are with the consistency of an axiom system and with
what can be said about axiom systems. In the second edition of his *Grundlagen
der Geometrie*, Hilbert said that an axiom system could be what he called
complete. By this he meant that there was a system of objects obeying the
axioms and there was no larger system of objects also obeying the axioms.
This was a different kind of axiom from the others, which described what
you could do with the objects. This was a statement about the nature of the
axiom system. It was intended to allow him to claim that the objects really
existed, as his friend Minkowski appreciated. He wrote in a letter to Hilbert
that "Your existence is as little to be doubted as in your $18 = 17 + 1$ axioms
for arithmetic."[1]

30.2.1 Frege versus Hilbert

But Hilbert's 18th axiom drew heavy criticism from Frege, then a struggling
philosopher of mathematics in the small University of Jena, but now commonly regarded as the greatest philosopher of mathematics since Aristotle and
the man who did more than anyone to analyse the idea that mathematics can
be deduced from logic. Frege had trained as a mathematician, and he possessed
a sharp appetite for criticism. He complained reasonably to Hilbert that the
18th axiom was like doing theology with an axiom that says God exists: "Axiom 3, there is at least one God".[2] Frege denied that completeness axioms can
be used to resolve questions of existence. Hilbert persisted, and asserted such an
axiom for the real numbers, thus obviating the need for explicit constructions in

---

[1] Minkowski to Hilbert 24 June 1899 [213, p. 116].

[2] Frege to Hilbert, 6 January 1900 [78, p. 46].

the manner of Cantor. Rather incautiously, in letters to Frege, he even claimed (Hilbert to Frege, 29 December 1899 [78, p. 39]) that "if the arbitrary given axioms do not contradict one another with all their consequences, then they are true and the things defined by them exist". The example of the parallel postulate should have commended "proved" (the word Hilbert preferred in Paris), rather than "true". Frege failed to get Hilbert to change his mind, so he eventually wrote most of his part of the correspondence as an article [79]. This probably did Hilbert little damage because Frege's manner was notoriously rude, and Hilbert had all the mathematicians on his side.

The polarity between Hilbert and Frege is crucial in modern mathematics. Hilbert was radically of the opinion that consistency implies existence. Consistent systems may admit multiple interpretations, different systems contradictory implications. One may have Euclidean geometry and non-Euclidean geometry in mathematics because each has a consistent set of axioms. Frege, contrarily, believed that existence was primarily a question of what objects there are in the world, and axiom systems without objects were void. In his view (most clearly in later, unpublished material) there is only one world and so only one geometry, and therefore non-Euclidean geometry, it would seem, is simply meaningless. This shows that Frege had no grasp at all of the relative consistency argument for, following Poincaré, these geometries had been shown to stand or fall together.

Hilbert was pioneering a distinction between axiom systems that codify and axiom systems that create. Codifying what is known is a traditional activity, almost certainly conducted by Euclid, and certainly conducted by Pasch for geometry. What Hilbert, and the Italians in a confused way, wanted to do, was to create mathematical objects by giving rules for their use. This was quite a novel approach, and it is not surprising that there was some doubt about how it can legitimately be done.

30.3 Relative consistency

Let me now turn to another contribution of Poincaré's, just mentioned briefly above: the idea that Euclidean geometry and non-Euclidean geometry stand or fall together. What would a self-contradiction in non-Euclidean geometry consist of? It would be an argument that something was simultaneously true and false in that geometry. The proof that something was true would consist of an argument, accompanied at each stage by a picture, let us say in the Poincaré disc. The contradiction would consist of claiming that the two pictures which resulted were contradictory. Now, every piece of the picture has a meaning in Euclidean geometry (see Poincaré's dictionary below) so the two arguments and their contradictory pictures are equally well arguments in Euclidean geometry leading to pictures that contradict each other in Euclidean geometry. So a contradiction in non-Euclidean geometry is also a contradiction in Euclidean geometry. Reverse the roles of Euclidean and non-Euclidean geometry, and the same argument shows that a contradiction in Euclidean geometry is also a contradiction in non-Euclidean geometry. The conclusion, which Poincaré was the first to draw explicitly, was that Euclidean and non-Euclidean geometry stand or fall together. This has the remarkable consequence that had anyone seeking to secure the absolute truth of Euclidean geometry by finding a contradiction in non-Euclidean geometry ever succeeded, they would have at that moment destroyed Euclidean geometry as well, the very thing they were hoping to make eternal. Like Samson, they would have pulled their own house down around them.

It is worth noting, in this context, that the German mathematician Otto Hölder, in the notes to the published version of his inaugural lecture in 1899 [122], observed that the usual descriptions of non-Euclidean geometry establish that it is free of contradictions on the assumption that Euclidean geometry is consistent. "This assumption, which before the discovery of non-Euclidean geometry was tacitly made everywhere, merely arises originally from the belief that Euclidean geometry is the true geometry for certain objective relations." Hölder then observed that, as Hilbert had shown in his *Grundlagen*, what Cartesian coordinate geometry does is establish that Euclidean geometry is relatively consistent. So one might conjecture that the relative consistency of Euclidean and non-Euclidean geometry was only apparent once the need to establish the consistency of Euclidean geometry had been felt, and met.

30.4 Poincaré on the relative consistency of Euclidean and non-Euclidean geometry

From "Non-Euclidean geometries", in *Science and hypothesis*

It would be easy to extend Beltrami's reasoning to three-dimensional geometries, and minds which do not recoil before space of four dimensions will see no difficulty in it; but such minds are few in number. I prefer, then, to proceed otherwise. Let us consider a certain plane, which I shall call the fundamental plane, and let us construct a kind of dictionary by making a double series of terms written in two columns, and corresponding each to each, just as in ordinary dictionaries the words in two languages which have the same signification correspond to one another:–

Space	The portion of space situated above the fundamental plane.
Plane	Sphere cutting orthogonally the fundamental plane.
Line.	Circle cutting orthogonally the fundamental plane.
Sphere	Sphere.
Circle	Circle.
Angle	Angle.
Distance between two points	Logarithm of the anharmonic ratio [cross-ratio] of these two points and of the intersection of the fundamental plane with the circle passing through these two points and cutting it orthogonally.
Etc.	Etc.

Let us now take Lobatschewsky's theorems and translate them by the aid of this dictionary, as we would translate a German text with the aid of a German–French dictionary. We shall then obtain the theorems of ordinary geometry. For instance, Lobatschewsky's theorem: "The sum of the angles of a triangle is less than two right angles," may be translated thus. "If a curvilinear triangle has for its sides arcs of circles which if produced would cut orthogonally the fundamental plane, the sum of the angles of this curvilinear triangle will be less than two right angles." Thus, however far the consequences of Lobatschewsky's hypotheses are carried, they will never lead to a contradiction; in fact, if two of Lobatschewsky's theorems were contradictory, the translations of these two theorems made by the aid of our dictionary would be contradictory also. But these translations are theorems of ordinary geometry, and no one doubts that ordinary geometry is exempt from contradiction. Whence is the certainty derived, and how far is it justified? That is a question upon which I cannot enter here, but it is a very interesting question, and I think not insoluble. Nothing, therefore, is left of the objection I formulated above. But this is not all. Lobatschewsky's geometry being susceptible of a concrete interpretation, ceases to be a useless logical exercise, and may be applied. [196, pp. 41–43]

31
On Writing the History of Geometry – 3

31.1 Assessment questions

The course at Warwick ended with a longer piece of work that students were to do over the Christmas holidays, in which they were asked to look back over the course and tease out the significance of various parts.

Here are two of the questions I used, together with the sort of advice I offered.

Question 31.1 (Essay Question 3)

Write a report on "Progress in geometry, 1860–1920".

Advice on interpreting the essay title: you may vary the dates (but not by much!) to suit your topic. You may suppose you are addressing an audience of mathematicians at around your terminal date, or a more general audience at that time, or you may decide to write a more straightforward historical essay aimed at a living audience. Do not pretend that you can see into the future (i.e. past 1920!). When you have taken these decisions, write a very short preamble stating how you have interpreted the question, stating your dates and intended audience. Failure to write such a preamble will result in a loss of marks.

The whole essay should be between 1,500 and 2,000 words.

J. Gray, *Worlds Out of Nothing*,
Springer Undergraduate Mathematics Series,
DOI 10.1007/978-0-85729-060-1_31, © Springer-Verlag London Limited 2011

Question 31.2 (Essay Question 4)

You should attempt both parts.

(a) Describe the main developments in geometry in the 19th century.

(b) Do you consider the changes in geometry in the 19th century amount to a revolution in geometry, or to be a process of incremental change?

The whole essay should be between 1,500 and 2,000 words. You might want to allocate more words to part (a) than part (b).

31.2 Advice on writing such essays

The topic of progress is a difficult one. It has strong overtones of improvement, and you might want to think about what constitutes improvement in a state of affairs. Is novelty improvement? Is diversity improvement? Is discovery improvement? You can find cases in this book where novelty, diversity, even discovery itself were regarded as making things worse. One way forward is to offer multiple perspectives. Another (and they are compatible) is to try to use the word "progress" as various people did at the time. They don't have to agree with each other, of course.

"Geometry" is a difficult word. There are many ideas about what geometry is, how it relates to other branches of mathematics, to science and to thought in general.

Decide what you would say were the main developments in geometry in the long 19th century. Try to pick some interesting periods (three or even four) when the nature of the subject changed, or the criteria for its importance shifted. You then have to say how they developed, and what is, or was, important about them. It may help to consider some of the key words people use on such occasions (novel, fundamental, profound, rigorous) and to consider what questions your chosen topics answer, or raise. Who found it important? Why? Give their reasons, not yours (or, if you prefer, give their reasons as well as yours). How numerous were these people: a large or small minority or majority? Don't forget the original authors. Were there dissenting voices? Factor them in if you think they colour the picture.

Where appropriate, your answer should show how and to what extent judgements of various kinds are grounded in the underlying mathematics. A wealth of mathematical detail is not necessary, but mathematical accuracy is. Try to avoid suggesting that the future could be predicted. Don't end up with a final paragraph full of superlatives that don't stand up to scrutiny.

For Question 31.2(b), you need to decide what constitutes a revolution in an academic subject. In the present case, there was no small group of revolutionary leaders who took power by force, but historians do talk of the Scientific Revolution, the Industrial Revolution, even the Neolithic Revolution. You need to decide if the 19th century saw a thorough turning upside down of cherished beliefs, or not. It might help to make a list of the main developments, the novelties and the continuities, and to decide if they were so dramatic as to merit the term "revolutionary". Try to imagine what the arguments for the contrary position are.

Finally, all manner of conclusions are admissible – if well argued! You might plump for revolution, or for steady progress, say there is some truth in both conflicting positions, or even that it's a false dichotomy. There are seldom simple right answers up a historian's sleeve, good essays make the reader sit up and think. But surely enough exciting things have happened in a century of mathematics to make the question worth discussing.

31.3 How the essays will be graded

It seems sensible to tell you what you are expected to do, and how well you are expected to do. I offer the following advice on obtaining a mark on each essay appropriate to the various classes of degree.

A First:

Quality of argument:	The argument is convincing, supported with relevant facts and well organised.
	The judgements reached are, when necessary, subtle and balanced.
	The coverage is broad, there are no potentially damaging omissions.
	There are no unnecessary digressions.
	Ideally, and without being cranky, the essay is original (in its emphases, or its conclusions).
	Good use is made of quotations.
Accuracy:	The historical facts are indeed correct and to the point.
	The mathematics is correct, clear and relevant.
	As a piece of prose, the essay is well written, and of the right length.

An Upper Second: Falls below the above in one or two significant ways.

A Lower Second:	Falls below what is required for an Upper Second in one or two significant ways. An unconvincing or desultory argument.
A Third:	Shows a bare knowledge of the topic, is poorly organised and/or at times inaccurate.
A Fail:	Does not demonstrate a knowledge of the topic.

Borderline distinctions are usually determined as follows. It can be hard to tell a First from a good Upper Second. Roughly speaking, a First-class essay is something you should be proud of and you could put with confidence in front of anyone, whereas varieties of Second go to good, and even very good, work. On any borderline, a good original point can push you up, a significant error can push you down. At the other end, a Third says "yes, you know some things you didn't know before but only just enough" and a Lower Second says that you are either generally, if intangibly, better than that, or at some identifiable point clearly better than that. A Fail mark goes to an essay that doesn't do more than recycle facts and lacks coherence or an argument, or fails to address the question of importance.

Plagiarism is the taking of information and arguments from sources you do not acknowledge and is theft. It will result in a Fail.

A

Von Staudt and his Influence

A.1 Von Staudt

The fundamental criticism of the work of Chasles and Möbius is that in it cross-ratio is defined as a product of two ratios, and so as an expression involving four lengths. This makes projective geometry, in their formulation, dependent on Euclidean geometry, and yet projective geometry is claimed to be more fundamental, because it does not involve the concept of distance at all. The way out of this apparent contradiction was pioneered by von Staudt, taken up by Felix Klein, and gradually made its way into the mainstream, culminating in the axiomatic treatments of projective geometry between 1890 and 1914.

That a contradiction was perceived is apparent from remarks Klein quotes in his *Zur nicht-Euklidischen Geometrie* [138] from Cayley and Ball.[1] Thus, from Cayley: "It must however be admitted that, in applying this theory of v. Staudt's to the theory of distance, there is at least the appearance of arguing in a circle." And from Ball: "In that theory [the non-Euclidean geometry] it seems as if we try to replace our ordinary notion of distance between two points by the logarithm of a certain anharmonic ratio. But this ratio itself involves the notion of distance measured in the ordinary way. How then can we supersede the old notion of distance by the non-Euclidean one, inasmuch as the very definition of the latter involves the former?"

The way forward was to define projective concepts entirely independently of Euclidean geometry. The way this was done was inevitably confused at first,

---

[1] In Klein, *Gesammelte mathematische Abhandlungen*, I [137, pp. 353–383], the quotations are on p. 354.

J. Gray, *Worlds Out of Nothing*,
Springer Undergraduate Mathematics Series,
DOI 10.1007/978-0-85729-060-1, © Springer-Verlag London Limited 2011

because it is a complicated process. An investigator has to decide what can be assumed, and what indeed is to be proved. Initially, the understanding was that the subject matter was that of real projective geometry in two and three dimensions – ideas about complex projective geometry were not at all those one would expect today. Then one has to decide how coordinates enter the picture: are they given in advance or to be derived from some logically antecedent structure? How are constructions related to transformations? What projective transformations are there? Two ideas in particular were to cause problems. One was continuity, the other the connection between coordinates and transformations. If the coordinates are to be real numbers, recall that Dedekind's rigorous ideas about them were published only in 1872, and if the transformations are to form a group, note that Jordan's major book on group theory came out only in 1870.

A.1.1 Von Staudt's *Geometrie der Lage*

The first mathematician to advance the study of projective geometry in its own right, independent of metrical considerations, was Karl Georg Christian von Staudt, who lived a quiet life working in the small University of Erlangen. It was a backwater, with few students, and his two major books crept almost unobserved onto the shelves, where they remained until after his death in 1867 and it became gradually clear that he had gone a long way to solve the problem of giving independent foundations to real projective geometry. Among the first to rescue him from obscurity was the young Felix Klein, at the time a student at Berlin, who was alerted by his friend Otto Stolz to the significance of von Staudt's work for questions he was interested in.

Von Staudt's first book, his *Geometrie der Lage* [228], is based on the idea that there are entities called points, lines and planes. Lines in the same plane may meet or be parallel – the presence of parallelism in his geometry is a complication that Klein was later to show can be written out of the theory. So one might say that von Staudt took over from Euclid all and only the non-metrical concepts of the *Elements*. He began by observing that if three points lie on a line then one is between the other two, and that if four points lie on a line then they form two separated pairs. Both statements are reasonable because he had not yet introduced points "at infinity". He noted that there is a unique line joining any two points. He could now define points at infinity in terms of a pencil of parallel lines in a plane, and he showed how to extend his earlier ideas to the new setting. He invoked the idea of figures in perspective as a typical transformation of figures, noting that a pencil of lines through a point can correspond to a pencil of parallel lines.

Now he introduced the idea of a reciprocity (his word) or duality between points and planes in space. Quite generally, von Staudt preferred to work on the geometry of three dimensions, deducing results about plane geometry as a consequence. So he stated his version of Desargues' theorem as a theorem about figures in two different planes, and used it to show [228, ch. 8] that given three distinct points on a line there is a point which is the fourth harmonic point with respect to these three, and moreover it is unique. Such a set of four points he called a harmonic set of points, and he showed that a harmonic set of points is mapped to another harmonic set of points by a perspectivity.

Von Staudt then introduced projective transformations, which Möbius and some later writers called collineations, as those 1–1 maps which send lines to lines (and, in three dimensions, planes to planes) and send sets of four harmonic points to sets of four harmonic points. A reciprocity may also be a projective transformation, if it sends a harmonic set of points to a harmonic set of planes (with the obvious definition). He showed by exhibiting a suitable sequence of perspectivities that any three distinct collinear points may be mapped to any three distinct collinear points by a collineation. Next he produced a peculiar argument [228, §106], much criticised by later writers, in support of the claim that a map sending three points on a line to three points on the same line extends to a map of the whole line. He argued that the claim is trivial if the point A is mapped to the point A', the point B to the point B', and the whole segment between A and B to the whole segment between A' and B', because then every point outside the segment is the harmonic conjugate of a point inside the segment and the map extends in an obvious way. If on the other hand the segment AB is not mapped in this fashion, then, he said, exactly one of the interior and the exterior of AB contains a point that is mapped to itself, but, by the theorem on the fourth harmonic point, this leads to a contradiction (the details are perforce omitted here). Klein was to argue that this requires a discussion of continuity.

Subsequent generations of mathematicians and historians of mathematics have been most impressed by von Staudt's insistence on duality. Von Staudt insisted on speaking of a figure and its dual simultaneously. For him, a duality (which he called a correlation) was a 1–1 correspondence between points and lines in a plane which sends harmonic sets of points to harmonic sets of lines and vice versa. For example, one might have a self-polar triangle in which, for each vertex, the line that corresponds to a vertex of the triangle is the corresponding side of the triangle. Given a polarity, it might be that a point P lies on the line ℓ to which it corresponds. This led von Staudt to his remarkable definition of a conic section as a locus of points each of which lies on its corresponding line. Indeed, for von Staudt, a conic was both its locus as a set of points and the corresponding dual locus of lines (such a conic, as he noted, may well be

an empty set, and here is an example: $x^2 + y^2 + z^2 = 0$). Möbius had noted
that in a space of even dimension the self-dual figures are conics, but in spaces
of odd dimension there are self-dual figures that are not conics – the so-called
null systems – and von Staudt did the same.

The upshot of all this work is that von Staudt showed in his *Geometrie der
Lage* [228] that the familiar, real projective geometry can be built up from the
non-metrical concepts of Euclidean geometry – or rather, and more precisely,
he had mapped out a way in which that might be done. However, many details
remained to be established properly.

He showed how one could as it were measure the cross-ratio of four points
(at least if they lie in a chain) by moving three of them into a standard position
and noting the coordinate of the fourth point. This shows that cross-ratios may
be used as lengths are in Euclidean geometry to give necessary and sufficient
conditions for one set of four points to be equivalent to another.[2]

He then showed how one could iterate the construction of the fourth harmonic point, to obtain what he called a chain of harmonic points on a line, and
to obtain a Möbius net from any four coplanar points (no three on a line).[3] The
Möbius net permits the introduction of coordinates which are rational multi-
ples of an arbitrary constant. He then assumed without discussion that a map
from a Möbius net in one plane onto a Möbius net in another plane extended
to a unique map of the one plane onto the other.

The difficulties with this work all lie beneath the surface. Some may even
strike the reader as artificial, and so they are if the aim is to establish real
projective geometry on its own terms, as von Staudt's was. But artificial or
not, the incidence axioms for plane projective geometry say things like this:
through any two distinct points there passes exactly one line; any two distinct
lines meet in exactly one point. They do not say that there are infinitely many
points on a line, or infinitely many lines through a point. They do not, for
example, guarantee that there are even four points on a line, and if there were
to be only three then the whole construction of the fourth harmonic point
would of course fail. (As we saw when discussing Fano's work, see page 272,
it is entirely possible to have a projective geometry with only three points on
each line, so there is something to do here.) Understandably, on occasions like
this, von Staudt assumed things that eventually later mathematicians felt the
need to prove, or to dispatch with an axiom.

The same is true of Desargues' theorem. Von Staudt was operating in a
context, not all of which he explicitly recognised, which permitted him to prove

---

[2] See also the discussion in Part II of the *Beiträge* [229] on sums, products, and
powers of transformations.

[3] A chain is a sequence of points $A_1, A_2, \ldots, A_j, \ldots$ such that for each $j = 1, 2, \ldots$
the fourth harmonic point of A_j, A_{j+1}, A_{j+2} is A_{j+3}.

Desargues' theorem in the plane. We shall return to this point later.

A.1.2 Klein's response to von Staudt

It is rather more understandable that von Staudt would slip into imprecision over the passage from Möbius nets in the projective plane, to collineations of the whole projective plane. Given a proper set of definitions, it is elementary to show that a continuous function defined on a dense set of points on the line extends to a continuous map on the closure of the dense set, but none of that body of theory was available to von Staudt, and even a rigorous definition of the real numbers had still to be given. Klein saw early on that it was not only possible to develop von Staudt's ideas without introducing the idea of parallel lines, but it was advisable to do so, because this opened the way to connections between non-Euclidean geometry and projective geometry. He was also of the opinion that something had to be done to establish the claim that the projective map sending three given distinct collinear points to three given distinct collinear points is unique (there is no problem, he agreed, in establishing its existence). Von Staudt had also shown that the sequence of fourth harmonic points established by a triple of points cannot suddenly stop (by closing up). But he did not show that it necessarily had points in every interval in the line. Klein therefore proposed in his article of 1873 [131] to insist that it did, whereupon Lüroth and Zeuthen wrote to him to say that some of his worries were unnecessary.

Klein replied in an article of 1874 [133]. He accepted Zeuthen's argument completely, even quoting it in his paper word for word in the original French. Zeuthen took four harmonic points A, B, C, D, where A and B separate C and D, and supposed there was a maximal segment FG on the line which the succession of fourth harmonic points obtained from A, B and C never entered. So if F is not a point of this interval, it is a limit of points in a chain of fourth harmonic points. He now argued by contradiction, as follows.

Let H be the fourth harmonic point of the points A, F and G, and let J be the point such that A and G harmonically separate F and J. Let B be a point of the chain suitably close to F and let K be the point such that A and H harmonically separate B and K. It is possible to choose B so that K is in the segment GJ and so close to G that KJ contains a point of the chain. Call this point C. Let L be the point such that A and L harmonically separate B and J, so L will be in HG. Now let D be the point such that A and D harmonically separate B and C. The point D lies not only in the segment HL but also in the segment FG, thus establishing the requisite contradiction.

This shows that a harmonic chain is a dense set of points on a projective line. Does it follow that a projective map defined on such a set extends to a unique map on the remaining points? Klein was now able to say that whatever it meant for a set of points on a line to be "continuous", the same applied to points on a projective line, because the matter had recently been clarified by Heine, Cantor and Dedekind. Since we now apply the adjective "continuous" to functions rather than sets of points, we must interpret this as concerning sets of points which are connected. But even so, he said, the answer was self-evidently "no", just as it was clear there was no way a function defined on the rational numbers could always be extended to a continuous function on the whole real line. It was necessary, he insisted, to add to von Staudt's definition of a collineation that it be continuous.

There the matter rested until 1880, when Darboux wrote to Klein (who promptly published the relevant part of the letter in *Mathematische Annalen*, a journal he now edited) [135]. Darboux said that while everyone had agreed with Klein, the only flaw in von Staudt's original presentation was with the method of proof, not the claim itself. In other words, collineations as defined by von Staudt were automatically continuous. Darboux's argument was very elegant. It was required to show that a map ϕ which maps three points to themselves is the identity map on all points. First, he said, suppose we are allowed metrical arguments. Then a simple argument from the information that $\phi(0) = 0$, $\phi(1) = 1$ and $\phi(\infty) = \infty$ shows that ϕ satisfies the functional equation $\phi(x) + \phi(y) = \phi(x + y)$. (There is no problem with the use of ∞, which merely simplifies the formulae.) Now, he said, this conclusion on its own would not show that $\phi(x) = x$ and therefore is continuous, as Cauchy had been the first to notice. But the conclusion would follow if ϕ satisfied some extra conditions, and in fact the functional equation had been derived without using all the properties of ϕ. A little more work showed that $\phi(x)$ was positive when x was positive, and this was enough to rule out pathological behaviour and show that indeed the function ϕ was continuous.

He then gave a non-metrical argument to the same conclusion, which invoked Zeuthen's result discussed above, and for good measure showed that some other theorems of a similar kind are true without the need for assumptions of continuity. For example, Möbius had shown that a continuous map of the plane sending circles to circles is an inversion or a sequence of inversions, but the assumption that the map be continuous was unnecessary.

The proof of Desargues' theorem that von Staudt offered also worried Klein. He noted that it was essentially an incidence proof, in which the key ingredients were that two "points" lie on a unique "line", two "planes" meet in a "line", and so forth, where the quotation marks are to indicate that it is the incidence properties that make the argument work, not any other properties of lines or planes. So one could imagine the theorem being true of figures drawn with the appropriate kinds of curved lines and planes: to be precise, curved surfaces which are determined by three distinct points and have the property that if two such surfaces meet in a curve, then any surface through two points on that curve contains the whole curve. Indeed, recall that the incidence proof of Desargues' theorem goes as follows. The lines OAA', OBB' and OCC' lie in a plane, and the lines AB and $A'B'$ in that plane meet in a point N. Similarly, the lines BC and $B'C'$ meet at the point L and the lines CA and $C'A'$ meet at the point M. The points L, M and N lie in the planes of the triangles ABC and $A'B'C'$, and so lie on the line common to these two planes. It is easy to see that the proof works for points, curves and surfaces subject to suitable restrictions; straightness and flatness are not involved.

What worried Klein was that Desargues' theorem in von Staudt's hands was the key to introducing coordinates in such a way that the surfaces involved had linear equations. This, Klein saw, invited an obvious generalisation down to two dimensions. One would discuss curves with the property that any two curves met in a point, and any two points determined a unique curve, and presumably deduce that Desargues' theorem allowed one to introduce coordinates in such a way that the curves were given by linear equations. But Klein knew that this could not be done, because Beltrami had shown that among the curves with that property in a disc-like region were geodesics with respect to a metric, and they could only be given linear equations if the metric had constant curvature.

This meant that von Staudt's trick of proving theorems in the projective geometry of two dimensions by passing to three dimensions could not be used. This suggested to Klein that projective geometry in three dimensions could be established more directly than projective geometry in only two dimensions, but he did not, as Enriques was later to suggest that Klein had done, conjecture that Desargues' theorem might even be false in two dimensions.

A.2 Non-orientability

In the course of all this work, a novel and unexpected topic emerged onto the mathematical scene: orientability. Both Möbius, who is usually credited with the discovery of non-orientable surfaces, because of the eponymous Möbius band, and Listing, seem to have been thinking of the band in 1858 – indeed Listing's unpublished note of that year [150] pre-dates Möbius's unpublished note [166] by a few months. Both men were connected to Gauss, who had died in 1855, and it might even be that the concept goes back to him. Be that as it may, the simple idea of orienting a surface is to imagine each point of the surface surrounded by a small disc. The boundary of each disc is a circle, and we can order the points on it by choosing three distinct points A, B, C say, and stating that they occur in that order. We say that the surface is orientable if all the discs can be oriented in a compatible way, and non-orientable otherwise. The cylinder is an example of an orientable surface, and the Möbius band an example of a non-orientable surface.[4]

The relationship of the real projective plane to the usual Euclidean plane was understood in many ways. For example, the projective plane can be thought of as the Euclidean plane with the addition of a line at infinity. By the early 1870s the work of several authors had promoted another consideration, that of algebraic topology. Möbius, Listing, Jordan and Riemann in various ways had produced an analysis of surfaces, including something like a classification of what, with later terminology and ideas could be called compact surfaces. These include the surfaces defined by complex algebraic curves, such as the sphere, the torus (defined, for example, by the equation $w^2 = z(z-1)(z-2)(z-3)$ and in this form familiar from the theory of elliptic functions) and others. Central to this approach was what Riemann called the order of connectivity of the surface, and which he defined, impressionistically, as the smallest number of closed curves that can be drawn on the surface without it falling into two pieces. The connectivity of the sphere is 0, of the torus 2, and so on.

In the early 1870s, Schläfli and Klein were independently interested in the surfaces that arise in projective geometry, and they noticed that more complicated behaviour can occur, and this imperilled the intuitive enumeration. In 1874 Schläfli wrote to Klein to say that order could be restored if one regarded the usual plane as a double plane or as the limiting case of a family of two-sheeted hyperboloids. Klein published his version of these ideas in the *Mathematische Annalen* in 1874 [133], but it must be said that they are a little obscure, which shows how unfamiliar this point must have been and how difficult to grasp. It seems better, therefore, to explain it without staying too close to the text of his paper.

Klein observed that Riemann's treatment of what happened out towards infinity had the effect of making infinity a point, and that this could be seen by stereographic projection. We might add that, topologically, this is the one-point compactification of the plane. However, in (three-dimensional) projective geometry one thought of there being a plane at infinity, and in plane projective geometry one supposed there was a line at infinity. The way forward was unclear to him, but in the article of 1874 [133] he can be seen groping for ideas like these. Consider the projective plane as the space of all lines through the origin in Euclidean three-space, and the Euclidean plane as the plane with equation

---

[4] Listing published his account of the band in 1861, Möbius only in 1865. What is at stake is the recognition of the mathematical significance of non-orientability; pictures of the band have been traced as far back as the 3rd century CE.

$z = -1$. Each Euclidean point gives rise to a sloping line through the origin (the line through the origin and the given point). The projective points correspond to the horizontal lines through the origin. Now, the space of all lines through the origin is an unpleasant thing to visualise, so represent each line through the origin by the two antipodal points it marks out on the unit sphere with centre the origin. We then immediately have a 2–1 map from the sphere to the real projective plane. The map sends a point on the sphere to the line through the origin and the given point. The map is 2–1 because antipodal points on the sphere define the same projective point.

Now, the sphere and the plane are essentially equivalent under stereographic projection. So one may think of the plane as a double cover of the projective plane. This is what Schläfli urged upon Klein. The double or Euclidean plane has an unexpected property: a line drawn upon one plane does not disconnect the double plane. This is easier to see if we switch over to the modern picture. The projective plane is, as noted, the image of the sphere under a 2–1 map. This allows us to see the projective plane as the northern hemisphere with antipodal points identified. Consider the effect of passing a plane through the origin. It cuts the sphere in a great circle, of course, but what can we say about the plane and the projective plane? We have a choice. It might be that we say that the great circle is mapped 2–1 onto its image, or we might merely look at the image. If we take the second alternative, we see a curve γ in the northern hemisphere that meets the equator at two antipodal points. This curve does not disconnect the projective plane, because of the identifications on the equator, and it lifts to a semicircle on the sphere which does not disconnect the plane. If we double the curve, however, we do disconnect the projective plane, and the image of the doubled curve is a whole line disconnecting the plane.

Klein went on to note that this strange property (there are curves which close up on doubling) was already visible in the Möbius band (which he did not call by that name). He did not observe that the cylinder is in the same relation to the Möbius band as the sphere is to the projective plane, and more tantalisingly he did not observe that the strange connectivity of the projective plane is connected to the fact that it contains a Möbius band. (Indeed, a thickened neighbourhood of the curve γ is a Möbius band.) In fact, the Möbius band is present in every drawing of a hyperbola and its asymptotes, once one knows to look for it. For example, consider the hyperbola with equation $x^2 - y^2 = 1$, and its asymptotes $x = y$, $x = -y$. For definiteness, consider the asymptote $x = y$. It goes off to infinity, as it were, north-east with the hyperbola on its right, and comes back (from the south-west) with the hyperbola on its left, showing that a thickened neighbourhood of the asymptote in the projective plane is a Möbius band, and that the asymptote has not cut the projective plane into two pieces (a north-west and a south-east part).

In terms of projective geometry, a straight (Euclidean) line extends to a closed curve, and a conic is also a closed curve in projective geometry. The difference between them is precisely that the straight projective line does not disconnect the projective plane, and the conic of course does. This observation also rippled through the community of projective geometers (it is visible in Zeuthen's article of 1876 [249], for example).

That Klein took these topological considerations to heart is noticeable in his little book of 1882, *Riemann's theory of algebraic functions and their integrals* [136], where in §23 the Klein bottle seems to make its first appearance. It is an amusing exercise to see the torus as a double cover of the Klein bottle.

A.3 Axiomatics – independence

In the years between 1899 and 1914 a number of mathematicians gave more or less definitive versions of axiomatic projective geometry, shorn of Pasch's links to the facts of experience. The Italians Pieri, Fano and Enriques were the first, followed in Germany by Hilbert and later Vahlen, in America by Veblen and Young, and then in England by Whitehead. In these years the Italians were widely appreciated, but for a variety of reasons they were eclipsed by Hilbert in the years after 1918, to the point where their achievements were almost forgotten, and they have had to be rediscovered by historians of mathematics.

What these mathematicians accomplished in various ways was the identification of projective geometry conceived analytically with a synthetic presentation given by axioms. By an analytic presentation is meant an account like this: projective space of dimension n consists of all the lines through the origin in an $(n + 1)$-dimensional space over the real numbers, the allowed transformations form the group $PSL(n + 1; \mathbb{R})$ and so forth. The question for all these investigators was: what should an appropriate axiom system be? Rather than pursue the historical development, let us jump to the end of the story and consider a set of suitable axioms for projective geometry. The treatment that follows is taken from Hartshorne's *Foundations of projective geometry* [108].

Four are entirely unproblematic:

A1. Two distinct points lie on exactly one line.

A2. Two distinct lines meet in at most one point.

A3. There are three non-collinear points.

A4. Every line contains at least three points.

It is clear that axiom A2 is equivalent to the assumption that two distinct lines meet in exactly one point, which is more obviously the dual version of A1. Axiom A3 says that the geometry is at least two-dimensional. Axiom A4 is needed to rule out the space consisting of three points and the three lines joining them in pairs as a projective space.

The next axioms are more substantial:

A5. Desargues' theorem holds.

A6. Pappus's theorem holds.

A7. (Fano's axiom): the diagonal points of a complete quadrilateral are not collinear.

It is striking that Desargues' theorem must be assumed. It is not a consequence of the first four axioms of projective geometry. This is all the more remarkable when one considers the incidence proof of it, and indeed if one writes down the obvious axioms for projective geometry in three or more dimensions then Desargues' theorem is a consequence of those axioms. But it is not a consequence of the axioms of plane projective geometry, and there are projective planes in which it is false.

It is also the case that Pappus's theorem implies Desargues' – a result known as Hessenberg's theorem after its discoverer, see Hessenberg [116]. So in any (necessarily plane) projective geometry in which Desargues' theorem does not hold, Pappus's theorem also fails. On the other hand, if Pappus's theorem (and therefore Desargues') is true and Fano's axiom holds, then one can prove the fundamental theorem of plane projective geometry: that there is a unique projective transformation taking any four points, no three of which are collinear, to any four points, no three of which are collinear.

What about the uniqueness of the fourth harmonic point? It doesn't hold in the Moulton plane. Moulton implies no fourth harmonic point, so the theorem of the fourth harmonic point implies that the plane is not a Moulton plane. What about Desargues' theorem in general?

It may be helpful to note that, in the presence of A1–A4, the only implication between axioms A5, A6 and A7 is that A6 implies A5 (Pappus implies Desargues). To establish the independence of the remaining axioms, examples must be given of geometries satisfying all the remaining possible combinations of axioms:

1. None of A5, A6 and A7 holds;

2. A5 holds, but not A6 or A7;

3. A6 holds (and therefore A5 holds) but not A7;

4. A7 holds, but not A5 or A6;

5. A5 and A7 hold, but not A6;

6. A5, A6, and A7 all hold.

These are all duly given in Hartshorne's book [108, ch. 6].

Another route, more in keeping with the Kleinian approach to geometry, is to accept the first four axioms and then to specify the existence of enough transformations. This is the approach of Artin in his *Geometric algebra* [4]. Artin preferred to study projective geometry via affine geometry, so he allowed himself the concept of parallel lines. (An affine plane is obtained from a projective plane by singling out a line (to be called the line at infinity) and restricting attention to transformations that map this line to itself. Two lines are said to be parallel if they meet in the line at infinity.) He defined a map to be a dilatation if it maps the points P and Q, say, to P' and Q' respectively, in such a way that the line through P' parallel to PQ passes through Q'. Degenerate cases aside, a dilatation maps a line to a parallel line. Artin called a dilatation a translation if it is either the identity map or has no fixed points.

Artin's axiom 4a asserts that given any two distinct points P and Q, there is a translation taking P to Q. His axiom 4b asserts that given three distinct collinear points P, Q and R, there is a dilation mapping P to itself and Q to R. The existence of translations imposes conditions on the group of projective transformations, and so, ultimately, on the coordinates (if any can be admitted) of points. For example, the group of all translations is a commutative group if translations exist with different directions. Artin observed that it can be the case that translations might be confined to a single direction, in which case it was not known if the corresponding group had to be commutative. The Moulton plane is such a space because the axis can only be mapped to itself, but here the corresponding group is commutative.

Artin confined his attention to what he called the "good" case, in which axioms 4a and 4b were satisfied, and he showed that in this case one can introduce coordinates for points and linear equations for lines. Naively, the idea is that one picks an origin O arbitrarily, and then picks distinct translations in different directions, say τ_1 and τ_2, and uses $\tau_1(O)$ and $\tau_2(O)$ as the units of length in these directions.

Artin then worked backwards, starting with a division ring, taking pairs of elements from the division ring as coordinates of points, thus obtaining an affine plane, and thence a coordinatised projective plane. He now assumed that the first three of his axioms applied in this setting, but not the fourth, and instead postulated either D_a or DP, that is, Desargues' theorem when the centre of perspective is either at infinity (D_a) or at a finite point (DP). He then established that D_a is true if and only if axiom 4a is true, and DP is true if and only if axiom 4b is true.

If a projective geometry satisfies modest extra symmetry requirements (see Chapter II of Artin's *Geometric algebra* [4]) then coordinates can be introduced that form a division ring. Facts about division rings include:

A finite division ring is a field (Wedderburn's theorem).

A weakly ordered division ring (other than 0, 1) is ordered (ordered means the additive subgroup is a union of three sets of the form $-P \cup \{0\} \cup P$, where P has the property that $P + P \subset P$, $P.P \subset P$).

There are ordered non-commutative division rings (one was constructed by Hilbert).

All Archimedean fields are subfields of the real numbers.

There is a unique ordering on the real numbers consistent with the ordering on the rational numbers.

Artin then established Hilbert's classic result that the division ring is commutative if and only if Pappus's theorem is true, and so, by Wedderburn's theorem, in a finite Desarguian plane Pappus's theorem is true – although, most intriguingly, no synthetic proof of that result was known.

Finally, by consideration of orderings that I have omitted, Artin showed that for an ordered geometry to come from a field which is isomorphic to a subfield of the field of real numbers with its natural ordering, it is necessary and sufficient that the Archimedean postulate holds. It follows that in an Archimedean field the theorem of Pappus holds and the field is necessarily commutative.

It seems that it is the introduction of non-Archimedean fields that provoked an attempt to eliminate continuity considerations from abstract projective geometry, say by the use of segment arithmetic (as done by Hilbert and again by Hölder). Hilbert showed that in plane geometry with congruence and parallels, but not continuity or an Archimedean axiom, Pappus's theorem can be proved. Also, Pappus's theorem cannot be proved in simple geometry without continuity or congruence.

A.4 References

Since this book was first published three valuable papers on von Staudt have appeared. They are:

Nabonnand, P. 2008 La théorie des *Würfe* de von Staudt – Une irruption de l'algebre dans la géométrie pure, *Archive for History of Exact Sciences* 62, 201–242.

Voelke, J.-D. 2008 Le théorème fondamental de la géométrie projective: evolution de sa prevue entre 1847 et 1900, *Archive for History of Exact Sciences* 62, 243–296.

Hartshorne, R. 2008 Publication history of von Staudt's *Geometrie der Lage*, *Archive for History of Exact Sciences* 62, 297–299.

Bibliography

[1] Andersen, K. 1991 *Brook Taylor's work on linear perspective: a study of Taylor's role in the history of perspective geometry. Including facsimiles of Taylor's two books*, Springer, London.

[2] Anderson, J.W. 2005 *Hyperbolic geometry*, 2nd edn., Springer Undergraduate Mathematics Series, Springer, London.

[3] Appell, P. 1925 *Henri Poincaré*, Plon, Paris.

[4] Artin, E. 1957 *Geometric algebra*, Wiley Interscience, New York.

[5] Avellone, M., Brigaglia, A., Zappulla, C. 2002 The foundations of projective geometry in Italy from De Paolis to Pieri, *Archive for History of Exact Sciences* 56(5), 363–425.

[6] Ball, R.S. 1879 The non-Euclidean geometry, *Hermathena* 3, 500–541.

[7] Baltzer, H.R. 1867 *Die Elemente der Mathematik*, 2nd edn., vol. 2, Teubner, Dresden.

[8] Balzac, H. de 2003 *Eugénie Grandet*, tr. S. Raphael, intro. C. Prendergast, World's Classics, Oxford University Press, Oxford.

[9] Beardon, A.F. 1995 *The geometry of discrete groups*, corrected reprint of 1983 original, Graduate Texts in Mathematics, Springer, New York.

[10] Belhoste, B. 1991 *Augustin-Louis Cauchy, a biography*, Springer, New York.

[11] Belhoste, B. 2003 *La formation d'une technocratie*, Belin, Paris.

[12] Bellivier, A. 1956 *Henri Poincaré ou la vocation souveraine*, Paris.

J. Gray, *Worlds Out of Nothing*,
Springer Undergraduate Mathematics Series,
DOI 10.1007/978-0-85729-060-1, © Springer-Verlag London Limited 2011

[13] Beltrami, E. 1868a Saggio di interpretazione della geometria non-euclidea, *Giornale di Matematiche* 6, 284–312, English translation in Stillwell [232, pp. 7–34].

[14] Beltrami, E. 1868b Teoria fondamentale degli spazii di curvatura costante, *Annali di Matematica Pura ed Applicata* 2(2), 232–255, English translation in Stillwell [232, pp. 41–62].

[15] Berger, M. 1987 *Geometry I*, tr. from French by M. Cole and S. Levy, Universitext, Springer, Berlin.

[16] Berger, M. 1987 *Geometry II*, tr. from French by M. Cole and S. Levy, Universitext, Springer, Berlin.

[17] Birkhoff, G., Bennett, M.K. 1988 Felix Klein and his "Erlangen Program", pp. 145–176, in *History and Philosophy of Modern Mathematics*, W. Aspray and P. Kitcher (eds.), Minnesota Studies in the Philosophy of Science, vol. XI.

[18] Blumenthal, O. 1935 Lebensgeschichte, pp. 388–429, in D. Hilbert, *Gesammelte Abhandlungen* 3.

[19] Bolyai, J. 1832 Appendix scientiam spatii absolute veram exhibens, in W. and J. Bolyai [20], tr. J. Houël, La science absolue de l'espace, *Mémoires de la Société des Sciences Physiques et Naturelles de Bordeaux* 5, 1867, pp. 189–248, tr. G. Battaglini, Sulla scienza della spazio assolutamente vera, *Giornale di* Matematiche 6, 1868, pp. 97–115, tr. G.B. Halsted, *Science absolute of space*, Appendix in Bonola [21].

[20] Bolyai, W. and J. 1832 *Tentamen juventutem studiosam in elementa matheosis purae...*, 2 vols., Maros-Vásérhely.

[21] Bonola, R. 1906 *La geometria non-euclidea*, Zanichelli, Bologna, tr. H.S. Carslaw 1912 *Non-Euclidean geometry. A critical and historical study of its development*, preface by F. Enriques, Open Court, Chicago, rep. Dover, New York, 1955.

[22] Bos, H.J.M., Kers, C., Oort, F., Raven, D.W. 1987 Poncelet's closure theorem, *Expositiones Mathematicae* 5(4), 289–364.

[23] Bottazzini, U. 2001 Italian geometers and the problem of foundations (1889–1899), *Bol. Unione Mat. Ital.* 4(2), 281–329.

[24] Bottazzini, U., Conte, A., Gario, F. (eds.) 1996 *Riposte armonie. Lettere di Federigo Enriques a Guido Castelnuovo*, Torino.

[25] Bottazzini, U., Tazzioli, R. 1995 Naturphilosophie and its role in Riemann's mathematics, *Revue d'Histoire des Mathématiques* 1, 3–38.

[26] Brannan, D.A., Esplen, M., Gray, J.J. 1999 *Geometry*, Cambridge University Press, Cambridge.

[27] Brianchon, C.J. 1806 Sur les courbes de deuxième degré, *Journal de l'École Polytechnique* 6, 92–121.

[28] Brieskorn, E., Knörrer, H. 1986 *Plane algebraic curves*, tr. J. Stillwell, Birkhäuser Verlag, Basel.

[29] Cayley, A. 1859 A sixth memoir on quantics, p. 592, in *Collected Mathematical Papers*, vol. 3, 1889.

[30] Cayley, A. 1878 On the geometrical representation of imaginary variables by a real correspondence of two planes, *Proceedings of the London Mathematical Society* 9, pp. 31–39, in *Collected Mathematical Papers* X(689), pp. 316–323.

[31] Chasles, M. 1837 *Aperçu historique sur l'origine et le développement des méthodes en géométrie... suivi d'un mémoire de géométrie... Mémoires sur les questions proposées par l'Académie Royale des Sciences et Belles-Lettres de Bruxelles*, tom. 11, Bruxelles.

[32] Chasles, M. 1852 *Traité de géométrie supérieure*, Paris.

[33] Chasles, M. 1865 *Traité des sections coniques*, Paris.

[34] Chateaubriand, F.A.-R. 1797 *Essais sur les révolutions*, London.

[35] Chateaubriand, F.A.-R. 1802 *Le génie du christianisme*, Chez Migneret, Paris.

[36] Chateaubriand, F.A.-R. 1849–50 *Mémoires d'outre-tombe*, Paris.

[37] Chemla, K. 2004 Euler's work in spherical trigonometry: contributions and applications, *Leonhardi Euleri Opera Omnia* (4) 10, Birkhäuser, Basel.

[38] Clifford, W.K. 1870 On the space theory of matter, *Proceedings of the Cambridge Philosophical Society* Feb. 21, in Clifford [39, pp. 21–22].

[39] Clifford, W.K. 1881 *Mathematical papers*, rep. Chelsea, New York, 1968.

[40] Codazzi, D. 1857 Intorno alle superficie le quali hanno costante il prodotto de due raggi di curvatura, *Ann. Sci. Mat. Fis.* = *Annali di Tortolini* 8, 346–355.

[41] Condorcet, Marie Jean Antoine Nicolas Caritat, Marquis de 1795 *Esquisse d'un tableau historique des progrès de l'esprit humain*, posth. ed. P.C.F. Daunou and the Marchioness de Condorcet, Paris.

[42] Cooke, R. 1984 *The mathematics of Sonya Kovalevskaya*, Springer, New York.

[43] Coolidge, J.L. 1940 *A history of geometrical methods*, Clarendon Press, Oxford, Dover reprint, 1955.

[44] Cremona, L. 1868 Mémoire de géométrie pure sur les surfaces du troisième ordre, *Journal für die reine und angewandte Mathematik* 68, 1–133.

[45] Cremona, L. 1873 *Elementi di geometria projettiva*, tr. C. Leusdorff as *Elements of projective geometry*, Clarendon Press, Oxford, 1885.

[46] Crilly, A. 2006 *Arthur Cayley: Mathematician laureate of the Victorian age*, Johns Hopkins University Press, Baltimore MA.

[47] Crowe, M.J. 1967 *A history of vector analysis: The evolution of the idea of a vectorial system*, University of Notre Dame Press, Dover reprint, 1994.

[48] Darboux, G. 1880 Sur le théorème fondamental de la géométrie projective (Extrait d'une lettre à M. Klein), *Mathematische Annalen* 17, 55–61.

[49] Dedekind, R. 1876 Riemann's *Lebenslauf*, pp. 571–590, in B. Riemann, *Gesammelte Werke*, 3rd edn., Springer, New York.

[50] Dehn, M. 1900 Die Legendre'schen Sätze über die Winkelsumme im Dreieck, *Mathematische Annalen* 53, 404–439.

[51] Desargues, G. 1864 *Oeuvres de Desargues réunies et analysées par M. Poudra*, 2 vols., Paris.

[52] Dhombres, J. and N. 1997 *Lazare Carnot*, Fayard, Paris.

[53] Digital Mathematics Library [online] http://www.wdml.org/, accessed 14/5/06.

[54] Dostoyevsky, F. 1993 *The brothers Karamazov*, tr. D. McDuff, Penguin Books, London.

[55] Dunnington, G.W. 2004 *Gauss – titan of science*, introduction and appendices J.J. Gray, Mathematical Association of America, Providence RI.

[56] Dupin, F.P.C. 1819 *Essai historique sur les services et les travaux scientifiques de G. Monge*, Paris.

[57] Dubrovin, B.A., Fomenko, A.T., Novikov, A.P. 1984 *Modern geometry – methods and applications*, 3 vols., Springer, New York.

[58] Ebbinghaus, H.-D., Hermes, H., Hirzebruch, F., Koecher, M., Mainzer, K., Neukirch, J., Prestel, A., Remmert, R. 1991 *Numbers*, Springer, New York.

[59] Einstein, A. 1905 Zur Elektrodynamik bewegter Körper, *Annalen der Physik*, vol. 17, 891–921, in *The Collected Papers of Albert Einstein* vol. 2, *The Swiss Years: Writings, 1900–1909*, pp. 414–427, English translation in A. Einstein, *The Principle of Relativity*, Dover, New York, 1952, pp. 35–65.

[60] Einstein, A. 1912a The speed of light and the statics of the gravitational field, English translation in *The Collected Papers of Albert Einstein* vol. 4, *The Swiss Years: Writings, 1912–1914*, pp. 95–106.

[61] Einstein, A. 1912b On the theory of the static gravitational field, English translation in *The Collected Papers of Albert Einstein* vol. 4, *The Swiss Years: Writings, 1912–1914*, pp. 107–120.

[62] Einstein, A. 1916 *Über die spezielle und allgemeine Relativitätstheorie*, Leipzig, English tr. R.W. Lawson, *Relativity; the special and the general theory, a popular exposition*, Methuen, London, 1920, many subsequent editions, rep. with an introduction by N. Calder, Penguin Classics, 2006.

[63] Encyclopaedia Britannica, http://www.britannica.com/.

[64] Engel, F., Stäckel, P. 1895 *Theorie der Parallellinien von Euklid bis auf Gauss*, Teubner, Leipzig.

[65] Engel, F. and Stäckel, P. 1913 *Urkunden zur Geschichte der Nichteuklidischen Geometrie, Wolfgang und Johann Bolyai*, Teubner, Leipzig and Berlin.

[66] Enriques, F. 1898 *Lezioni di geometria proiettiva*, Zanichelli, Bologna.

[67] Enriques, F. 1906 *Problemi della scienza*, tr. K. Royce as *Problems of science*, Open Court, Chicago, 1914.

[68] Enriques, F. 1907 Prinzipien der Geometrie, *Encyklopädie der Mathematischen Wissenschaften*, III.I.1, 1–129.

[69] Epple, M. 2002 From quaternions to cosmology: spaces of constant curvature, c. 1873–1925 *Proceedings of the International Congress of Mathematicians*, Beijing, vol. III, pp. 935–945, Higher Education Press, Beijing, 2002.

[70] Euclid 1956 *The thirteen books of Euclid*, ed. and tr. Sir T.L. Heath, Cambridge University Press, 3 vols., Dover reprint, New York, 1956.

[71] Euler, L. 1770 *Vollständige Anleitung zur Algebra*, St Petersburg.

[72] Fano, G. 1892 Sui postulati fondamentali della geometria proiettiva, *Giornale di Matematiche* 30, 106–131.

[73] Fauvel, J.G., Flood, R., Wilson, R.J. 1993 *Möbius and his band: Mathematics and astronomy in 19th-century Germany*, Oxford University Press, Oxford.

[74] Fauvel, J.G., Gray, J.J. 1987 *The history of mathematics; a reader*, Macmillan, London.

[75] Field, J.V., Gray, J.J. 1986 *The geometrical work of Girard Desargues*, Springer, New York.

[76] Fischer, G. 2001 *Plane algebraic curves*, tr. L. Kay, Students' Mathematical Library 15, American Mathematical Society, Providence RI.

[77] Fowler, D.H. 1998 *The mathematics of Plato's Academy: a new reconstruction*, Oxford University Press, New York.

[78] Frege, G. 1980 *Philosophical and mathematical correspondence*, G. Gabriel, H. Hermes, F. Kambartel, C. Thiel and A. Veraart (eds. of German edn.), abridged from German edn. by B. McGuinness, tr. H. Kraal, Blackwell, Oxford.

[79] Frege, G. 1903, 1906 On the Foundations of Geometry, in Frege [80, pp. 273–285, 293–340], original German papers in *Jahrsbericht der Deutschen Mathematiker-Vereinigung* 12, 319–324, 368–375 and 15, 293–309, 377–403, 423–430.

[80] Frege, G. 1984 *Collected papers on mathematics, logic, and philosophy*, B. McGuinness (ed.), Blackwell, Oxford.

[81] Freudenthal, H. 1957 Zur Geschichte der Grundlagen der Geometrie, *Nieuw Archief voor Wiskunde* 5(4), 105–142.

[82] Freudenthal, H. 1962 The main trends in the foundations of geometry in the 19th century, pp. 613–621, in *Logic, Methodology and Philosophy of Science*, E. Nagel, F. Suppes and A. Tarski (eds.), Stanford University Press, Stanford CA.

[83] Gauss, C.F. 1801 *Disquisitiones arithmeticae*, Leipzig, rep. in *Werke* vol. I.

[84] Gauss, C.F. 1828 *Disquisitiones generales circa superficies curvas* rep. in *Werke* vol. IV, pp. 217–258, ed. P. Dombrowski, in *Astérisque* 62, 1978, Latin original, with a reprint of the English translation by A. Hiltebeitel and J. Morehead, 1902, and as *General investigations of curved surfaces*, ed. P. Pesic, Dover Books, New York, 2005.

[85] Gauss, C.F. 1860–65. *Briefwechsel zwischen C.F. Gauss and H.C. Schumacher*, 6 vols., Altona.

[86] Gauss, C.F. 1870 *Werke* vol. I, Teubner, Leipzig.

[87] Gauss, C.F. 1880 *Werke* vol. IV, Teubner, Leipzig.

[88] Gauss, C.F. 1900 *Werke* vol. VIII, Teubner, Leipzig.

[89] Gergonne, J.D. 1827 Géométrie de situation, *Annales de Mathématiques Pures et Appliquées* 18, 125–216.

[90] Gillispie, C.C. 1981 *Dictionary of scientific biography*, American Society of Learned Societies/Scribner, New York.

[91] Gillispie, C.C. 1997 *Pierre-Simon Laplace, 1749–1827; a life in exact science*, Princeton University Press, Princeton NJ.

[92] Gow, R. 1997 George Salmon 1819–1904: his mathematical work and influence, *Irish Mathematical Society Bulletin* 39, 26–76.

[93] Grassmann, H.G. 1844 *Die Lineale Ausdehnungslehre...*, Leipzig, 2nd edn., *Die Ausdehnungslehre von 1844, oder die lineale Ausdehnungslehre...*, Leipzig, 1878.

[94] Grassmann, H.G. 1862 *Die Ausdehnungslehre...*, T.C.F. Enslin, Berlin.

[95] Grassmann, H.G. 1844 *Die Lineale Ausdehnungslehre*, Leipzig, English translation L.C. Kannenberg as *A new branch of mathematics: The "Ausdehnungslehre" of 1844 and other works*, Open Court, Chicago, 1995.

[96] Grassmann, H.G. 1862 *Die Ausdehnungslehre, Vollständig und in strenger Form*, T.C.F. Enslin, Berlin, English translation L.C. Kannenberg as *Extension theory*, American and London Mathematical Societies, Providence RI, 2000.

[97] Grattan-Guinness, I. 1990 *Convolutions in French mathematics, 1800–1840*, 3 vols., Birkhäuser Verlag, Basel.

[98] Grattan-Guinness, I. 1996 Number, magnitudes, ratios, and proportions, in Euclid's *Elements*: how did he handle them?, *Historia Mathematica* 23(4), 355–375.

[99] Gray, J.J., Tilling, L. 1978 Johann Heinrich Lambert, mathematician and scientist, 1728–1777, *Historia Mathematica* 5, 13–41.

[100] Gray, J.J. 1989 *Ideas of space, Euclidean, non-Euclidean and relativistic*, Oxford University Press, Oxford.

[101] Gray, J.J. 2000a *Linear differential equations and group theory from Riemann to Poincaré*, 2nd edn., Birkhäuser, Boston and Basel, 2003.

[102] Gray, J.J. 2000b *The Hilbert challenge*, Oxford University Press.

[103] Gray, J.J. 2004 *János Bolyai, non-Euclidean geometry and the nature of space*, Burndy Library, M.I.T. Press, 2004.

[104] Gray, J.J. 2005 A history of prizes in mathematics, pp. 3–27, in *The Millennium Prize Problems*, J. Carlson, A. Jaffe and A. Wiles (eds.), Clay Mathematics Institute and American Mathematics Society, Cambridge MA.

[105] Greenberg, M.J. 2008 *Euclidean and non-Euclidean geometries: development and history*, 4th edn., New York, W.H. Freeman, Basingstoke, Palgrave.

[106] Greenberg, M.J. 2010 Old and new results in the foundations of elementary plane Euclidean and non-Euclidean geometries, *American Mathematical Monthly*, March PAGES.

[107] Hallett, M., Majer, U. 2004 *David Hilbert's lectures on the foundations of geometry, 1891–1902*, Springer, Berlin.

[108] Hartshorne, R. 1967 *Foundations of projective geometry*, Benjamin, New York.

[109] Hartshorne, R. 2000 *Geometry: Euclid and beyond*, Springer, New York.

[110] Hawkins, T. 2000 *Emergence of the theory of Lie groups*, Springer, New York.

[111] Helmholtz, H. von 1868 Über die tatsächlichen Grundlagen der Geometrie, *Nachr. König. Ges. Wiss. zu Göttingen*, vol. 15, 193–221, in *Abhandlungen*, 2, 1883, 618–639, tr. On the facts underlying geometry, in *Hermann von Helmholtz, Epistemological Writings*, P. Hertz and M. Schlick (eds.), Boston Studies in the Philosophy of Science, Reidel, Dordrecht and Boston, 1977, pp. 39–71.

[112] Helmholtz, H. von 1870 Über den Ursprung und die Bedeutung der geometrischen Axiome, in *Populäre wissenschaftliche Vorträge, Braunschwieg*, 2, tr. M.F. Lowe On the origin and significance of the axioms of geometry, in *Hermann von Helmholtz, Epistemological Writings*, P. Hertz and M. Schlick (eds.), Boston Studies in the Philosophy of Science, 37, Reidel, Dordrecht and Boston, 1977, pp. 1–38.

[113] Henderson, L.D. 1983 *The fourth dimension and non-Euclidean geometry in modern art*, Princeton University Press, Princeton NJ.

[114] Hesse, L.O. 1848, in *Gesammelte Werke*, 1st pub. 1897, rep. Chelsea, 1972.

[115] Hesse, L.O. 1897 *Gesammelte Werke*, rep. Chelsea, New York, 1972.

[116] Hessenberg, G. 1905 Beweis des Desarguesschen Satzes aus dem Pascalschen, *Mathematische Annalen* 61, 161–172.

[117] Hilbert, D. 1894 *Zahlbericht (Report on the theory of numbers)*, Deutsche *Mathematiker-Vereinigung*, tr. I. Adamson as *The theory of algebraic number fields*, Springer, New York, 1998.

[118] Hilbert, D. 1899 *Grundlagen der Geometrie*, tr. E.J. Townsend as *The foundations of geometry*, Open Court, Chicago, 1902.

[119] Hilbert, D. 1899 *Grundlagen der Geometrie*, many subsequent editions, tr. L. Unger as *Foundations of geometry*, 10th English edn. of the 2nd German edn., Open Court, Illinois, 1971.

[120] Hilbert, D. 1902 Über die Grundlagen der Geometrie, *Mathematische Annalen* 56, 81–422, rep. in *Grundlagen der Geometrie*, 7th edn., Teubner, Leipzig, 1930, pp. 178–230.

[121] Hilbert, D. 1903 Neue Begrundung der Bolyai-Lobatschefskyschen Geometrie Über die Grundlagen der Geometrie, *Mathematische Annalen* 57, 137–150, rep. in *Grundlagen der Geometrie*, 7th edn., Teubner, Leipzig, 1930, pp. 159–177.

[122] Hölder, O. 1900 *Anschauung und Denken in der Geometrie*, Teubner, Leipzig.

[123] Hoüel, J. 1863 Essai d'une exposition rationelle des principes fondamentaux de la géométrie élémentaire, *Archiv der Mathematik und Physik* 40, 171–211.

[124] Huntington, E.V. 1924 A new set of postulates for betweenness, with proof of complete independence, *Transactions of the American Mathematical Society* 26, 257–283.

[125] Jacobi, C.G.J. 1850 Beweis des Satzes, dass eine Curve $n$ten Grades im Allgemeinen $\frac{1}{2}n(n-2)(n^2-9)$ Doppeltangenten hat, *Journal für die reine und angewandte Mathematik* 40, 237–260, in *Gesammelte Werke*, Berlin: Verlag von G. Reimer, 1881, III, 517–542, rep. Chelsea, New York, 1969.

[126] Kagan, V.F. 1957 *N. Lobachevskii and his contribution to science*, Foreign Languages Publishing House, Moscow.

[127] Killing, W. 1885 *Die nicht-Euklidischen Raumformen*, Teubner, Leipzig.

[128] Killing, W. 1893–1898 *Einführung in die Grundlagen der Geometrie*, 2 vols., Paderborn.

[129] Klein, C.F. 1871 Über die sogenannte nicht-Euklidische geometrie I, *Mathematische Annalen* 4, 573–625, in F. Klein, *Ges. Math. Abh.*, vol. 1, 254–305.

[130] Klein, C.F. 1872 Vergleichende Betrachtungen über neuere geometrische Forschungen (Erlanger Programm), 1st pub. Deichert, Erlangen, in *Ges. Math. Abh.*, vol. 1 (no. XXVII), 460–497.

[131] Klein, C.F. 1873 Über die sogenannte nicht-Euklidische Geometrie II, *Mathematische Annalen* 6, in F. Klein, *Ges. Math. Abh.*, vol. 1, 311–343.

[132] Klein, C.F. 1874a Nachtrag zu dem "zweiten Aufsatz über nicht-Euclidische Geometrie", *Mathematische Annalen* 7, 531–537, in *Ges. Math. Abh*, vol. 1, 344–350.

[133] Klein, C.F. 1874b Bemerkungen über den Zusammenhang der Flächen, *Mathematische Annalen* 7, 549–557, and 9, 476–482, in *Ges. Math. Abh*, vol. 2, 63–77.

[134] Klein, C.F. 1879 Über die Transformation siebenter Ordnung der elliptischen Funktionen, *Mathematische Annalen* 14, in *Ges. Math. Abh.*, vol. 3 (no. LXXXIV), 90, 134.

[135] Klein, C.F. 1880 Über die geometrische Definition der Projectivität auf den Grundgebilden der ersten Stufe, *Mathematische Annalen* 17, 52–54, in *Ges. Math. Abh.*, vol. 1, (no. XXVII), 351–352.

[136] Klein, C.F. 1882 *Über Riemanns Theorie der algebraischen Funktionen und ihrer Integrale*, Teubner, Leipzig, in *Ges. Math. Abh.* III (no. XCIX), 499–573, tr. F. Hardcastle as *Riemann's theory of algebraic functions and their integrals*, Macmillan and Bowes, Cambridge, 1893.

[137] Klein, C.F. 1921–1923 *Gesammelte mathematische Abhandlungen*, 3 vols., R. Fricke and A. Ostrowski (eds.), Springer, Berlin.

[138] Klein, C.F. 1899 Zur nicht-Euklidischen Geometrie, in C.F. Klein, *Ges. Math. Abh.*, vol. 1, 353–383.

[139] Klein, C.F. (ed.) various dates *Encyklopädie der mathematischen Wissenschaften*, 23 vols.

[140] Klein, C.F. 1926 *Vorlesungen über die Entwicklung der Mathematik im 19. Jahrhundert*, vol. 1, Teubner, ed. R. Courant, O. Neugebauer, re-edition, Chelsea, New York, 1967.

[141] Klein, C.F. 1923 Autobiography, *Göttingen Mitteilungen des Universitäts Bundes Göttingen* 5(1).

[142] Kline, M. 1972 *Mathematical thought from ancient to modern times*, Oxford University Press, Oxford.

[143] Klügel, A.G. 1763 *Conatuum praecipuorum theoriam parallelarum demonstrandi recensio...*, Göttingen.

[144] La Hire, P. de 1685 *Sectiones conicae, in novem libros distributae*, Paris.

[145] Lambert, J.H. 1759 *J.H.L.'s freye Perspektive, oder Anweisung, jeden perspektivischen Aufriss von freyen Stücken und ohne Grundriss zu verfertigen*, 2nd edn., 1774.

[146] Lambert, J.H. 1761 *Cosmologische Briefe über die Einrichtung des Weltbaues*, Klett, Augsburg, tr. with introduction and notes by Stanley L. Jaki, *Cosmological letters on the arrangement of the world-edifice*, Scottish Academic Press, Edinburgh, 1976.

[147] Lambert, J.H. 1786 Theorie der Parallellinien, in F. Engel and P. Stäckel [64].

[148] Laugwitz, D. 1999 *Bernhard Riemann, 1826-1866: Turning points in the conception of mathematics*, tr. Abe Shenitzer with the editorial assistance of the author, Hardy Grant and Sarah Shenitzer, Birkhäuser Verlag.

[149] Legendre, A.M. 1794 *Éléments de géométrie*, Paris, with many subsequent editions, e.g. 12th edn., 1823, Paris.

[150] Listing, J.B. 1861 Der Census räumlicher Complexe, *Abh. K. Ges. Wiss. Göttingen, Math-Phys Cl.* 10, 97–182.

[151] Lobachevskii, N.I. 1836, 1837, 1838 *New elements of geometry, with a complete theory of parallels* (in Russian), *Gelehrten Schriften der Universität Kasan*, German translation in Lobatschefskij [155].

[152] Lobachevskii, N.I. 1837 Géométrie imaginaire, *Journal für die reine und angewandte Mathematik* 17, 295–320.

[153] Lobachevskii, N.I. 1840 *Geometrische Untersuchungen*, Berlin, rep. Mayer and Müller, 1887, tr. J. Hoüel as Études géométriques sur la théorie des parallèles, *Mémoires de la Société des Sciences Physiques et Naturelles de Bordeaux*, 4, 1867, pp. 83–128, rep. Gauthier-Villars, Paris, 1866, tr. G.B. Halsted as *Geometrical researches on the theory of parallels*, Open Court, Chicago, 1914. Appendix in Bonola [21].

[154] Lobachevskii, N.I. 1856 *Pangéométrie, ou précis de géométrie fondée sur une théorie générale des parallèles*, Kasan, *Pangéométrie*, tr. and ed. H. Liebmann, Leipzig, Engelmann, 1912.

[155] Lobatschefskij, N.I. 1899 *Zwei geometrische Abhandlungen*, tr. F. Engel, Teubner, Leipzig.

[156] Lobachevskii, N.I. 2010 *Pangeometry*, ed. and tr. Athanase Papadopoulos, Herit. Eur. Math., European Math. Soc. Publ. House, Zürich.

[157] Loria, G. 1904 Luigi Cremona et son oeuvre mathématique, *Bibliotheca Mathematica* 3(5), 125–195.

[158] Lützen, J. 1990 *Joseph Liouville, 1809–1882: Master of pure and applied mathematics*, Studies in the History of Mathematics and Physical Sciences 15, Springer, New York.

[159] Martin, G.E. 1996 *The foundations of geometry and the non-Euclidean plane*, corrected 3rd printing of 1975 original, Undergraduate Texts in Mathematics, Springer, New York.

[160] Marx, K. 1852 *The eighteenth Brumaire of Louis Napoleon*, German edition 1869, online version at Marxists internet archive http://marxists.org/, accessed 7/6/06, 1995, 1999.

[161] Meschkowski, H. 1964 *Non-Euclidean geometry*, tr. A. Shenitzer, Academic Press, New York.

[162] Minding, H.F. 1839 Wie sich entscheiden lässt, ob zwei gegebener Krummen Flächen..., *Journal für die reine und angewandte Mathematik* 19, 370–387.

[163] Minkowski, H. 1908 Raum und Zeit, in *Jahrsbericht den Deutschen mathematiker Vereinigung* 1908, English translation in *The Principle of Relativity*, A. Einstein et al., Dover, New York, 1952.

[164] Mittler, E. (ed.) 2005 *Wie der Blitz einschlägt, hat sich das Rätsel gelöst – Carl Friedrich Gauß in Göttingen*, Göttinger Biblioteksschriften Nr. 30. Available at http://www.sub.uni-goettingen.de/ebene_1/shop/schriften.html

[165] Möbius, A.F. 1827 *Der barycentrische Calcul*, Hirzel, Leipzig.

[166] Möbius, A.F. 1865 Über die Bestimmung des Inhalts eines Polyëders, *Berichte der Sächsische Ges. Wiss., Leipzig, Math-Phys Cl.* 17, 31–68, in *Werke*, 2, pp. 473–512.

[167] Möbius, A.F. 1885–1887 *Gesammelte Werke*, R. Baltzer and W. Scheibner (eds.), Hirzel, Leipzig.

[168] Monge, G. 1811 *Traité de géométrie descriptive*, Paris.

[169] Monge, G. 1850 *Application de l'analyse à la géométrie...*, 5th edn., revised, corrected and annotated by M. Liouville, Paris.

[170] Moore, E.H. 1902 On the projective axioms of geometry, *Transactions of the American Mathematical Society* 3, 142–158.

[171] Morgan, F. 1998 *Riemannian geometry: A beginner's guide*, 2nd edn., A K Peters Ltd., Wellesley MA.

[172] Moulton, R.F. 1902 A simple non-Desarguesian plane geometry, *Transactions of the American Mathematical Society* 3, 192–195.

[173] Mumford, D., Series, C., Wright, D. 2002 *Indra's pearls: The vision of Felix Klein*, Cambridge University Press, Cambridge.

[174] Nagel, E. 1939 The formation of modern conceptions of formal logic in the development of geometry, *Osiris*, 7, 142–224.

[175] Neuenschwander, E. 1981 Lettres de Bernhard Riemann à sa famille, *Cahiers du Séminaire d'Histoire des Mathématiques* 2, 85–131.

[176] Newton, I. 1968 *The mathematical papers of Isaac Newton vol. II*: 1667–1670, ed. D.T. Whiteside, with the assistance in publication of M.A. Hoskin, Cambridge University Press, London–New York.

[177] Newton, I. 1969 *The mathematical papers of Isaac Newton vol. III*: 1670–1673, ed. D.T. Whiteside, with the assistance in publication of M.A. Hoskin and A. Prag, Cambridge University Press, London–New York.

[178] O'Connor, J.J., Robertson, E.F. 1996 Non-Euclidean geometry [online] http://www-groups.dcs.st-and.ac.uk/~history/HistTopics/NonEuclidean_geometry.html accessed 3/6/2006.

[179] Olesko, K.M. 1991 *Physics as a calling: Discipline and practice in the Königsberg seminar for physics*, Cornell University Press, Ithaca NY.

[180] O'Neill, B. 1966 *Elementary differential geometry*, Academic Press, New York.

[181] Ore, O. 1974 *Niels Henrik Abel; mathematician extraordinary*, Chelsea, New York.

[182] Padoa, A. 1902 Un nouveau système de définitions pour la géométrie euclidienne, *Compte Rendu du Deuxième Congrès International des Mathématiciens*, Paris.

[183] Parshall, K.H. 2006 *James Joseph Sylvester: Jewish mathematician in a Victorian world*, History of Science and Technology Series, Johns Hopkins University Press, Baltimore MA.

[184] Pasch, M. 1882 *Vorlesungen über neuere Geometrie*, Teubner, Leipzig.

[185] Peano, G. 1891 Osservazione del Direttore, *Rivista di Matematica* 1, 66–69.

[186] Peano, G. 1894 Sui fondamenti della geometria, *Rivista di Matematica* 4, 73.

[187] Peano, G. 1889 *I principii di geometria logicamente espositi*, Turin, rep. in *Opere scelte* 2, Rome, 1958, pp. 56–91.

[188] Pesic, P. 2007 *Beyond geometry: Classic papers from Riemann to Einstein*, Dover, New York.

[189] Peters, C.A.F. (ed.) 1860–1863 *Briefwechsel zwischen C.F. Gauss und H.C. Schumacher*, 6 vols., Esch, Altona.

[190] Pieri, M. 1895 Sui principi che reggiono la geometria di posizione, *Atti Accademia Torino* 30, 54–108.

[191] Pieri, M. 1899 I principii della geometria di posizione, composti in sistema logico deduttivo, *Memorie della Reale Accademia delle Scienze di Torino* 48(2), 1–62.

[192] Plücker, J. 1828, 1831 *Analytisch-Geometrische Entwicklungen*, 2 vols., Baedecker, Essen.

[193] Plücker, J. 1830 Über ein neues Coordinatensystem, *Journal für die reine und angewandte Mathematik* 5, 1–36.

[194] Plücker, J. 1835 *System der analytischen Geometrie*, Berlin.

[195] Plücker, J. 1839 *Theorie der algebraischen Curven*, Bonn.

[196] Poincaré, H. 1903 *La science et l'hypothèse*, Paris, tr. W.J. Greenstreet as *Science and hypothesis*, Walter Scott Publishing, London, 1905, rep. Dover, 1952.

[197] Poincaré, H. 1905 Non-Euclidean geometries, pp. 35–50, in *Science and Hypothesis*, rep. Dover, 1952.

[198] Poincaré, H. 1909 *Science et méthode*, Flammarion, Paris.

[199] Poincaré, H. 1916–1954 *Oeuvres*, 11 vols., Paris.

[200] Poincaré, H. 1985 *Papers on Fuchsian functions*, Springer, Heidelberg and New York.

[201] Poincaré, H. 1997 *Three supplementary essays on the discovery of Fuchsian functions*, ed. J.J. Gray and S. Walter, with an introductory essay, Akademie Verlag, Berlin and Blanchard, Paris.

[202] Poncelet, J.-V. 1822 *Traité des propriétés projectives des figures*, Gauthier-Villars, Paris.

[203] Poncelet, J.-V. 1832 Analyse des transversales..., *Journal für die Reine und Angewandte Mathematik* 4, 38–158.

[204] Poncelet, J.-V. 1834 *Notice analytique sur les travaux de M. Poncelet*, Bachelier, Paris.

[205] Poncelet, J.-V. 1862–1864 *Applications d'analyse et de géométrie*, 2 vols., Mallet-Bachelier, Paris.

[206] Pont, J.-C. 1986 *L'aventure des parallèles*, Lang, Berne.

[207] Reid, C. 1970 *Hilbert*, Springer, New York.

[208] Richards, J.L. 1988 *Mathematical visions: The pursuit of geometry in Victorian England*, Academic Press, Boston.

[209] Riemann, B. 1854 *Habilitatationsvortrag*, tr. in Clifford [39, pp. 55–71], originally in *Nature*, 1873.

[210] Riemann, B. 1990 *Bernhard Riemann's gesammelte mathematische Werke und wissenschaftlicher Nachlass*, R. Dedekind and H. Weber, with Nachträge, ed. M. Noether and W. Wirtinger, 3rd edn. R. Narasimhan (ed.), Springer, New York, 1st edn. R. Dedekind, H. Weber (eds.), Leipzig, 1876.

[211] Rosenfeld, B.A. 1988 *A history of non-Euclidean geometry: Evolution of the concept of a geometric space*, Springer, New York.

[212] Rousseau, J.J. 1762 *Émile: ou de l'éducation*, numerous editions, e.g. La Haye, chez Jean Néaulme; [i.e. Paris: Duchesne], 1762 and Amsterdam [i.e. Lyon] : chez Jean Néaulme, 1762.

[213] Rüdenberg, L. and Zassenhaus H. (eds.) 1973 *Hermann Minkowski – briefe an David Hilbert*, Springer, Berlin and New York.

[214] Saccheri, G. 1733 *Euclides ab omni naevo vindicatus*, Milan, ed. and tr. G.B. Halsted, *Girolamo Saccheri's Euclides vindicatus*, Open Court, Chicago, 1920.

[215] Salmon, G. 1852 *A treatise on the higher plane curves*, Hodges, Foster and Figgis, Dublin, 3rd edn., 1879.

[216] Samuel, P. 1988 *Projective geometry*, Springer, New York.

[217] Sartorius, W. von Waltershausen 1856 *Gauss zum Gedächtnis*, Hirzel, Leipzig, rep. Martin Sandig oHG, 1965.

[218] Scholz, E. 1982 Herbart's influence on Bernhard Riemann, *Historia Mathematica* 9(4), 413–440.

[219] Scholz, E. 2004 C.F. Gauss' Präzisionsmessungen terrestricher Dreiecke und seine Überlegungen zur empirischen Fundierung der Geometrie in den 1820er Jahren, pp. 355–380, in *Form, Zahl, Ordung; Studien zur Wissenschafts- und Technikgeschichte*, R. Seising, M. Folkerts and U. Hashagen (eds.), Franz Steiner Verlag.

[220] Schur, F. 1899 Über die Fundamentalsatz der projectiven Geometrie, *Mathematische Annalen* 51, 401–409.

[221] Schweitzer, A.R. 1909 A theory of geometrical relations, *American Journal of Mathematics* 31, 365–410.

[222] Segre, C. 1884 Studio sulle quadriche in uno spazio lineare a un numero qualunque di dimensioni, *Memorie della Reale Accademia delle Scienze di Torino* 36(2), 3–86.

[223] Segre, M. 1994 Peano's axioms in their historical context, *Archive for History of Exact Sciences* 48, 201–342.

[224] Silvester, J.R. 2001 *Geometry, ancient and modern*, Oxford University Press, Oxford.

[225] Smith, H.J.S. 1877 On the present state of some branches of pure mathematics, *Proceedings of the London Mathematical Society* 8, 6–29.

[226] Stachel, J. 1989 The rigidly rotating disc as the "missing link" in the history of general relativity, pp. 48–62, in *Einstein and the History of General Relativity*, D. Howard and J. Stachel (eds.) Birkhäuser, Boston and Basel.

[227] Stäckel, P. 1913 *Wolfgang und Johann Bolyai, geometrische Untersuchungen, Leben und Schriften der beiden Bolyai*, Teubner, Leipzig and Berlin.

[228] Staudt, K.G.C. von 1847 *Geometrie der Lage*, Bauer and Raspe, Nürnberg.

[229] Staudt, K.G.C. von 1856 *Beiträge zur Geometrie der Lage*, Bauer and Raspe, Nürnberg.

[230] Struik, D.J. 1988 *Lectures on classical differential geometry*, 2nd edn., Dover Publications Inc., New York.

[231] Steiner, J. 1832 *Systematische Entwicklung der Abhängigkeit geometrischer Gestalten von einander*, Leipzig.

[232] Stillwell, J. 1996 *Sources of hyperbolic geometry*, American and London Mathematical Societies.

[233] Sylvester, J.J. 1869 "A Plea for the Mathematician" Presidential address to the British Association for the Advancement of Science Section on Mathematics and Physics, pp. 650–661, in vol. 2, *The Collected Mathematical Papers of James Joseph Sylvester*, H.F. Baker (ed.), 4 vols., 1904–1912.

[234] Taton, R. 1951 *L'oeuvre mathématique de G. Desargues*, 2nd edn., 1988, Vrin, Paris.

[235] Taton, R. 1951 *L'oeuvre scientifique de Monge*, Paris.

[236] Toepell, M.-M. 1986 *Über die Entstehung von David Hilberts "Grundlagen der Geometrie"*, Vandenhoeck & Ruprecht, Göttingen.

[237] Toth, I. 1969 Non-Euclidean geometry before Euclid, *Scientific American* 221(5), 87–98.

[238] Toth, I. 2000 *Palimpseste: Propos avant un triangle*, Presses Universitaire de France, Paris.

[239] Tribout de Morembert, H. 1936 *Un grand savant. Le général Jean-Victor Poncelet, 1788–1867*, p. 225. Paris. 8o. [With a portrait.]

[240] Vassilief, A. 1896 *Éloge historique de Nicolas-J. Lobatchevsky*, Hermann, Paris.

[241] Veblen, O. 1904 A system of axioms for geometry, *Transactions of the American Mathematical Society* 5, 343–384.

[242] Veblen, O. 1905 *Princeton lectures on the foundations of geometry*, The University of Chicago Press, Chicago and London.

[243] Veronese, G. 1882 Behandlung der projectivischen Verhältnisse der Räume von verschiedenen Dimensionen durch das Princip des Projicirens und Schneidens, *Mathematische Annalen* 19, 161–234.

[244] Voelke, J.-D. 2005 *Renaissance de la géométrie non euclidienne entre 1860 et 1900*, Peter Lang, Bern.

[245] Walker, R.J. 1978 *Algebraic curves*, rep. of 1950 edn., Springer, New York.

[246] Wallis, J. 1693 De postulato quinto et definitione lib. 6 Euclidis deceptatio geometrica, pp. 665–678 in *Operum Mathematicorum*, vol. 2.

[247] Weyl, H.K.H. 1944 David Hilbert and his mathematical work, *Gesammelte Abhandlungen* 4, Springer, Berlin.

[248] Wiener, H.L.G. 1890 Über Grundlagen und Aufbau der Geometrie, *Jahresbericht den Deutschen Mathematiker-Vereinigung* 1, 45–48.

[249] Zeuthen, H.G. 1876 Note sur les singularités des courbes planes, *Mathematische Annalen* 10, 210–220.

Some Geometers

Ampère, André-Marie 1775–1836

Arago, François Jean Dominique 1786–1853

Artin, Emil 1898–1962

Ball, Robert 1840–1913

Bartels, Martin 1769–1836

Beltrami, Eugenio 1835–1900

Bertrand, Joseph 1822–1900

Bessel, Wilhelm 1784–1836

Blumenthal, Otto 1876–1944

Bolyai, Wolfgang Farkas 1775–1856

Bolyai, János 1802–1860

Brianchon, Charles Julien 1783–1864

Cantor, Georg Ferdinand Ludwig Philipp 1845–1918

Carnot, Lazare Nicolas Marguérite 1753–1823

Cauchy, Augustin Louis 1789–1857

Cayley, Arthur 1821–1895

Chasles, Michel 1793–1880

Clebsch, Rudolf Friedrich Alfred 1833–1872

Clifford, William Kingdon 1845–1879

Codazzi, Delfino 1824–1873

Crelle, August Leopold 1780–1855

Cremona, Antonio Luigi Gaudenzio Giuseppe 1830–1903

Darboux, Jean Gaston 1842–1917

Dedekind, Julius Wilhelm Richard 1831–1916

Dehn, Max 1878–1952

Desargues, Girard 1591–1661

Descartes, René 1596–1650

Dirichlet, Johann Peter Gustav Lejeune 1805–1859

Dupin, Pierre Charles François 1784–1873

Einstein, Albert 1879–1955

Enriques, Federigo 1871–1946

Fano, Gino 1871–1952

Frege, Friedrich Ludwig Gottlob 1848–1925

Fuss, Nicolaus 1755–1826

Gauss, Johann Carl Friedrich 1777–1855

Geiser, Karl Friedrich 1843–1934

Gergonne, Joseph Diaz 1771–1859

Gerling, Christian Ludwig 1788–1864

Grassmann, Hermann 1809–1877

Hankel, Hermann 1839–1873

Helmholtz, Hermann Ludwig Ferdinand von 1821–1894

Herbart, Johann Friedrich 1776–1841

Hermite, Charles 1822–1901

Hesse, Ludwig Otto 1811–1874

Hessenberg, Gerhard 1874–1925

Hilbert, David 1862–1943

Hölder, Otto Ludwig 1859–1937

Hoüel, Guillaume Jules 1823–1886

Hurwitz, Adolf 1859–1919

Jacobi, Carl Gustav Jacob 1804–1851

Jordan, Marie Ennemond Camille 1838–1922

Kaestner, Abraham Gotthelf 1719–1800

Kant, Immanuel 1724–1804

J. Gray, *Worlds Out of Nothing*,
Springer Undergraduate Mathematics Series,
DOI 10.1007/978-0-85729-060-1, © Springer-Verlag London Limited 2011

Killing, Wilhelm Karl Joseph
  1847–1923
Klein, Felix Christian 1849–1925
Klügel, Georg Simon 1739–1812
Lacroix, Sylvestre François 1765–1843
Lagrange, Joseph Louis 1736–1813
Lambert, Johann Heinrich 1728–1777
Laplace, Pierre Simon 1749–1827
Legendre, Adrien Marie 1752–1833
Lie, Marius Sophus 1842–1899
Lindemann, Carl Louis Ferdinand
  von 1852–1939
Liouville, Joseph 1809–1882
Lipschitz, Rudolf Otto Sigismund
  1832–1903
Listing, Johann Benedict 1708–1882
Lobachevskii, Nikolai Ivanovich
  1792–1856
Lotze, Rudolf Hermann 1817–1881
Lüroth, Jacob 1844–1910
Minding, Ernst Ferdinand Adolf
  1806–1885
Minkowski, Hermann 1864–1909
Möbius, August Ferdinand 1790–1868
Monge, Gaspard 1746–1818
Moore, Eliakim Hastings 1862–1932
Moulton, Forest Ray 1872–1952
Olbers, Heinrich Wilhelm Matthäus
  1758–1840
Ostrogradskii, Mikhail Vasilevich
  1801–1862
Pascal, Blaise 1623–1662
Pasch, Moritz 1843–1930
Peano, Giuseppe 1858–1932
Pieri, Mario 1860–1913
Playfair, John 1748–1819
Plücker, Julius 1801–1868
Poincaré, Henri 1854–1912

Poncelet, Jean-Victor 1788–1867
Ricci-Curbastro, Gregorio 1853–1925
Riemann, Georg Friedrich Bernhard
  1826–1866
Russell, Bertrand Arthur William,
  Earl 1872–1970
Saccheri, Giovanni Girolamo
  1667–1733
Salmon, George 1819–1904
Sartorius, Wolfgang von Walter-
  shausen 1809–1876
Schering, Ernst Christian Julius
  1833–1897
Schläfli, Ludwig 1814–1895
Schubert, Hermann Cäsar Hannibal
  1848–1911
Schumacher, Heinrich Christian
  1780–1850
Schur, Friedrich Heinrich 1856–1932
Schweikart, Ferdinand Karl 1780–
  1859
Segre, Corrado 1863–1924
Smith, Henry John Stephen 1826–
  1883
Staudt, Karl Georg Christian von
  1798–1867
Steiner, Jakob 1796–1863
Stolz, Otto 1842–1905
Sylvester, James Joseph 1814–1897
Taurinus, Franz Adolph 1794–1874
Veblen, Oswald 1880–1960
Veronese, Giuseppe 1854–1917
Weber, Heinrich Martin 1842–1913
Weierstrass, Karl Theodor Wilhelm
  1815–1897
Weyl, Hermann Klaus Hugo
  1885–1955
Wiener, Hermann 1857–1939

Index

Abel, Niels Henrik, 194
absolute measure of length, 86, 88
affine geometry, 235, 356
Ampère, André-Marie, 65
angle of parallelism, 109–110, 125, 224
Apollonius of Perga, 14, 16, 21, 23, 44
Arago, François Jean Dominique, 7, 8,
  47, 48, 66, 75
Archimedes of Syracuse, 208
Aristophanes, 112
Aristotle, 80, 336
Aronhold, Siegfried, 170
Artin, Emil, 356, 357
astral geometry, 94, 120, 126, 133
axiom, 19, 81, 227, 257, 266, 302, 317,
  336, 348
– Archimedean, 262, 263, 357
– Artin's, 356
– Fano's, 355
– Hilbert's, 263
– Hilbert's 18th, 336
– Pasch's, 255, 256
– Playfair's, 80
– system, 201, 266, 310, 336, 337

Ball, Robert, 248, 249, 345
Baltzer, Heinrich Richard, 217
Balzac, Honoré de, 8
Bartels, Martin, 91, 116, 132
barycentric coordinates, 150, 154
Belidor, Bernard Forest de, 12
Beltrami, Eugenio, vi, 202, 215–225,
  227–229, 234, 239, 246, 257, 285, 286,
  310, 338, 351
– his Saggio, 215, 219
– his Teoria, 217, 227
Beltrami disc, 222, 225
Bernoulli, Daniel, 12
Bernoulli, Johann (III), 85
Bertrand, Joseph, 283
Bessel, Wilhelm, 97, 134, 169, 170, 299
Bézout, Etienne, 12, 176, 186, 193
Bézout's theorem, 176, 186, 193
bitangent, 164–166, 173–175, 177, 178,
  193
Blumenthal, Otto, 260, 262
Bobillier, Etienne, 170
Bolyai, Wolfgang Farkas, 92, 93, 101,
  130–132, 288
Bolyai, János, v, 100–107, 112, 115, 120,
  131, 132, 134, 135, 143, 144, 147, 198,
  199, 229, 231, 234, 310, 326, 328, 334
Borda, Jean Charles de, 12
Brianchon, Charles Julien, 13, 21, 23,
  51, 53–55, 68, 74
Brianchon's theorem, 55
Brioschi, Francesco, 249
Bunsen, Robert Wilhelm, 170

Cantor, Georg Ferdinand Ludwig
  Philipp, 261, 337, 350
Carnot, Lazare Nicolas Marguérite, 13,
  14, 52
Castelnuovo, Guido, 273
cathode rays, 167

J. Gray, *Worlds Out of Nothing*,
Springer Undergraduate Mathematics Series,
DOI 10.1007/978-0-85729-060-1, © Springer-Verlag London Limited 2011

Cauchy, Augustin Louis, v, 47–51, 63,
    65–67, 156, 198, 350
Cayley, Arthur, 168, 192, 231, 233, 234,
    236, 247–249, 254, 345
Cayley metric, 231, 233
central projection, 17, 69, 251
Ceva, Giovanni, 28
Chasles, Michel, vii, xi, 50–52, 55, 63,
    67–74, 156, 170, 243, 251, 254, 345
Chateaubriand, François René de, 13
Clebsch, Rudolf Friedrich Alfred, 156,
    170, 176, 193, 194, 197, 231, 232, 248
Clifford, William Kingdon, 200, 202,
    204, 258
coaxial circles, 10
Codazzi, Delfino, 216, 227
collinearity, 35, 73, 156, 256
Columbus, Christopher, 135
complex curve, 192, 193, 352
concurrence, 35, 73, 156, 256
conic, 16, 17, 23, 28, 33, 34, 47, 49, 71,
    155, 184, 243, 314, 347, 348, 354
– conic part, 184
– construction of, 72, 73
– definition of, 71, 72, 347
– pair of, 40
conjugate diameters, 44, 45, 52, 253
conventionalism, vi, 300
coordinate, 150, 151, 154, 212, 219, 239,
    263, 273, 323, 327, 338, 346, 348, 351,
    356
– affine, 165, 184
– barycentric, 150–154, 157, 309
– Cartesian, 152–154, 330, 331
– complex, 45, 192
– geodesic, 212
– homogeneous, 39, 161, 167, 168, 182,
    187, 188
– line, 155, 158, 165
– non-Euclidean, 237, 239
– plane, 157
– polar, 212, 220, 239, 240, 296
– projective, 154, 155, 157, 159, 165,
    263, 271, 279
– spherical, 211, 212
Copernicus, Nicolaus, 115, 135
Cornu, Alfred, 282
Coulomb, Charles Augustin de, 12
Couturat, Louis, 276
Crelle, August Leopold, 67, 119, 144,
    167, 194
Cremona, Antonio Luigi Gaudenzio
    Giuseppe, xi, 244, 249–252, 254, 310

cross-ratio, 33, 36–40, 68–73, 77, 159,
    233, 234, 236, 243, 244, 251, 256, 339,
    348
– called anharmonic ratio, 71
– definition of, 36, 70, 77, 243, 256, 345
– in non-Euclidean geometry, 233, 310
– invariance of, 36, 68, 70, 159, 234, 235,
    244, 253
cubic curve, 56, 57, 60, 68, 163–165, 174,
    181, 185, 186, 245, 278
curve, 60, 164, 174, 175, 193, 244, 245
– algebraic, 161, 164, 165, 168, 173, 176,
    193, 247, 309
– closed, 352
– complex, 191, 193, 309, 352
– dual of, 56, 57, 60, 164, 173, 174, 176
– non-singular, 179, 190, 270, 278
– singular, 251
– space-filling, 270
cusp, 163–165, 173, 174, 176, 181, 185,
    186, 216, 244, 278, 309

Darboux, Jean Gaston, 282, 350
Dedekind, Julius Wilhelm Richard,
    196–198, 260, 346, 350
Dehn, Max, 264, 265
Desargues, Girard, 23, 25, 26, 28, 29, 32,
    35, 38, 51, 53, 73, 242, 262–264, 310,
    347–351, 355, 356
Desargues' theorem, 23, 26, 28, 32, 38,
    78, 263, 264, 348, 355
Descartes, René, 73, 168, 176, 186, 245,
    321
descriptive geometry, 5, 7–9, 46, 48, 51,
    55, 66, 69, 73–75, 141, 142, 232, 247,
    273, 282
differential geometry, 7, 65, 114, 198,
    199, 217, 285, 295
Dirichlet, Johann Peter Gustav Lejeune,
    196
Dostoyevsky, Fyodor Mikhailovich, 135
double point, 163, 164, 173, 174, 176,
    180, 181, 183, 189, 278, 309
duality, xi, 22, 23, 35, 44, 53–57, 60,
    61, 63, 67, 73, 155–158, 174, 251–253,
    256, 274, 309, 313, 347
duality paradox, vi, 161, 163, 164, 173,
    176, 177
dual theorem, 53
Dupin, Pierre Charles François, 5, 51
Durége, Heinrich, 170

Einstein, Albert, vi, 115, 134, 202, 300, 322, 323, 326–328, 330, 331
elliptic function, 169, 193, 194, 352
Encke, Johann Franz, 133
Enriques, Federigo, 273, 275, 277, 301–303, 311, 351, 354
Erlangen Program, 232, 235–237
Esperanto, 270
Euclid of Alexandria, 69, 79–82, 94–96, 108, 115, 118, 119, 124, 135, 198, 201, 203, 207, 229, 257, 266, 305–307, 326, 337, 346
Euclid's *Elements*, 18, 52, 79–82, 101, 105, 225, 251, 255, 288, 334
Euler, Leonhard, v, 12, 13, 84, 101, 188, 189, 193

Fano, Gino, 269, 272, 273, 311, 354, 355
First Fundamental Form, 211, 212
first polar, 183, 184, 189, 190
Flamm, 328
Fourier, Jean Baptiste Joseph, v, 48
fourth harmonic point, 26–28, 33, 38, 39, 54, 71, 273, 347–349, 355
Frege, Friedrich Ludwig Gottlob, 336, 337
Fresnel, Augustin Jean, 322
Freudenthal, Hans, 269
Fuchsian functions, 283, 284
Fuss, Nicolaus, 117

Galileo Galilei, 14, 208
Galois, Evariste, 92, 197
Gauss, Johann Carl Friedrich, vi, vii, 91–93, 95–101, 107, 116, 120, 130–134, 138, 139, 144, 146, 147, 156, 169, 194, 196, 198–202, 217, 228, 229, 235, 248, 260, 262, 265, 282, 299, 302, 309, 327, 331, 334, 351
– on the parallel postulate, 93
Gaussian curvature, 98, 99
Geiser, Karl Friedrich, 327
general relativity, 322, 327
geodesic, 98, 129, 199, 206, 208, 212, 214, 219–224, 227, 239, 286, 294, 296, 310, 351
geodetic projection, 212
Gergonne, Joseph Diaz, 14, 41, 51, 55–57, 59–61, 63, 67, 68, 74, 166, 167
Gerling, Christian Ludwig, 94, 95, 131
Gordan, Paul Albert, 260
Grassmann, Hermann, viii, 150
great circle, 24, 119, 199, 213, 214, 353
Grossmann, Marcel, 327

group, viii, 2, 193, 194, 235–238, 251, 271, 284, 287, 343, 346, 354, 356
group of transformations, 236
group theory, 236, 237, 287, 346
Grundsatz, 255, 256

HAA, 81, 82, 85–87, 94
Habilitation, 169, 196–198, 260
Hachette, Jean Nicolas Pierre, 51
Halphen, George Henri, 283
Hamilton, William Rowan, 150
Hankel, Hermann, 197
Hattendorff, Karl, 197
Heath, Sir Thomas Little, 79
Heiberg, Johan Ludwig, 79
Heine, Heinrich Eduard, 350
Helmholtz, Hermann Ludwig Ferdinand von, 170, 248, 256, 257
Herbart, Johann Friedrich, 197, 229
Hermite, Charles, 283
Hesse, Ludwig Otto, 161, 165, 168–171, 193, 244, 260
Hessenberg, Gerhard, 263, 355
Hessian, 169, 176, 185–187, 190
higher dimensions, 158, 270, 277
higher plane curve, vi, 15–17, 56, 59, 60, 161, 164, 179
Hilbert, David, vi, vii, 115, 228, 259–266, 269, 270, 275, 277, 303, 310, 314, 315, 326, 336–338, 354, 357
HOA, 81, 82, 85, 88
Hölder, Otto Ludwig, 338, 357
homography, 70, 71
horocycle, 120, 289
horosphere, 120
Hoüel, Guillaume Jules, 229
HRA, 81
Huntington, Edward V., 317
Hurwitz, Adolf, 260
Huxley, Thomas, 248
Huygens, Christiaan, 13
hyperbolic functions, 96, 224
hyperbolic geometry, 219
hyperbolic trigonometry, 216
hyperboloid, 237–239, 285

ideal points, 18, 69, 256
imaginary points, 51
inflection points, 161, 164, 165, 168, 169, 173–176, 185, 186, 190
inversion, 235, 289–292, 350
involution, 242, 243, 254

Jackson, Nicholas, xii

Jacobi, Carl Gustav Jacob, 165, 169,
    170, 194, 196, 260
Jordan, Marie Ennemond Camille, 237,
    282, 346, 352

Kaestner, Abraham Gotthelf, 85, 92
Kant, Immanuel, 84, 85, 229, 257, 259,
    335
Killing, Wilhelm Karl Joseph, 237–239,
    258, 262
Klein, Felix Christian, vi, 147, 156, 187,
    188, 198, 216, 218, 219, 230–237, 248,
    249, 254, 256, 258, 260–262, 269, 270,
    275, 277, 278, 287, 288, 297, 302, 310,
    345–347, 349–354, 368
Klügel, Georg Simon, 85
Knorre, 133
Kotelnikov, Petr Ivanovich, 132
Kovalevskaya, Sofia Vasilyevna, 283
Kronecker, Leopold, 197
Kummer, Ernst Eduard, 231

Lacroix, Sylvestre François, 5–7
Lagrange, Joseph Louis, v, 5, 6, 13, 14,
    16, 48
Lambert, Johann Heinrich, 79, 84–86,
    94–96
Lancret, Michel Ange, 51
Laplace, Pierre Simon, v, xii, 5–8, 13,
    14, 16, 48
Legendre, Adrien Marie, v, 5–7, 13, 16,
    48, 79, 86, 92, 124, 198–200, 203, 225
Lehrsatz, 256
Leibniz, Gottfried Wilhelm, 84
Levi-Civita, Tullio, 327
Lie, Marius Sophus, 231, 237
Lindemann, Carl Louis Ferdinand von,
    248, 260
line at infinity, 33, 35, 46, 47, 184, 309,
    352, 356
Liouville, Joseph, 216, 217
Lipschitz, Rudolf Otto Sigismund, 170,
    321
Listing, Johann Benedict, 351, 352
Lobachevskii, Nikolai Ivanovich, v, x–xii,
    82, 100, 115–120, 123, 130, 132–135,
    143, 144, 147, 198, 199, 218, 227, 229,
    231, 234, 246, 284, 288, 289, 300,
    305–307, 310, 326, 334, 339
log-spherical geometry, 96
Lotze, Rudolf Hermann, 198
Louis XVIII, 5
Lüroth, Jacob, 349

MacLaurin, Colin, 164
Magnitsky, 116–118
Malus, Etienne Louis, 51
Marat, Jean-Paul, 8
Maxwell, James Clerk, 250
Mendelssohn-Bartholdy, Felix, 196
Menelaus of Alexandria, 28, 37
Mentovich, Franz, 130
Mercator, Girardus, 112, 113
Miliradowitch, Field-Marshal, 12
Minding, Ernst Ferdinand Adolf, 129,
    130, 215–217
Minding's surface, 129, 130, 215
Minkowski, Hermann, 260, 261, 326, 327
Möbius, August Ferdinand, v, 149–156,
    158, 159, 177, 233, 235, 253, 256, 309,
    345, 347–353
Möbius net, 348, 349
Monge, Gaspard, v, vii, 4–10, 13, 14, 16,
    46, 48, 51, 54, 55, 63, 66, 69, 73–75,
    141, 142, 217
Moore, Eliakim Hastings, 265, 270, 317
Moulton, Forest Ray, 264, 310, 355, 356
Moulton's lines, 264, 355, 356

Nachlass of Gauss, 134
Nagel, Ernest, 313, 314
Napoleon Bonaparte, 2, 3, 5–7, 13, 17,
    141, 142, 146
negative curvature, 100, 129, 130, 200,
    215–217, 222, 224, 296
Nelson, Horatio, 3
Neumann, Carl Gottfried, 170
Neumann, Franz Ernst, 169, 260
Newton, Isaac, 5, 13, 134, 161, 165, 169,
    208, 209, 245, 321, 329
Ney, Michel, 11
Noether, Emmy Amalie, 314
non-Euclidean geometry, v–vii, 86, 96,
    107, 108, 111–113, 117, 123, 130, 134,
    135, 143, 147, 198–202, 215, 216, 218,
    219, 225, 227–231, 233, 234, 236, 237,
    239, 246, 248, 256, 257, 266, 271, 281,
    283–288, 299, 300, 302, 305, 310, 311,
    314, 321, 322, 326, 328, 334, 337, 338,
    349
non-metrical geometry, 69, 70, 346, 348,
    350
non-orientable, 351, 352
non-singular point, 179, 180
null system, 348

Olbers, Heinrich Wilhelm Matthäus, 93,
    96

Olivier, Théodore, 51
Orban, Rosalie von, 132
Ostrogradskii, Mikhail Vasilevich, 132

Pappus of Alexandria, 25, 73, 263, 355,
    357
Pappus's theorem, 25, 31, 32, 35, 73,
    262, 263, 355, 357
paradox, v, 18, 63
parallel, 20, 29–35, 44, 52, 80–82, 86, 89,
    102, 104, 105, 107–109, 112, 113, 119,
    120, 124, 125, 133, 152, 153, 168, 220,
    222, 223, 227, 238, 252, 253, 256, 264,
    266, 289, 334, 346, 349, 356
– asymptotic, 21
parallel postulate, 79–82, 85–87, 93,
    95–97, 102, 105–107, 109–111, 119,
    225, 288, 334, 335, 337
Pascal, Blaise, 25, 27, 34, 35, 44, 47, 54,
    262, 263
Pascal's theorem, 35, 44, 47
Pasch, Moritz, 255, 256, 262, 271, 276,
    310, 335, 337
Peano, Giuseppe, 269–273, 275–277, 311
perspectivity, 70, 71, 347
Piazzi, Giuseppe, 92
Pieri, Mario, vii, 276, 311, 354
Playfair, John, 80
Plücker, Julius, v, vi, xii, 60, 61, 156,
    161–168, 170, 173–178, 186, 189, 231,
    244, 309, 310
Plücker formulae, 173, 175, 176, 309
Poincaré, Henri, vi, xi, 147, 217–219,
    233, 237, 246, 257, 260, 281–289, 296,
    300–302, 304, 310, 322, 326, 327, 337,
    338
Poincaré disc, xi, 246, 288, 289, 296, 337
Poisson, Siméon Denis, v, 47, 48
polar, 11, 15, 17, 21–24, 34, 35, 43, 44,
    47, 53, 55, 58, 67, 68, 73, 155, 158,
    183, 184, 189, 190, 212, 220, 239, 240,
    243, 253
pole, 11, 15, 17, 21–24, 34, 35, 43, 47, 53,
    55, 56, 58, 68, 73, 74, 155, 200, 253
Poncelet, Jean-Victor, v–vii, x, 13–18,
    21–23, 25, 28, 31, 35, 40, 41, 43–57,
    59–61, 63, 64, 66–70, 74, 140, 142,
    143, 147, 165–167, 170, 244, 248
Popper, Sir Karl Raimund, 303
porism, 14, 16, 35, 40, 41, 244
Proclus Diadochus, 80
projective geometry, v–vii, 25, 28, 30,
    35, 43, 51, 52, 64, 66–70, 73, 140,
    142, 149, 218, 231, 233–235, 237, 242,

247–252, 254–256, 260–262, 264, 269,
    271, 273–276, 278, 309, 310, 313, 334,
    335, 345, 346, 348, 351, 352, 354–357
projective transformation, 43, 46, 68,
    154, 157–159, 234, 236, 243, 256, 309,
    346, 347, 355, 356
Prony, Gaspard Clair François Marie
    Riche de, 51
pseudogeometry, 284
pseudosphere, 129, 130, 215–217
Ptolemy, Claudius, 80, 135
Pythagoras of Samos, 80, 111, 212

quartic curve, xi, 165, 174–177, 244, 309

radical axis, 10
Reid, Constance, 259
Ricci-Curbastro, Gregorio, 327
Richelot, Friedrich, 169
Riemann, Georg Friedrich Bernhard, vi,
    113, 176, 188, 193, 195–202, 204, 217,
    218, 228, 229, 248, 256, 257, 260, 271,
    287, 288, 305, 309, 310, 321, 327, 331,
    352
Riemann surface, 193
Roch, Gustav, 197
rotating disc, 326–328, 330
Rousseau, Jean Jacques, 101
Russell, Bertrand Arthur William, Earl,
    276

Saccheri, Giovanni Girolamo, 81–85,
    130, 199
Salmon, George, xi, 176, 182, 244, 245,
    254
Sartorius, Wolfgang von Waltershausen,
    302
Schering, Ernst Christian Julius, 197,
    256
Schiller, Johann Christoph Friedrich
    von, 101
Schläfli, Ludwig, 352, 353
Schroeter, Heinrich Eduard, 170
Schubert, Hermann Cäsar Hannibal, 68
Schumacher, Heinrich Christian, 133,
    156, 229
Schur, Friedrich Heinrich, 262, 265
Schweikart, Ferdinand Karl, 91, 93–96,
    133
Schweitzer, Arthur Richard, 317
Segre, Corrado, 270–273, 276, 311
Shakespeare, William, 101
singular curve, 173, 174, 278

singular point, 129, 168, 173, 179, 183, 244, 245, 270, 309
Smith, Henry John Stephen, 248
Sommerfeld, Arnold Johannes Wilhelm, 326
special relativity, 325–327
spherical trigonometry, 82, 95, 100, 134, 224, 237, 238
Staudt, Karl Georg Christian von, vii, 254, 256, 346–351
Steiner, Jakob, vii, 70, 72, 156, 167, 170, 171, 243, 251, 254, 278
Stolz, Otto, 231, 346
Sturm, Jacques Charles François, 251
supplementaries, 45
supplementary hyperbola, 46
Sylvester, James Joseph, 167, 168, 170, 247–249
synthetic geometry, 17, 18, 167, 255
Szász, Otto, 102, 104

tangent, 14, 16, 17, 21–23, 34, 40, 43, 44, 52, 55–59, 72, 94, 158, 161–166, 173, 176, 179–184, 186, 188, 189, 193, 200, 244, 245, 278, 309, 310, 327
Taurinus, Franz Adolph, 91, 93, 95–97
Taylor, Brook, 181
tessellation, 188, 288

tractrix, 129, 130, 215, 216
truth, vi, 16, 18, 20, 29, 69, 75, 95, 102, 103, 112, 138, 140, 144, 228, 257, 326, 338

Vahlen, Theodor, 354
Vauban, Sébastien Le Prestre, Marquis de, 12
Veblen, Oswald, 265, 317, 354
Veronese, Giuseppe, 270, 277–279
Victor Emmanuel II, 250

Wachter, Friedrich Ludwig, 96, 97, 107
Wald, Abraham, 317
Wallis, John, 81, 86
Weber, Heinrich Martin, 197, 260
Weber, Wilhelm, 196–198, 231, 262
Wedderburn, Joseph Henry Maclagen, 357
Weierstrass, Karl Theodor Wilhelm, 197, 231, 237, 239
Weyl, Hermann Klaus Hugo, 259, 261
Whitehead, Alfred North, 354
Wiener, Hermann, 262
Wolf, Johann Rudolf, 88
Wordsworth, William, 3

Young, Thomas, 322, 354